Advanced Rail Geotechnology – Ballasted Track

T0239782

Advanced Rail Geotechnology – Ballasted Track

Buddhima Indraratna, PhD, FIEAust, FASCE, FGS

Professor and Head, School of Civil, Mining & Environmental Engineering,
Director, Centre for Geomechanics & Railway Engineering,
Faculty of Engineering, University of Wollongong,
Wollongong City, NSW 2522, Australia

Wadud Salim, PhD, MIEAust

Senior Geotechnical Engineer,
Planning, Environment and Transport Directorate,
Gold Coast City Council, Queensland 9729, Australia
(Adjunct Research Fellow, Centre for Geomechanics and
Railway Engineering, University of Wollongong)

Cholachat Rujikiatkamjorn, PhD

Senior Lecturer,
Centre for Geomechanics and Railway Engineering,
School of Civil, Mining & Environmental Engineering,
University of Wollongong, Wollongong City, NSW 2522, Australia

CRC Press
Taylor & Francis Group
Boca Raton London New York

CRC Press is an imprint of the
Taylor & Francis Group, an **informa** business

A BALKEMA BOOK

CRC Press
Taylor & Francis Group
6000 Broken Sound Parkway NW, Suite 300
Boca Raton, FL 33487-2742

First issued in paperback 2018

CRC Press/Balkema is an imprint of the Taylor & Francis Group,
an informa business

© 2011 by Taylor & Francis Group, LLC

No claim to original U.S. Government works

ISBN-13: 978-0-415-66957-3 (hbk)
ISBN-13: 978-1-138-07289-3 (pbk)

Typeset by MPS Limited, a Macmillan Company, Chennai, India

This book contains information obtained from authentic and highly regarded sources. Reasonable efforts have been made to publish reliable data and information, but the author and publisher
cannot assume responsibility for the validity of all materials or the consequences of their use. The
authors and publishers have attempted to trace the copyright holders of all material reproduced in
this publication and apologize to copyright holders if permission to publish in this form has not
been obtained. If any copyright material has not been acknowledged please write and let us know so
we may rectify in any future reprint.

Except as permitted under U.S. Copyright Law, no part of this book may be reprinted, reproduced,
transmitted, or utilized in any form by any electronic, mechanical, or other means, now known or
hereafter invented, including photocopying, microfilming, and recording, or in any information
storage or retrieval system, without written permission from the publishers.

For permission to photocopy or use material electronically from this work, please access www.
copyright.com (http://www.copyright.com/) or contact the Copyright Clearance Center, Inc.
(CCC), 222 Rosewood Drive, Danvers, MA 01923, 978-750-8400. CCC is a not-for-profit organiza-
tion that provides licenses and registration for a variety of users. For organizations that have been
granted a photocopy license by the CCC, a separate system of payment has been arranged.

Trademark Notice: Product or corporate names may be trademarks or registered trademarks, and
are used only for identification and explanation without intent to infringe.

Library of Congress Cataloging-in-Publication Data
Indraratna, Buddhima.
  Advanced rail geotechnology–ballasted track / Buddhima Indraratna,
Wadud Salim, Cholachat Rujikiatkamjorn.
    p. cm.
  Includes bibliographical references and index.
  ISBN 978-0-415-66957-3 (hardback) – ISBN 978-0-203-81577-9 (ebook)
  1. Ballast (Railroads) I. Salim, Wadud. II. Rujikiatkamjorn, Cholachat.
III. Title.

  TF250.I53 2011
  625.1′41—dc22

                                                    2010054313

Published by: CRC Press/Balkema
               P.O. Box 447, 2300 AK Leiden, The Netherlands
               e-mail: Pub.NL@taylorandfrancis.com
               www.crcpress.com – www.taylorandfrancis.co.uk – www.balkema.nl

**Visit the Taylor Francis Web site at
http:www.taylorandfrancis.com**

**and the CRC Press Web site at
http:www.crcpress.com**

Dedication

This Book is dedicated to those who have perished in fatal train derailments all over the world.

Contents

Preface

For several hundred years, the design of railway tracks has practically remained unchanged, even though the carrying capacity and speeds of both passenger and freight trains have increased. Essentially, the rail track is a layered foundation consisting of a compacted sub-ballast or capping layer placed above the formation soil, and a coarse granular medium (usually hard rock ballast) placed above the sub-ballast. The steel rails are laid on either timber or concrete sleepers that transmit the stress to the ballast layer which is the main load bearing stratum. Only a minimum amount of confining pressure is applied from the shoulder ballast on the sides and crib ballast between sleepers to reduce lateral spread of the ballast during the passage of trains. Against the common knowledge of the mechanics of rockfill, the ballast has remained practically to be an unconfined load bearing layer.

The high lateral movement of ballast in the absence of sufficient confinement, fouling of ballast by dust, slurried (pumped) formation soils (soft clays and silts liquefied under saturated conditions) and coal from freight trains as well as ballast degradation (fine particles then migrating downwards) has been the cause for unacceptably high maintenance costs in railways. Quarrying for fresh ballast in spite of stringent environment controls, stockpiling of used ballast with little demand for recycling and routine interruption of traffic for track repairs have been instrumental in the allocation of significant research funds for the improvement of ballasted rail tracks in Australia, North America and Western Europe. Finding means of reducing the maintenance costs and reducing the frequency of regular repair cycles have been a priority for most railway organisations running busy traffic schedules. In this Book, the authors have also highlighted the role of geosynthetics in the improvement of recycled ballast. Naturally it is expected that the use of geosynthetics will encourage the re-use of discarded ballast from stockpiles, reducing the need for further quarrying and getting rid of the unsightly spoil tips often occupying valuable land areas in the metropolitan areas.

Although the amount of research conducted on sand, road base and rockfill (for dams) has been extensive, limited research has been conducted on the behaviour of ballast under monotonic loading. Under cyclic loading, the available literature on ballast is even more limited. For many decades, the ballast layer has been considered to be 'elastic' in design by railway (structural) engineers. It is only since recently, that the behaviour of ballast under high train axle loading has been considered to be initially elasto-plastic, and then fully-plastic under conditions of significant degradation including breakage. Observations of removed ballast during maintenance indicate clearly the change in particle sizes due to degradation. The associated track settlement and lateral displacements are the blatant tell-tale signs of the need to evaluate ballast as a material that encounters plastic deformation after several thousands of loading cycles.

In this Book, the authors have attempted to describe the behaviour of ballast through extensive large-scale equipment, namely, the cylindrical and prismoidal triaxial tests and impact chambers. These experimental studies conducted in large testing rigs under both static and cyclic loads are unique, as very few research institutions have designed and built such facilities for the purpose of ballast testing. The authors have proposed various constitutive models to describe the ballast behaviour under both monotonic and cyclic loads. The mathematical formulations and numerical model are validated by experimental evidence from the above mentioned tests and also by field trials where warranted. The book also provides an extensive description of the use of geosynthetics in track design, and provides a fresh insight to design and performance of tracks capturing particle degradation, fouling and drainage. Non-destructive testing is described to monitor the track condition. The benefits of subsurface drainage to stabilise rail tracks are discussed and demonstrated using a case study. In terms of practical specifications, a more appropriate ballast gradation with a less uniform particle size distribution is presented for modern tracks carrying heavier and faster trains.

The writing of this Book would not have been possible without the encouragement and support of various individuals and organisations. Firstly, the authors are most grateful for the continuous support and invaluable advice of David Christie, Senior Geotechnical Consultant, Rail Infrastructure Corporation (NSW). The support from the former Cooperative Research Centre for Railway Engineering and Technologies and the current Cooperative Research Centre for Rail Innovation (Rail-CRC) during the past 10 years is gratefully appreciated. During the past 8–10 years, the funds for various research projects were provided by the Australian Research Council and the Rail-CRC. The research efforts of former PhD students, Dr Dominic Trani, Dr Daniela Ionescu, Dr Behzad Fatahi, Dr Joanne Lackenby and Dr Pramod Thakur are gratefully acknowledged. The efforts of Dr. Hadi Khabbaz and Dr. Mohamed Shahin (former Research Fellows) have been significant. Continuing support of Julian Gerbino (Polyfabrics Australia Pty Ltd) is appreciated. Assistance of George Fannelli (formerly of BP-Amoco Chemicals Pty Ltd, Australia) is also acknowledged. The dedicated laboratory assistance and the workmanship of Alan Grant, Ian Bridge and Ian Laird of University of Wollongong and the technical staff of the former Rail Services Australia (RSA) Workshop are gratefully appreciated. Special thanks to Dr Anisha Sachdeva for her assistance to the authors through speedy editing efforts during her short stay at University of Wollongong. Most Chapters have been copy edited and proofread by Manori Indraratna and Bill Clayton.

Selected technical data presented in numerous Figures, Tables and some technical discussions have been reproduced with the kind permission of various publishers. In particular, the authors wish to acknowledge:

Prof. Coenraad Esveld: author of *Modern Railway Track*, MRT Productions, Netherlands, 2001.

Thomas Telford Ltd. (UK): permission granted to reproduce selected data from the book, *Track Geotechnology and Substructure Management*, E. T. Selig and J. M. Waters, 1994; and authors' previous publication in *Geotechnical Engineering*, Proc. of the Institution of Civil Engineers (UK).

Elsevier Science Publishers Ltd: permission granted to reproduce several Figures from the book, *Geotextiles and Geomembranes Manual*, T. S. Ingold, 1994.

Dr. Akke S. J. Suiker, author of *The Mechanical Behaviour of Ballasted Tracks*, Delft University Press, Netherlands, 2002.

Canadian Geotechnical Journal, International Journal of Geotechnical and Geoenvironmental Engineering, International Journal of Geomechanics, ASTM Geotechnical Testing Journal and Geomechanics and Geoengineering: An International Journal permission granted to reproduce technical contents of the authors' previous publications.

SPECIAL ACKNOWLEDGMENT

Numerous contributions from Dr Sanjay Nimbalkar, Dr Jayan Vinod and Dr Lijun Su of the Centre for Geomechanics and Railway Engineering (GRE) at University of Wollongong are gratefully acknowledged. Their assistance for including salient outcomes through various research projects conducted at the GRE Centre has made this book comprehensive in a track geotechnology perspective.

Buddhima Indraratna
Wadud Salim
Cholachat Rujikiatkamjorn

Foreword

Railways around the world are undergoing a renaissance in all sectors including urban rail operations, High Speed Rail, Heavy Haul and Intermodal Freight. This second book on the design of ballast tracks will help underpin the revitalization of rail by providing practical means of reducing maintenance costs and improving track availability. It particularly focuses on the use of geosynthetics to improve drainage and extend life. It also focuses on how to recycle ballast in order to reduce the demand for further quarrying. In so doing it also provides tools for improving and modernizing railway track performance.

Buddhima Indraratna, Wadud Salim and Cholochat Rujikiatkamjorn have led the way in finding innovative solutions in ballast design. The University of Wollongong (UoW), Australia has pushed the frontiers of knowledge in track geotechnology. UOW is one of the founding Universities in the CRC (Cooperative Research Centre) for Rail Innovation. The CRC, together with others, has contributed to the funding and leadership of this research.

This book is based on the knowledge acquired through years of painstaking observations and studies of track under both static and cyclic loading. The research has used state-of-the-art laboratory testing and the use of non destructive ballast testing to monitor track condition. This book examines the benefits of sub surface drainage to stabilize rail tracks and the role of subballast of various characteristics. Field instrumentation for track performance and verification, together with modeling of ballast and track have resulted in a practical specification of appropriate ballast gradations for modern track and faster and/or heavier trains.

Australia is playing a leading role in the development of heavy haul railway operations. This includes the operation of 3 kilometer long trains, payloads greater than 40,000 tonnes operating with wagon axle loads of 40 tonnes in extreme weather conditions. As such having access to world leading knowledge on track structure and ballast is critical to sustaining such operations over the longer term.

This book is not only a comprehensive study of mechanics and behavior of ballast, but the research also includes the role that geosynthetic reinforcing materials can play in strengthening ballast and improving track drainage. It provides a pictorial guide for track instability assessment and performance verification through modern track instrumentation.

This world leading work is designed to provide support for practicing railway engineers. It introduces new specifications for ballast gradations which take into account ballast response to train loadings including degradation and deformation particularly important in today's more demanding operating environments.

This book is an excellent example of collaborative research, delivering competitive ground breaking solutions for the railway industry. It represents a major contribution to railway track knowledge for researchers, students and practicing engineers. A culmination of this work will be the much anticipated smart tool software expected to be commercialized by the CRC for Rail Innovation in conjunction with the University of Wollongong in the near future, to guide field engineers in the management of ballast.

It is important and innovative work that the CRC (Cooperative Research Centre) for Rail Innovation is delighted to have financed and sponsored.

David George
CEO, Cooperative Research Centre for Rail Innovation,
Australia

About the Authors

Professor Buddhima Indraratna (FIEAust, FASCE, FGS, CEng, CPEng) is an internationally acclaimed geotechnical researcher and consultant. After graduating in Civil Engineering from Imperial College, University of London he obtained a Masters in Soil Mechanics also from Imperial College, and subsequently earned a PhD in Geotechnical Engineering from the University of Alberta, Canada. He is currently Professor and Head, School of Civil, Mining & Environmental Engineering, University of Wollongong, Australia. In 2009, he was appointed as Honorary Professor in Civil Engineering, University of Shanghai for Science and Technology. His outstanding professional contributions encompass innovations in railway geotechnology, soft clay engineering, ground improvement, environmental geotechnology and geo-hydraulics, with applications to transport infrastructure and dam engineering. Under his leadership, the Centre for Geomechanics & Railway Engineering at the University of Wollongong has evolved to be a world class institution in ground improvement and transport geomechanics, undertaking national and international research and consulting jobs.

Recognition of his efforts is reflected by numerous prestigious Awards, such as: *2009 EH Davis* Memorial Lecture by the Australian Geomechanics Society for outstanding contributions to the theory and practice of geomechanics and Australian Commonwealth government hosted 2009 Business-Higher Education Round Table award for Rail Track Innovations, among others. He is the author of 4 other books and over 350 publications in international journals and conferences, including more than 30 invited keynote lectures worldwide. In the past, several of his publications have received outstanding contribution awards from the International Association for Computer Methods and Advances in Geomechanics (IACMAG), the Canadian Geotechnical Society and the Swedish Geotechnical Society.

Dr Wadud Salim is a Senior Geotechnical Engineer in the Planning, Environment & Transport Directorate of Gold Coast City Council in Queensland, Australia. After graduating from Bangladesh University of Engineering & Technology, he obtained a Masters in geotechnical engineering from the Asian Institute of Technology, Thailand, and a PhD in geotechnical engineering from the University of Wollongong. He is an Adjunct Research Fellow of the Centre for Geomechanics and Railway Engineering, University of Wollongong, and a former Geotechnical Engineer at RailCorp, Sydney. He is the co-author of a previous book on rail track geotechnics and numerous technical papers in various international journals and conferences in the area of rail track modernisation.

 **Dr Cholachat Rujikiatkamjorn** is a Senior Lecturer in Civil Engineering at the University of Wollongong. He is a Civil Engineering graduate from Khonkaen University, Thailand (BEng) with a Masters (Meng) from the Asian Institute of Technology, Thailand. He obtained his PhD in Geotechnical Engineering from the University of Wollongong. His key areas of expertise include ground improvement for transport infrastructure and soft soil engineering. In 2009, he received an award from the International Association for Computer Methods and Advances in Geomechanics (IACMAG) for an outstanding paper by an early career researcher, and the 2006 Wollongong Trailblazer Award for innovations in soft soil stabilisation for transport infrastructure. He has published over 50 articles in international journals and conferences.

Chapter 1

Introduction

Rail track network forms an essential part of the transportation system of a country and plays a vital role in its economy. It is responsible for transporting freight and bulk commodities between major cities, ports and numerous mineral and agricultural industries, apart from carrying passengers in busy urban networks. In recent years, the continual competition with road, air and water transport in terms of speed, carrying capacity and cost have substantially increased the frequency and axle load of the trains with faster operational speeds. On one hand this implies continuous upgrading of track, and on the other, this imparts inevitable pressure for adopting innovative technology to minimise construction and maintenance costs. Hundreds of millions of dollars are spent each year for the construction and maintenance of rail tracks in many countries including USA, Canada, China, India and Australia. The efficient and optimum use of these funds is a challenging task which demands innovative and cutting edge technologies in railway engineering.

Traditional rail foundations or track substructures consisting of one or two granular layers overlying soil subgrade have become increasingly overloaded due to the utilisation of faster and heavier trains. Rail tracks built over areas with adverse geotechnical conditions that are coupled with substructures not built to counteract greater design requirements demand more frequent maintenance cycles. Finding economical and practical techniques to enhance the stability and safety of the substructure is vital for securing long term viability of the rail industry, and to ensure sufficient capacity to support further increases in load.

In the past, most attention was paid to the superstructure (sleepers, fasteners and rails) of the track, and less consideration was given to the substructure components, namely ballast, subballast and subgrade layers. Many researchers have indicated that the major portion of any track maintenance budget is spent on substructure [1, 2]. Economic studies by Wheat and Smith [3] into British rail infrastructure showed that more than a third of the total maintenance expenditure for all railway networks that operate on ballasted track goes into substructure. Railway authorities in the USA spend tens of millions of dollars annually for ballast and related maintenance [4], while the Canadian railroads have reported an annual expenditure of about 1 billion dollars, where most of which includes track replacement and upkeep costs [5]. Fast train lines such as the Shinkansen Line (Japan) and the TGV-Sud-Est Line (France) face even higher maintenance costs. In Australia, the cost of public funding would exceed $2.1 billion per year to maintain operations above and below rail [6]. The huge cost involved in substructure maintenance can be significantly reduced if thorough understanding of the physical and mechanical characteristics of the rail substructure and of the ballast layer in particular is obtained.

1.1 NATURE OF TRACK SUBSTRUCTURE

The properties of the substructure elements are highly variable and more difficult to determine than those of the superstructure components [7]. Some key stability problems of paramount importance are issues related to ballast. The ballast layer is a key component of the conventional track structure. Its importance has grown with increasing axle loads and train speeds. Ballast is defined as the selected crushed granular material placed as the top layer of the substructure in which the sleepers are embedded to support the rails. It is usually comprised of hard and strong angular particles derived from high strength unweathered rocks. However, ballast undergoes gradual and continuing degradation due to cyclic rail loadings (Fig. 1.1). Minimising ballast degradation is imperative to sustain its primary functions and overall working of the substructure.

In conventional track design, ballast degradation and associated plastic deformations are generally ignored. This problem stems from a lack of understanding of complex ballast breakage mechanisms and the absence of realistic stress-strain constitutive models that include plastic deformation and particle breakage under a large number of load cycles, typically a few millions. This limited understanding results in oversimplified empirical design and/or technological inadequacies in the construction of track substructure, inevitably requiring frequent remedial measures and costly maintenance. In order to reduce high maintenance costs, the predominant problems in rail track substructure need to be well understood in view of cause and effect. The main issues with track substructure are identified and discussed below.

1.1.1 Fouling

Ballast fouling is used to indicate contamination by fines. Fouling of ballast is one of the primary reasons for the deterioration of the track geometry. The sharp edges and corners break due to high stress concentrations at the contact points between adjacent particles reducing the angularity and the angle of internal friction of ballast (hence, shear strength). This process is continuous, and the fines generated add to those resulting from the expected weathering of ballast grains under harsh field environment. Fouling occurs by upward intrusion of the slurried subgrade, air/water borne debris and spillage from freight traffic such as coal and other mineral ore. Extensive field and laboratory studies conducted in North America [8] have concluded that ballast breakdown is the main source of track fouling (Fig. 1.2). This finding is contrary to the popular belief by the railroad industry that mud on the ballast surface is mostly derived from the fine subgrade soil underlying the ballast [8]. The fouling of ballast usually increases track settlement due to a reduction in the friction angle, and may also cause differential settlement (Fig. 1.3). In severe cases, fouled ballast needs to be cleaned or replaced to maintain desired track stiffness (resiliency), bearing capacity, alignment and level of safety. Examples of contaminated ballast are shown in Figure 1.4.

Figure 1.1 Degradation and fouling of ballast in track in New South Wales Australia.

1.1.2 Drainage

The layers of subballast and subgrade generally contain some moisture at any given time. They perform best under cyclic load when sustaining an intermediate moisture

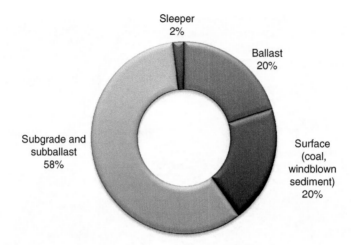

Sleeper
2%

Ballast
20%

Subgrade and
subballast
58%

Surface
(coal,
windblown
sediment)
20%

Figure 1.2 Comparison of different sources of ballast fouling from coal fouled, low-lying tracks.

Figure 1.3 Differential settlement in rail track causing significant risk to trains (after Suiker, [9]).

state (between dry and saturated state) [7]. Under gravitational forces, the fines generated by ballast breakdown migrate downwards and fill pore spaces between the particles. The fines decrease the void volume and retain moisture, thereby assisting further abrasion with time. As the pores get filled, the ballast loses its ability to drain the track superstructure attributed to reduced permeability (Fig. 1.5). Excess substructure water, particularly when it creates a saturated state similar to the situation in Figure 1.6, causes a significant increase in the cost of track maintenance. Trapped moisture leads to increased pore water pressure and subsequent loss of shear strength

Figure 1.4 Fouled ballast.

and stiffness (resiliency). Such conditions lead to a reduction in track stability and continued deterioration of track components over time.

1.1.3 Subgrade Instability

Where subballast is not in use, excessive load transfer may occur to the subgrade from the overlying ballast affecting the stability at the ballast-subgrade surface. In low lying

Figure 1.5 Insufficient drainage along a railway line in New South Wales, Australia.

Figure 1.6 Ponding water in the load bearing ballast along a Sydney metropolitan line in Australia.

coastal areas where the subgrade is generally saturated, the presence of water and its softening effect can result in the formation of slurry (liquefaction) at the interface. In the absence of a suitable separation layer, cyclic loading from passing traffic can cause this slurry to be pumped up to the ballast surface initiating pumping failure [1, 7].

The fine particles resulting from clay pumping or ballast degradation form a thin layer coating the larger grains thereby increasing overall compressibility. The fine particles also fill the void spaces between larger aggregates and reduce the drainage potential of the ballast bed. With time, this clay pumping phenomenon may clog the ballast bed

Figure 1.7 Clay pumping along a railway line in the state of Victoria, Australia.

and promote undrained shear failure [10]. Figure 1.7 shows an occurrence of subgrade pumping along one of the railway lines in Victoria, Australia.

1.1.4 Hydraulic degradation of ballast and sleepers

A particularly severe problem of ballast and sleeper degradation has been documented and studied by British Railways [11]. This problem seems to be most commonly associated with limestone ballast, for two reasons:

(a)   limestone abrades more readily than other rock aggregates, and
(b)   limestone particles tend to adhere so that they remain in a zone around the sleeper where they trap water, restrict drainage, and form an abrasive slurry that pumps up with high velocity.

The sleeper degradation as well as most of the ballast attrition are also believed to be associated with the high hydraulic gradients generated beneath the sleepers. In this situation, the speed of loading can be more critical than the magnitude of axle load. Due to traffic loading, the sleeper will settle giving rise to high fluid pressures within the substructure. This excess fluid pressure dissipates itself by jetting sideways and upwards. Naturally, high speed traffic loading induces much higher substructure water pressures, and that is why this type of undrained failure is seldom associated with low speed lines [7]. This problem of hydraulic erosion can also begin from other sources of fouling, which cause the ballast around the sleeper to become impermeable, resulting in the pooling of highly polluted water in the absence of constant maintenance.

Furthermore, the jetting action can displace ballast particles from the vicinity of the sleeper, thereby reducing the lateral resistance offered by the ballast to the sleeper. The jetted material is highly abrasive and in extreme cases can degrade the concrete sleepers to the point of exposing their prestressing wires. When dried, the eroded fines can

Figure 1.8 Dried slurry deposited around the sleepers.

drastically reduce the hydraulic conductivity around the sleeper and further exacerbate the abrasiveness of the jetted liquid. Figure 1.8 shows an example of dried slurry deposited around the sleepers. The factors common to this type of failure are (12):

(a)   poor drainage;
(b)   concrete sleeper giving high contact stresses on particles;
(c)   low wear resistant ballast material; and
(d)   void under sleeper resulting in adverse impact and hydraulic action.

1.1.5   Lateral confinement

Lateral buckling of rail track is usually observed in hot weather where degraded ballast is not able to provide sufficient lateral confinement to maintain track stability. As a result, the continuous welded track buckles with the formation of large lateral misalignments as shown in Figure 1.9 (top). Radial widening and track buckling (Fig. 1.9, bottom) can also occur on curves when the lateral confinement is reduced due to movement of ballast down slopes.

In order to reduce the maintenance costs caused by the above-mentioned problems, a proper understanding of how the ballast performs its tasks is imperative. Also, the exact role of the geotechnical parameters that contribute for optimising lateral confining pressure in track needs thorough examination. Only limited efforts have been made in the past to understand the role of track confinement through detailed laboratory testing and field experimentation. This has been quite a challenging task as the engineering behaviour of ballast is affected by various physical factors such as: particle mineralogy, grain size and shape, particle size distribution, porosity and moisture content, together with other variables often difficult to quantify including weathering

Figure 1.9 Track buckles due to insufficient lateral confinement.

effects and chemical attack. The use of geosynthetics to increase the lateral confinement in track is now proven to be a promising technique as vividly described later.

1.1.6 Aspects of load-deformation

Until today, the vast majority of railway engineers have regarded ballast as a quasi-elastic medium. Although the accumulation of plastic deformation under cyclic traffic loading is evident, most researchers have concentrated their studies on modelling the dynamic resilient modulus of ballast. Only limited research has been conducted on analytical modelling of ballast considering plastic deformation associated with cyclic loading. In the past, many researchers have attempted to simulate the plastic deformation of ballast empirically. Despite spending a large annual sum for the construction and maintenance of railways, track design is still predominantly empirical in nature, often using a trial and error basis for decision making [9].

The load-deformation behaviour of substructure elements under cyclic train loading is also not understood well, and often difficult to predict with reasonable accuracy. Modern ballast testing is usually focused on the actual track loading and boundary conditions which should be represented as closely as possible in laboratory model studies [13, 14]. Trains impart a quasi-static load [14], which incorporates a combination of static and dynamic loads superimposed onto the static load [7]. Raymond and Davies [15] pointed out that vertical stresses under static wheel loads are in the order of $140\,kN/m^2$, and trains on a stiff track running at high speed can increase this stress more than three times. Therefore, the importance of dynamic (cyclic) testing to evaluate ballast behaviour cannot be underestimated. Lama and Vutukuri [16] indicated that repeated loads can cause failure at stresses much lower than the static strength by the process commonly known as mechanical fatigue.

Selig and Waters [7] and Ionescu et al. [2] indicated that the behaviour of coarse granular aggregates under repeated loading is non-linear and stress-state dependent, and it is very different from that under static (monotonic) loading. Selig and Waters [7] also pointed out that failure of ballast under cyclic loading is progressive and occurs at smaller stress levels than under static loading. Raymond and Williams [17] reported that the volumetric strain of ballast under repeated loading is twice the magnitude of that under static loading.

It is well known that all carriages are not of the same weight, and trains do not travel at the same speed. Apart from testing ballast under dynamic loading, it is vital to vary the cyclic stress levels, instead of applying constant cyclic load amplitude. This is because the ballast behaviour under these two different load amplitudes can be very different [18]. Indraratna et al. [14, 19] reported similar findings reiterating that ballast settlement is significantly influenced by the loading pattern.

Apart from the above considerations, the current usage of geosynthetics to control track deformation needs further exploration. Geosynthetics are proven to be effective reinforcement for the rail embankment, including the ballast bed and at the subgrade and subballast levels. Geogrids with suitable apertures can reduce lateral displacement and associated particle degradation. In addition, the application of geosynthetics has grown recently with the increased utilisation of recycled ballast after the implementation of strict regulations by environmental regulatory authorities on the disposal of fouled ballast. Recycled ballast usually has reduced angularity and may show significantly higher settlement and lateral deformation than fresh ballast. Therefore, to improve the performance of recycled ballast, the inclusion of geosynthetics is regarded as a suitable and attractive alternative. However, the degree of improvement and associated implications are still far from being advanced given the complex particle-aperture interactions and the load distribution mechanisms within a composite (layered) track medium.

Ultimately, there is a greater need to identify and develop new analytical and numerical models that can account for complex cyclic loading and associated degradation mechanisms of track elements. Advanced computational tools need to be developed to provide detailed insight into important short term and long term load-deformation processes in track substructure. Furthermore, sophisticated constitutive models will have to be formulated and calibrated by comparing their predictions with the laboratory observations. Also, attention must be focused on the changes in track response due to the increased train velocity and frequency of operation, the effect of which serves as an important criterion for the design of high-speed railway lines.

1.2 CARBON FOOTPRINT AND IMPLICATIONS

According to United Nations statistical data centre, Australia has one of the highest overall greenhouse gas (GHG) emission rates per capita in the world [20]. One of the major contributors for GHG emission is the transportation industry, which according to the Federal Department of Climate Change and Energy Efficiency accounted for 13.2% of Australia's domestic emissions in 2007 [21]. Emissions from road transport alone was 87% (68.5 million tonnes of carbon dioxide equivalent) of total transport emissions, while the contributions from other forms of transport were nominal (civil aviation – 6.8%, domestic shipping – 3.7% and railways – 2.5 %) [21]. This to a certain extent is attributed to high dependence on light vehicles, buses and trucks for transport. The Australian Government intends to reduce at least 60% of GHG emissions by 2050 from the year 2000 levels [22]. To achieve these goals, sharp reductions in transport emissions are essential and will require going beyond emissions trading to a new generation of transport policies [23].

Bureau of Transport and Regional Economics had projected that due to population growth and increased trade, Australia's freight movements would double by 2020, and triple by 2050 from 2006 levels [24]. This would result in increased traffic congestions and energy prices. To keep up with various policies to reduce GHG emissions and to accommodate future freight movements, a modal shift to rail transport system can be a favourable option. In addition to providing significant cost savings to the Australian economy, this can also provide significant social and environmental benefits. Rail transportation generates up to 10 times less emissions than road freight and is also 10 times more fuel efficient [20]. Furthermore, rail freight network holds the key to improvement of road congestion (one freight train removes about 150 trucks off the road). In recognition of this, Australian rail organisations particularly in the states of New South Wales and Queensland are currently spending hundreds of millions of dollars for making rail more competitive by track modernisation and upgrading various rail corridors across the country. However, the challenges posed by future developments in demographics and trade need to be addressed by the larger community. Historically, Australian government has seemingly underspent on railway infrastructure when compared to road transport, with rail infrastructure only receiving between 20–30% of the combined investment in road and rail. In year 2007–08, the Australian Government invested approximately 13.1 billion on road infrastructure whereas it was only about 2.1 billion dollars on rail. Such scenarios are true for various other countries too, where investments on rail infrastructure had fallen behind proactive road infrastructure development. To promote greater economies of scale and transport efficiency, more serious consideration should be given to investment in large scale rail systems in populated large countries, with the implementation of novel applied research in the light of long term socio-economic returns.

1.3 SCOPE

This book presents creative and innovative solutions to rail industry worldwide, and is the result of extensive research in track geotechnology conducted at the University of Wollongong, Australia. Keeping the critical issues of track substructure in mind, the authors present the current state of research concentrating on the factors governing the stress-strain behaviour of ballast, its strength and degradation characteristics based on detailed laboratory experiments, the effectiveness of various geosynthetics in minimising ballast breakage and controlling track settlement, and the role of constitutive modelling of ballast under cyclic loading. The authors hope that this book would generate further interest among both researchers and practicing engineers in the wide field of rail track geotechnology and promote much needed track design modifications. The ultimate goal is to provide better understanding of this complex subject, improvement in the design and maintenance of track substructure and the speedy adoption of cutting edge technologies to minimise maintenance costs while promoting resilient high speed tracks.

Chapter 2 describes various types of rail tracks used in current practice, different components of track structure, and the various forces to which a track is typically subjected to. Chapter 3 describes the key factors governing ballast behaviour. The details of the state-of-the-art laboratory testing of ballast are presented in Chapter 4. The general stress-strain responses, and quantified strength, degradation and deformation behaviour of ballast with and without geosynthetics under static and dynamic (cyclic) loadings are discussed in Chapter 5. An overview of the currently available ballast deformation models is presented in Chapter 6. A new stress-strain constitutive model for ballast incorporating particle breakage is presented in Chapter 7. The drainage aspects in rail tracks and the application of geosynthetics in track are discussed in Chapter 8. Chapter 9 investigates the role of sub-ballast as a filtration medium apart from its load distribution function. Chapter 10 describes the field instrumentation for monitoring and verifying track performance. Chapter 11 describes in detail the distinct element modelling (DEM) of ballast densification and degradation, while numerical modelling of tracks and its applications to case studies are presented and discussed in Chapter 12. Chapter 13 focuses on non-destructive testing and track condition assessment. The different sources of ballast fouling and the various equipment, machinery and techniques employed in track maintenance schemes are described in Chapter 14. A new range of ballast gradations has been recommended in Chapter 15 based on various research outcomes described in earlier Chapters. Finally, bio-engineering for track stabilization is discussed with the application to selected case studies in Chapter 16.

REFERENCES

1. Indraratna, B., Salim, W. and Christie, D.: Improvement of recycled ballast using geosynthetics. *Proc. 7th International Conference on Geosynthetics*, Nice, France, 2002, pp. 1177–1182.
2. Ionescu, D., Indraratna, B. and Christie, H. D.: Behaviour of railway ballast under dynamic loads. *Proc. 13th Southeast Asian Geotechnical Conference*, Taipei, 1998, pp. 69–74.

3. Wheat, P., and Smith, A.: Assessing the marginal infrastructure maintenance wear and tear costs of Britain's railway network. *Journal of Transport Economics and Policy*, Vol. 42, 2008, pp. 189–224.

4. Chrismer, S. M.: Considerations of Factors Affecting Ballast Performance. *AREA Bulletin*, AAR Research and Test Department Report No. WP-110, 1985, pp. 118–150.

5. Raymond, G. P.: Research on Railroad Ballast Specification and Evaluation. *Transportation Research Record* 1006-Track Design and Construction, 1985, pp. 1–8.

6. BITRE: Rail Infrastructure Pricing: Principles and Practice. Bureau of Infrastructure, *Transport and Regional Economics*, Report 109, 2003.

7. Selig, E. T. and Waters, J. M.: *Track Technology and Substructure Management*. Thomas Telford, London, 1994.

8. Selig, E. T. and DelloRusso, V.: Sources and causes of ballast fouling. Bulletin No. 731, *American Railway Engineering Association*, Vol. 92, 1991, pp. 145–457.

9. Suiker, A. S. J.: *The mechanical behaviour of ballasted railway tracks*. PhD Thesis, Delft University of Technology, The Netherlands, 2002.

10. Indraratna, B., Balasubramaniam, A. and Balachandran, S.: Performance of test embankment constructed to failure on soft marine clay. *Journal of Geotechnical Engineering*, Vol. 1181, 1992, pp. 12–33.

11. Johnson, D. M.: A Reappraisal of the BR Ballast Specification, *British Rail Research Technical Memorandum*, TM TD 1, 1982.

12. Burks, M. E., Robson, J. D. and Shenton, M. J.: Comparison of Robel Supermat and Plasser 07-16 Track Maintenance Machines. Technical note SM 139, British Railways Board, R and D Division, 1975.

13. Indraratna, B., Salim, W. and Christie, D.: Cyclic loading response of recycled ballast stabilized with geosynthetics. *Proc. 12th Panamerican Conference on Soil Mechanics and Foundation Engineering*, Cambridge, USA, 2003, pp. 1751–1758.

14. Indraratna, B., Ionescu, D., and Christie, H. D.: State-of-the-Art Large Scale Testing of Ballast. *Conference on Railway Engineering (CORE 2000)*, Adelaide, 2000, pp. 24.1–24.13.

15. Raymond, G. P. and Davies, J. R.: Triaxial tests on dolomite railroad ballast. *Journal of the Geotechnical Engineering. Division*, ASCE, 1978, Vol. 104, No. GT6, pp. 737–751.

16. Lama, R.D. and Vutukuri, V. S.: Handbook on Mechanical Properties of Rocks, Vol. III, *Trans Tech Publications*, Clausthat, Germany, 1978.

17. Raymond, G.P. and Williams, D.R.: Repeated load triaxial tests on dolomite ballast. *Journal of the Geotechnical Engineering Division*, ASCE, Vol. 104 (GT 7), 1978, pp. 1013–1029.

18. Diyaljee, V. A.: Effects of stress history on ballast deformation. *Journal of the Geotechnical Engineering*, ASCE, Vol. 113, No. 8, 1987, pp. 909–914.

19. Indraratna, B., Ionescu, D., Christie, H. D. and Chowdhury, R. N.: Compression and degradation of railway ballast under one-dimensional loading. *Australian Geomechanics*, December, 1997, pp. 48–61.

20. The Australian Railway Association: Towards 2050-National freight strategy, The role of rail, 2010, pp. 1–42.

21. Department of Climate Change: Australia's national green house accounts – National greenhouse gas inventory, accounting for the Kyoto target, 2009, pp. 1–23.

22. Department of Climate Change and Energy Efficiency: Carbon Pollution Reduction Scheme: *Australia's Low Pollution Future*, Vol. 1, 2008, pp. 1.1–11.32.

23. CRC for Rail Innovation: Transforming Rail: A Key Element in *Australia's Low Pollution Future Final Report. Brisbane*, Queensland, Australia, 2008, pp. 1–44.

24. BITRE: Freight Measurement and Modelling in Australia. *Bureau of Infrastructure, Transport and Regional Economics*, Report 112, Canberra, 2006.

Chapter 2

Track Structure and Rail Load

The purpose of a railway track structure is to provide a stable, safe and efficient guided platform for the train wheels to run at various speeds with different axle loadings. To achieve these objectives, the vertical and lateral alignments of track must be maintained and each component of the structure must perform its desired functions satisfactorily under various axle loads, speeds, environmental and operational conditions.

This Chapter describes the types of track structure used in practice, various components of a conventional track structure, different types of loading imposed on a track system during its predicted life cycle and the load transfer mechanism.

2.1 TYPES OF TRACK STRUCTURE

Currently, the two types of rail tracks commonly used are conventional ballasted track and slab track. Most rail tracks are of the traditional ballasted type, however, there are some recent applications of non-ballasted slab tracks depending on the load-deformation characteristics of the subgrade (Fig. 2.1). Recent studies indicate that slab tracks may be more cost effective than conventional ballasted tracks when appropriate considerations are given to their life cycles, maintenance cost, and the extent of traffic disruption during maintenance [1]. However, rigid platforms do not perform as well as flexible, self-adjusting tracks where differential settlement can pose serious instability.

2.1.1 Ballasted track

These tracks are widely used throughout the world. In this conventional type of track, rails are supported on sleepers, which are embedded on a compacted ballast layer up to 350 mm thick. A common problem with this type of track is the progressive deterioration of ballast with increasing traffic passage (number of load cycles). The breakage of sharp corners, repeated grinding and wearing of aggregates, and crushing of weak particles under heavy cyclic loading may cause differential track settlement and unevenness of the surface. To maintain the desired safety level, design speed and passenger comfort, routine maintenance is imperative in a ballasted track.

The following are the main advantages of a ballasted track:

- Relatively low construction cost and use of indigenous materials,
- Ease of maintenance works,

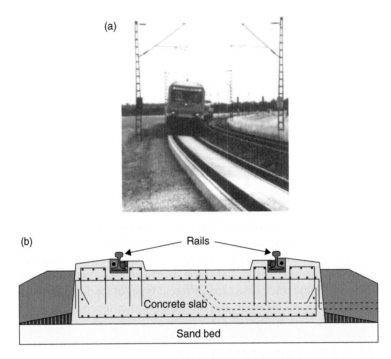

Figure 2.1 (a) Slab track and (b) cross-section of a slab track (modified after Esveld, [1]).

- High hydraulic conductivity of track structure, and
- Simplicity in design and construction.

The disadvantages are significant and are as follows:

- Degradation and fouling of ballast, requiring frequent track maintenance and routine checks,
- Disruption of traffic during maintenance operations,
- Reduction in hydraulic conductivity due to the clogging of voids by crushed particles and infiltrated fines from the subgrade,
- Pumping of subgrade clay- and silt-size particles (clay pumping) to the top of ballast layer particularly in areas of saturated and soft subgrade,
- Emission of dust from ballast resulting from high speed trains,
- Substructure becomes relatively thicker and heavier, which requires a stronger bridge and viaduct construction [1].

The mechanical behaviour of ballast and the other key aspects of a ballasted track are discussed in the following Sections and Chapters.

2.1.2 Slab track

Slab tracks are more suitable to high-speed and high-intensity traffic lines where lengthy routine maintenance and repairs are difficult. Since ballasted tracks are more maintenance intensive, causing frequent disruptions to traffic schedules, there is an increasing demand for low-maintenance tracks. The construction of slab tracks offers an attractive solution and is gaining popularity amongst rail track designers [1].

The main advantages of a slab track are:

- Almost maintenance free,
- Minimal disruption of traffic,
- Long service life,
- Reduced height and weight of substructure, and
- No emission of dust from the track, thus maintains a cleaner environment.

The disadvantages are:

- Higher initial construction cost,
- In case of structural damage or derailment, repair works are more time consuming and costly,
- Subgrade requires additional preparation and treatment, and
- Design and constructions are relatively more complex.

High initial construction cost still limits the widespread use of slab tracks, which is why conventional ballasted tracks are still popular.

2.2 COMPONENTS OF A BALLASTED TRACK

A ballasted track system typically consists of the following components: (a) steel rail, (b) fastening system, (c) timber or concrete sleepers or ties, (d) natural rock aggregates (ballast), (e) subballast and (f) subgrade. Figure 2.2 shows a typical track section and its different components. Although the principle of a ballasted track structure has not changed substantially, important improvements were put forward after the Second World War. As a result, a traditional ballasted superstructure can still satisfy the high demands of the Train à Grande Vitesse (TGV), the high speed trains in France. The track components may be classified into two main categories: (a) superstructure, and (b) substructure. The superstructure consists of rails, fastening system and sleepers. The substructure comprises ballast, subballast and subgrade. The superstructure is separated from the substructure by the sleeper-ballast interface, which is the most important element of track governing load distribution to the deeper track section.

2.2.1 Rails

Rails are longitudinal steel members that guide and support the train wheels and transfer concentrated wheel loads to the supporting sleepers (timber or pre-cast concrete),

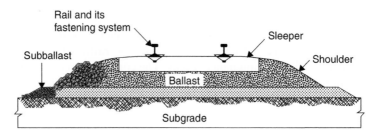

Figure 2.2 Typical section of a ballasted rail track.

which are evenly spaced along the length of track. Rails must be stiff enough to support train loading without excessive deflection between the sleepers and may also serve as electric signal conductors and ground lines for electric power trains [2].

The vertical and lateral profiles of the track assembly and the wheel-rail interaction govern the smoothness of traffic movement as the wheels roll over the track. Consequently, any appreciable defect on the rail or wheel surface can cause an excessive magnitude of stress concentration (dynamic) on the track structure when the trains are running fast. Excessive dynamic loads caused by rail or wheel surface imperfections are detrimental to other components of the track structure, because design for imperfections is difficult to incorporate.

Rail sections may be connected by bolted joints or welding. With bolted joints, the rails are connected with drilled plates called 'fishplates'. The inevitable discontinuity resulting from this type of joint can cause vibration and additional dynamic load, which apart from reducing passenger comfort may cause accelerated failure around the joint. The combination of impact load and reduced rail stiffness at the joints produces extremely high stresses on the ballast and subgrade layers which exacerbates the rate of ballast degradation, subsequent fouling and track settlement. Numerous track problems are found at bolted rail joints where frequent maintenance is required. Therefore, in most important passenger and heavily used freight lines, bolted joints are now being replaced by continuously welded rail (CWR), as described by Selig and Waters [2]. CWR has several advantages, including substantial savings in maintenance due to the elimination of joint wear and batter, improved riding quality, reduced wear and tear on rolling stock, and reduction in substructure damage [2].

2.2.2 Fastening system

Steel fasteners are used to hold the rails firmly on top of the sleepers to ensure they do not move vertically, longitudinally, or laterally [2]. Various types of steel fastening systems are used by different railway component manufacturers throughout the world, depending on the type of sleeper (concrete vs timber) and geometry of rail section.

The main components of a fastening system commonly include coach screws to hold the baseplate to the sleeper, clip bolts, rigid sleeper clips, and spring washers and nuts [1]. In addition, rail pads are often employed on top of the sleepers to dampen the dynamic forces generated by high-speed traffic movements. Fastening systems are categorised into two groups, namely, direct and indirect fastening. With direct fastening, the rail and baseplate are connected to the sleeper using the same fastener, but in indirect fastening, the rail is connected to the baseplate with one fastener while the baseplate is attached to the sleeper by a different unit. The indirect fastening system enables a rail to be removed from the track without removing the baseplate from the sleeper and allows the baseplate to be attached to the sleeper before being placed on the track.

2.2.3 Sleeper

Sleepers (or ties) provide a resilient, even and flat platform for holding the rails, and form the basis of a rail fastening system. The rail-sleeper assembly maintains the designed rail gauge. Sleepers are laid on top of the compacted ballast layer a specific distance apart. During the passage of trains, the sleepers receive concentrated vertical, lateral and longitudinal forces (described later in the Chapter) from the rails, and these forces are distributed by the sleepers over a wider area to decrease the stress at the sleeper/ballast interface to an acceptable level.

Sleepers can be typically made of timber, concrete or even steel. Timber sleepers (Fig. 2.3a) are still commonly observed worldwide in older tracks including in Australia and South Asia, but mainly due to environmental preservation as well as higher rate of degradation, mass scale production of concrete sleepers has become a more attractive financial option. The problems with wood are the tendency to rot, particularly around the fastenings used to hold the rails to them. Steel sleepers are considerably more expensive and are used only in very special situations. Concrete has now become the most popular type of sleepers. Concrete sleepers are much heavier than wooden ones, so they resist movement better but they have the disadvantage that they cannot be cut to size for turnouts and special track work. They work well under most conditions but under the high cyclic and impact loads of heavy haul freight trains, fracturing of concrete sleepers has caused concern.

In recent times, prestressed concrete sleepers (Fig. 2.3b) are becoming the primary choice as they are economical in various countries due to mass production in pre-cast yards. Pre-stressed concrete sleepers are potentially more durable, stronger, heavier, and more rigid than their timber counterparts. A main advantage is that the geometry of the concrete sleepers can be easily modified to extend the support area beneath the rails (Fig. 2.4). The extended support area decreases the ballast/sleeper contact stress, hence minimising track settlement and particle breakage. Concrete sleepers can provide an overall stiffer track, which may enhance fuel consumption benefits, although some researchers indicate that timber sleepers are more resilient and less abraded by the surrounding ballast than concrete sleepers [3].

Recently, a number of companies have started to offer sleepers manufactured of recycled plastic materials. They can be used in harsh climatic conditions and are more environmental friendly. These sleepers are said to outlast the classical wooden sleepers as they are impervious, but otherwise exhibit the same properties as their wooden counterparts with respect to damping of impact loads, lateral stability, and sound absorption. These products have gained limited acceptance, mainly because of the speed of mass production of stronger concrete sleepers.

Figure 2.3 (a) Timber sleepers, and (b) concrete sleepers used in track (site near Wollongong city, Australia).

2.2.4 Ballast

Essentially, the term 'ballast' used in railway engineering means coarse aggregates placed above subballast (finer grained) or subgrade (formation) to act as a load-bearing platform to support the track superstructure (sleepers, rails etc.). The sleepers

Figure 2.4 Concrete-frame sleeper used in track (Courtesy RailCorp).

(or ties) are embedded into a ballast layer that is typically 250–350 mm thick (measured from lower side of the sleeper). Ballast is usually composed of blasted (quarried) rock aggregates originating from high quality igneous or metamorphic rock quarries. For lighter passenger trains, well-cemented sedimentary rocks may also serve the purpose. Traditionally, crushed angular hard stones and rock aggregates having a uniform gradation and free of dust have been considered as acceptable ballast materials [2].

The source of ballast (parent rock) varies from country to country depending on the quality and availability of rock, environmental regulations, and economic considerations. No universal specification of ballast for its index characteristics such as size, shape, hardness, friction, texture, abrasion resistance and mineral composition that will provide the optimum track performance under all types of loading, subsoil and track environments can ever be established. Therefore, a wide variety of materials (e.g. basalt, limestone, granite, dolomite, rheolite, gneiss and quartzite) are used as ballast throughout the world. Aggregates that often fail to perform as ballast would include various types of sandstones mainly because of softening upon wetting and the inability to withstand high cyclic loads. Certain types of waste materials such as blast furnace slag have also been considered, but their load carrying capacity cannot be compared to freshly quarried natural rock aggregates.

2.2.4.1 Functions of ballast

Ideally, ballast should perform the following functions [4]:

- Provide a stable load-bearing platform and support the sleepers uniformly,
- Transmit high imposed stress at the sleeper/ballast interface to the subgrade layer at a reduced and acceptable stress level,
- Provide acceptable stability to the sleepers against vertical, longitudinal and lateral forces generated by typical train speeds,

Table 2.1 Ballast size and gradation [5].

Sieve size (mm)	% passing by weight (Nominal ballast size = 60 mm)
63.0	100
53.0	85–100
37.5	20–65
26.5	0–20
19.0	0–5
13.2	0–2
9.50	–
4.75	0–1
1.18	–
0.075	0–1

Table 2.2 Minimum ballast strength and maximum strength variation [5].

Minimum wet strength (kN)	Wet/dry strength variation (%)
175	≤25

- Provide required degree of elasticity and dynamic resiliency for the entire track,
- Provide adequate resistance against crushing, attrition, bio-chemical and mechanical degradation and weathering,
- Provide minimal plastic deformation to the track structure during typical maintenance cycles,
- Provide sufficient permeability for drainage,
- Facilitate maintenance operations,
- Inhibit weed growth by reducing fouling,
- Absorb noise, and
- Provide adequate electrical resistance.

2.2.4.2 Properties of ballast

In order to fulfil the above functions satisfactorily, ballast must conform to certain characteristics such as particle size, shape, gradation, surface roughness, particle density, bulk density, strength, durability, hardness, toughness, resistance to attrition and weathering, as discussed below.

Various standards and specifications have been made by different railway organisations throughout the world to meet their design requirements. In general, ballast must be angular, uniformly graded, strong, hard and durable under anticipated traffic loads and tough environmental conditions. Australian Standard AS 2758.7 [5] states the general requirements and specifications of ballast, and the recommended grain size distribution is given in Table 2.1. It also specifies the minimum wet strength and the wet/dry strength variation of the ballast particles (in accordance with AS 1141.22 [6]) for the fraction of aggregates passing 26.5 mm sieve and retained on 19.0 mm sieve, as shown in Table 2.2.

Table 2.3 Ballast Specifications in Australia, USA and Canada [10].

Ballast property	Australia	USA	Canada
Aggregate Crushing Value	<25%		
LAA	<25%	<40%	<20%
Flakiness Index	<30%		
Misshapen Particles	<30%		<25%
Sodium Sulphate Soundness		<10%	<5%
Magnesium Sulphate Soundness			<10%
Soft and Friable Pieces		<5%	<5%
Fines (<No. 200 sieve)		<1%	<1%
Clay Lumps		<0.5%	<0.5%
Bulk Unit Weight kg/m^3	>1200	>1120	
Particle Specific Gravity	>2.5		>2.6

The durability of ballast is usually assessed by conducting several standard tests such as Los Angeles Abrasion (LAA) test (AS 1141.23 [7]), Aggregate Crushing test (AS 1141.21 [8]), Wet Attrition test (AS 1141.27 [9]) etc. Indraratna et al. [10] gives a comparison between the specifications of ballast used in Australia [5], USA [11] and Canada [11, 12], as given in Table 2.3.

In order to effectively design the track substructure, it is essential to know the magnitude of sleeper/ballast contact stress and the distribution of stresses with depth through the ballast, sub-ballast and subgrade layers. The ballast thickness required for a track structure should depend on maximum stress intensity at the sleeper/ballast interface, acceptable bearing pressure of the underlying layer (subballast or subgrade) and stress distribution within the ballast body. Various methods, including simplified mathematical models, semi-empirical and empirical solutions are used in practice to evaluate the distribution of vertical stress through the ballast layer [13]. These methods are based on calculating stress under a uniformly loaded strip of infinite length and circular loaded area, in accordance with elastic theory.

Under cyclic loading imposed by repeatedly passing wheel loads, ballast undergoes irrecoverable plastic deformation and particle degradation, in addition to recoverable elastic strains. Accumulated plastic deformation may become significantly high after a few million load cycles. The continuous degradation process causes the originally sharp angular particles into relatively less angular or semi-rounded grains, thereby reducing inter-particle friction. This reduction in frictional resistance leads to a further increase in plastic strains. Ballast degradation and associated plastic deformation have been ignored in conventional design and analysis of track substructure. Traditionally, when the plastic deformation exceeds a certain tolerance level and/or ballast becomes excessively fouled by degradation or pumping of formation soils, these shortcomings in design and analysis are compensated for by frequent and costly maintenance operations, which disrupt traffic. Where timely maintenance is not carried out, devastating accidents including loss of lives and properties can result.

In order to design a more efficient track structure and minimise maintenance cost, ballast degradation and plastic track deformation must be examined and studied in detail. Moreover, the effects of particle breakage must be included in the constitutive stress-strain formulation so that a more appropriate and rational analysis and design method can be employed. With the advent of geosynthetic technology, the degradation and deformation of ballast may be minimised. These innovative ideas and techniques will be discussed in detail in the following Sections and Chapters.

2.2.5 Subballast

Subballast is the layer of aggregates placed between the ballast layer and the subgrade. This is usually comprised of well-graded crushed rock or a sandy gravel mixture. The subballast layer should be designed to prevent the penetration of coarse ballast grains into the subgrade and upward migration of subgrade particles (fines) into the ballast layer. Subballast therefore, acts as a filter and separating layer in the track substructure, transmits and distributes stress from the ballast layer down to the subgrade over a wider area, and also acts as a drainage medium to dissipate cyclic pore water pressures. In current design approaches, the main role of sub-ballast is to protect a soft subgrade soil (e.g. compressible estuarine clay) from being excessively loaded. In other words, the sub-ballast layer is compacted to a much higher stiffness than the natural soil formation, such that the load distribution to the underlying subgrade is significantly reduced as well as being uniform.

When designing the subballast layer, attention must also be given to its drainage and filtering functions. Therefore, it is usually composed of broadly graded materials, where empirical filter design methods often govern its particle size distribution [2]. Where there is no subballast or where poorly designed subballast is used, saturated subgrade clay and silt-size particles can become slurried or liquefied with infiltrated water. The slurried soil may subsequently pump upwards to foul the ballast under high cyclic loading, a phenomenon commonly known as clay pumping. In low lying coastal areas where rail tracks are founded on soft soils, ballast fouling by clay pumping is commonly observed during and after heavy rainfall. Use of geosynthetics in track substructure may prevent or minimise ballast fouling, and this aspect will be discussed further in later Chapters.

In Australia and in some other countries, the sub-ballast is often replaced by the term *capping*, to distinguish clearly its broadly graded nature and higher compaction that distinctly separates the subgrade from the overlying ballast layer. This subballast layer may vary from about 100–150 mm. In Australia, the capping layer thickness is often considered as a variable in the design approach depending on the subgrade properties, while the ballast bed is maintained at a constant thickness of 300 mm. On rail bridges, the ballast thickness is often reduced to less than 250 mm. The particle size distribution of sub-ballast is usually in the range of well graded medium to coarse grained granular fills (i.e. typically fine sand to fine gravels) as described further in Chapter 9.

The main functions of the sub-ballast or capping layer have been elucidated by Selig and Waters [2] and only a summary is given below:

- reducing the traffic induced stress at the bottom of the ballast layer to a tolerable level for the top of subgrade;
- extending the subgrade frost protection;
- preventing interpenetration of subgrade and ballast (separation function);

- preventing upward migration of fine material emanating from the subgrade;
- preventing subgrade attrition by ballast, which in the presence of water, leads to slurry formation, and hence prevent this source of pumping;
- shedding water, i.e., intercepts water coming from the ballast and directs it away from the subgrade to ditches at either side of the track; and
- permits drainage of water that might be flowing upward from the subgrade.

2.2.6 Subgrade

Subgrade is the ground (formation) on which the rail track structure is built. It may be naturally deposited soil or artificially placed fill material, e.g. rail embankment. The subgrade must have adequate stiffness and bearing capacity to resist traffic induced stresses at the subballast/subgrade interface. Subgrade soils are subjected to lower stresses than the overlying ballast and subballast layer. This stress decreases with depth, and the controlling subgrade stress is usually at the top zone unless unusual conditions such as a layered subgrade of sharply varying water contents or densities changes the location of the controlling stress. An investigation of the soil prior to design should check for these conditions. Instability or failure of subgrade will inevitably result in unacceptable distortion of track geometry and alignment, even with the placement of high quality ballast and subballast layers. If a track is to be constructed on soft soil (e.g. estuarine floodplain), the subgrade may be stabilised by one or more of the several ground improvement techniques, for example, installation of prefabricated vertical drains (PVD), lime-cement columns, deep cement/lime grouting, vibratory (pneumatic) compaction among other techniques.

2.3 TRACK FORCES

In order to analyze and design a resilient track substructure, the type and magnitude of loads that may be imposed on the ballast bed during its lifetime must be quantified. As discussed earlier, these loads are exerted by the sleepers onto the ballast bed by standing or running trains (wheel-rail-sleeper interactions), and are a complex combination of 'moving' static loads and dynamic forces.

The requirements for the bearing strength and overall quality of the track depend largely on the vertical load per axle, tonnage borne as the sum of the axle loads and the running speed. The static axle load level, to which the dynamic increment is added as a function of speed, determines the required load carrying capacity of the track. The accumulated tonnage determines the deterioration of the track quality and provides an indication of when maintenance and renewal are necessary. The dynamic load component, which depends on speed and horizontal and vertical track geometry, also plays an essential role. The maximum speed on a specific section is expressed usually in km/hour. In many European countries and some parts of Southeast Asia, freight trains are allowed to run at maximum speeds of 100 to 120 km/h whereas passenger trains on main lines run at 160 to 200 km/h. At the extreme end, high speed trains may travel at 250 to 300 km/h. It is known that the world record for high-speed rail is still held by French Railways National Company's (Société Nationale des Chemins de fer Français) Train à Grande Vitesse (TGV), which set a rail speed record approaching almost 575 km/h [14].

2.3.1 Vertical forces

As discussed by Esveld [1], the total vertical wheel load on a rail may be classified into two groups: quasi-static load and dynamic load. The quasi-static load is composed of three components, as given below:

$$Q_{total} = Q_{quasi\text{-}static} + Q_{dyamic} \tag{2.1}$$

$$Q_{quasi\text{-}static} = Q_{static} + Q_{centrifugal} + Q_{wind} \tag{2.2}$$

where, Q_{static} = static wheel load, $Q_{centrifugal}$ = increase in wheel load on the outer rail in curves due to non-compensated centrifugal force, Q_{wind} = increase in wheel load due to wind, $Q_{dynamic}$ = dynamic wheel load component resulting from sprung mass, unsprung mass, corrugations, welds, wheel flats etc.

The static load on each wheel is equal to half of static axle load, thus:

$$Q_{static} = \frac{G}{2} \tag{2.3}$$

where, G = weight of vehicle per axle. Considering the limit equilibrium of forces acting on a vehicle, as shown in Figure 2.5, Esveld [1] proposed the following expressions for the centrifugal and wind forces:

$$Q_{centrifugal} + Q_{wind} = G\frac{p_c h_d}{s^2} + H_w \frac{p_w}{s} \tag{2.4}$$

$$h_d = \frac{sV^2}{gR} - h \tag{2.5}$$

where, H_w = cross wind force, s = track width, V = speed, g = acceleration due to gravity, R = radius of curved track, h = cant (or superelevation), p_c = distance between center of rails and center of gravity of vehicle, and p_w = vertical distance of resultant wind force from center of rails. The maximum wheel load usually occurs at the outer rail ($h_d > 0$), thus combining Equations 2.3 and 2.4, we get:

$$Q_{e_{max}} \approx \frac{G}{2} + G\frac{p_c h_d}{s^2} + H_w \frac{p_w}{s} \tag{2.6}$$

The most uncertain part of the wheel load is the dynamic component, $Q_{dynamic}$. In order to obtain an approximate rough estimate of $Q_{dynamic}$, the static wheel load may be multiplied by a dynamic amplification factor (otherwise known as the impact factor), in lieu of conducting a purely cyclic load analysis [1]. The major factors affecting the magnitude of dynamic load component are [15]:

- Speed of train,
- Static wheel load and wheel diameter,
- Vehicle unsprung mass and vehicle condition,

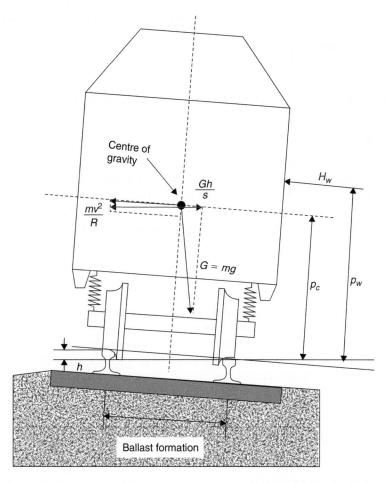

Figure 2.5 Quasi-static vehicle forces on a curve track (modified after Esveld, [1]).

- Track condition (including track joints, track geometry and track modulus), and
- Track construction aspects and properties of ballast and subballast.

A range of empirical formulae has been used by different railway organizations for determining the design vertical wheel load. It is usually expressed empirically as a function of the static wheel load. Various types of expressions developed are presented in the following sections.

2.3.1.1 AREA method

For the purpose of track design, Li and Selig [16] proposed the following simple expression for the computation of design wheel load based on the recommendation by the American Railway Engineering Association (AREA):

$$P_d = \phi P_s \tag{2.7}$$

where, P_d = design wheel load (kN) incorporating dynamic effects, P_s = static wheel load (kN), and ϕ = dimensionless impact factor (>1.0) and is given by Equation 2.8.

$$\phi = \left(1 + \frac{0.0052V}{D_w}\right) \tag{2.8}$$

where, D_w = diameter of the wheel (m), and V = velocity of the train (km/h).

2.3.1.2 ORE method

A comprehensive method of determining the impact factor has been developed by the Office of Research and Experiments (ORE) of the International Union of Railways [15, 17]. In this method, Equation 2.7 remains the same but the impact factor is entirely based on the measured track forces [17]. This impact factor is defined in terms of dimensionless speed coefficients, namely, α', β' and γ', as given by the following equation:

$$\phi = 1 + \alpha' + \beta' + \gamma' \tag{2.9}$$

where, α' and β' are related to the mean value of the impact factor, and γ' is related to the standard deviation of the impact factor.

The coefficient α' depends on track irregularities, vehicle suspension and vehicle speed. Although, it is difficult to correlate α' with track irregularities, it has been empirically found that for the poorest case, α' increases with the cubic function of speed, hence:

$$\alpha' = 0.04\left(\frac{V}{100}\right)^3 \tag{2.10}$$

where, V = vehicle speed (km/h).

The coefficient β' is the contribution to the impact factor due to the wheel load shift in curves [17], and may be expressed by either Equation 2.11 or 2.12:

$$\beta' = \frac{2d \cdot h}{G_h^2} \tag{2.11}$$

$$\beta' = \frac{V^2(2h + c)}{127Rg} - \frac{2c \cdot h}{G_h^2} \tag{2.12}$$

where, G_h = horizontal distance between rail centerlines (m), h = vertical distance from rail top to vehicle center of mass (m), d = super-elevation deficiency (m), c = super-elevation (m), g = acceleration due to gravity (m/sec²), R = radius of curve (m), and V = vehicle speed (km/h).

The last coefficient, γ', depends on the vehicle speed, track condition (e.g. age, hanging sleepers etc.), vehicle design, and maintenance condition of the

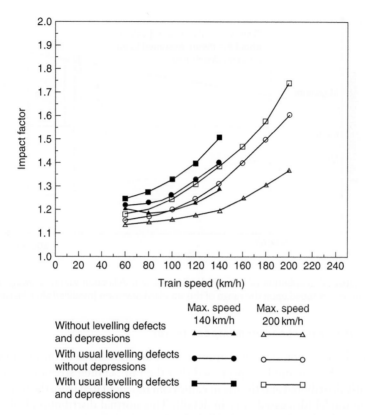

Figure 2.6 Impact factor in track design (data from Jeffs and Tew, [15]).

locomotives [17]. It was found that the coefficient, γ', increases with the speed, and can be approximated by the following algebraic expression:

$$\gamma' = 0.10 + 0.017\left(\frac{V}{100}\right)^3 \tag{2.13}$$

The ORE impact factor (ϕ) for different train speeds and various standards of tangent track has been plotted graphically, as shown in Figure 2.6 [15].

2.3.1.3 Equivalent dynamic wheel load

Atalar et al. [18] proposed the following simple equation to compute the equivalent dynamic wheel load:

$$P'_w = P_w\left(1 + \frac{V}{100}\right)(1 + C) \tag{2.14}$$

where, P'_w = equivalent dynamic wheel load for design, P_w = static wheel load, V = maximum velocity (km/hour), and C = a coefficient ≈ 0.3.

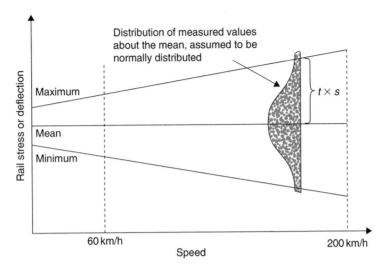

Figure 2.7 Statistical distribution of measured rail stress and deflection values, showing the effect of increased speed upon the range of the standard deviation (modified after Eisenmann, [19]).

2.3.1.4 Rail stress, speed and impact factor

This method is based on statistical approach for determination of the magnitude of the impact factor. Eisenmann [19] proposed that the rail bending stresses and deflections are normally distributed and the mean values can be calculated from the beam on elastic foundation model (discussed later in detail). This normal distribution is illustrated in Figure 2.7 for both rail stress and rail deflection values.

The mean rail stress and its corresponding deviation are represented by the expression:

$$s = x \cdot \delta \cdot \eta \tag{2.15}$$

where, $x =$ mean rail stress, $s =$ corresponding standard deviation of mean rail stress, $\delta =$ factor dependent upon the track condition (0.1 for very good condition track, 0.2 for good condition track and 0.3 for poor track condition) and $\eta =$ factor dependent on speed of the vehicle V (km/h) and the following values have been suggested for use:

$$\eta = 1 \qquad\qquad \text{if } V < 60 \text{ km/h} \tag{2.16}$$

$$\eta = \left(1 + \frac{V - 60}{140}\right) \quad \text{if } 60 \leq V \leq 200 \text{ km/h} \tag{2.17}$$

The corresponding maximum applied load (rail deflection) is given by [15]:

$$X = x + s \cdot t \tag{2.18}$$

where, $X =$ maximum applied load or deflection, $t =$ value depending upon the upper confidence limit (UCL) defining the probability that the maximum applied load will not be exceeded (0 for 50% UCL, 1 for 84.1% UCL, 2 for 97.7% UCL and 3 for 99.9% UCL).

Attributed to the assumed linearity between the applied load and rail stress, Equation 2.7 can be rewritten as:

$$X = \phi \cdot x \tag{2.19}$$

Combining Equations 2.15 and 2.18 and comparing with Equation 2.19, one can obtain an expression for impact factor as follows:

$$\phi = 1 + \delta \cdot \eta \cdot t \tag{2.20}$$

The four types of methods discussed for determining the vertical wheel load are not specifically interrelated, but a general observation can be made on the predicted magnitude of the impact factor. The envelope defined by Eisenmann's curve of impact factor for very good and good track conditions, incorporates both AREA and ORE impact factor curves that have been derived for probable average track conditions [15].

2.3.2 Lateral forces

Lateral loads in tracks are far more complex than vertical loads and are less understood [3]. Selig and Waters [2] indicated that there are two principal sources of lateral loads: (a) lateral wheel force, and (b) buckling reaction force. Lateral wheel forces are initiated by the lateral force component of friction between the wheel and rail, plus the lateral force applied by the wheel flange on the rail. Buckling reaction forces in the lateral direction are developed due to the high compressive stresses caused by high rail temperatures.

Similar to vertical force (Eq. 2.1), lateral force exerted by the wheel on outer rail is also equal to the sum of the quasi-static and dynamic loads, thus,

$$Y_{total} = Y_{quasi\text{-}static} + Y_{dyamic} \tag{2.21}$$

$$Y_{quasi\text{-}static} = Y_{flange} + Q_{centrifugal} + Q_{wind} \tag{2.22}$$

where, Y_{flange} = lateral force in curve caused by flanging against the outer rail, $Y_{centrifugal}$ = lateral force due to non-compensated centrifugal force, Y_{wind} = increase in lateral force due to cross wind, $Y_{dynamic}$ = dynamic lateral force component.

Now, if an assumption is made that the centrifugal and wind lateral forces act entirely on the outer rail, then the lateral equilibrium equation obtained from Figure 2.5 will be as follows:

$$Y_{e_{max}} \approx G\frac{h_d}{s} + H_w \tag{2.23}$$

Similarly, as in vertical force estimation, to account for dynamic component of lateral force, the static component of the force can be multiplied by the dynamic amplification factor (DAF), thus,

$$H = DAF\left(G\frac{h_d}{s} + H_w\right) \tag{2.24}$$

The Office of Research and Experiments (ORE) in Netherlands [17, 20] also carried out test programs for train speeds up to 200 km/hour, in order to assess lateral forces in track. These studies found that the lateral track force is dependent only on the radius of curvature, and the following empirical expression was proposed:

$$H = 35 + \frac{7400}{R} \tag{2.25}$$

where, H = lateral force at curved track (kN), and R = radius of curve (m).

A similar empirical formula for the lateral rail force is used in France, where the lateral track force is considered to increase with the traffic load [3], and given by:

$$H_s > 10 + \frac{P}{3} \tag{2.26}$$

where, H_s = force (kN) required to initiate lateral displacement, and P = Axle load (kN).

2.3.3 Longitudinal forces

The longitudinal force imposed on the rail head can be due to any change in length of the released rail occurring as a result of a significant change in temperature This is insignificant in fixed rails because the resistance is produced by friction forces between rails and sleepers and between sleepers and ballast. Other phenomenon causing longitudinal forces include, track creep, accelerating and braking of the vehicle, and shrinkage stresses caused by rail welding [1].

2.3.4 Impact forces

Rail track structures are often subjected to the impact loads due to abnormalities in either a wheel or a rail. The magnitude of these impact loads is very high within the very short impulse duration (frequency range upto 2000 Hz) and usually depends on the nature of wheel or rail irregularities, as well as on the dynamic response of the track [21, 22].

Impact loads are caused by wheel or rail abnormalities such as wheel-flat, wheel-shells, dipped rails, turnouts, crossings, insulated joints, expansion gap between two rail segments, imperfect rail welds and rail corrugations etc. [1, 23–25]. A diagrammatic representation of these typical sources of impact is shown in Figure 2.8. A wheel-flat is formed on the wheel of a vehicle becoming locked during braking, and sliding along the track and may be typically 50–100 mm long. A wheel-shell is caused by micro-cracks initiated by high wheel-rail contact forces. The geometry of a rail joint can be characterised by the gap width (typically 5–20 mm) and the height difference (typically 0.5–2 mm) in the two sides of a gap. These discontinuities on the wheel and rail can generate large impact forces between the wheel and track when wheels with flats and/or shells subsequently rotate or wheels roll over a rail joint [24]. At railway turnout or crossings, a large wheel impact force is generated due to traversing of wheel over the rail discontinuity [23]. Besides, at bridge approaches, road crossings and track transitions such as concrete slab track merging to ballast track, the abrupt

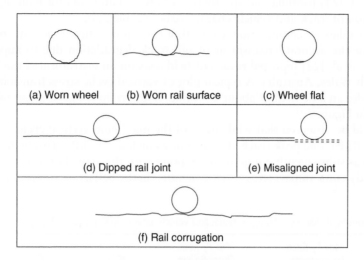

Figure 2.8 Various sources of impact loads in rail tracks.

change in track stiffness gives rise to high impact forces leading to accelerated track degradation [26].

Two types of distinct force peaks are observed during impact loading, i.e. an instantaneous sharp peak and a much longer duration gradual peak of smaller magnitude. The British Rail researchers termed these peak forces as P_1 and P_2 respectively, the universal terminology that is now being widely used by track engineers [21]. The impact force P_1 is due to the inertia of the rail and sleepers resisting the downward motion of the wheel and leads to compression of the contact zone between the wheel and rail. Its effects are mostly filtered out by the rail and sleepers and therefore, do not directly affect the ballast or the subgrade. The force P_2, lesser in magnitude compared to P_1, prevails over a longer duration and its occurrence is attributed to the mechanical resistance of the track substructure leading to its significant compression [28]. Since, P_2 forces are of greater importance in the assessment of track degradation, Jenkins et al. [21] proposed a simplified formula to calculate P_2 forces for a vehicle negotiating a vertical ramp discontinuity in rail top profile, equivalent to a dipped rail joint, at its maximum design operating velocity:

$$P_2 = P_0 + 2\alpha V_m \sqrt{\frac{M_u}{M_u + M_t}} \cdot \left[1 - \frac{C_t \pi}{4\sqrt{K_t (M_u + M_t)}} \right] \cdot \sqrt{K_t M_u} \qquad (2.27)$$

where, P_0 = maximum static wheel load, M_u = vehicle unsprung mass per wheel (kg), 2α = total dip angle (radians), V_m = maximum normal operating velocity (m/s), M_t = equivalent vertical rail mass per wheel (kg), K_t = equivalent vertical rail stiffness per wheel (N/m), C_t = equivalent vertical rail damping per wheel (Ns/m).

UK Railway group standards [29, 30] suggest that for the safety of the track, the P_2 force should not exceed 322 kN when a vehicle, with class 55 Deltic locomotive, negotiates a vertical ramp discontinuity at its maximum design operating velocity of 160 km/h. Australian standards [31–33] recommend the calculation of P_2 forces using

Jenkins et al. [21] formula and specify the following guidelines for limiting P_2 forces as a function of track and vehicle characteristics (Table 2.4).

Field studies in combination with laboratory tests often represent an efficient strategy for the accurate assessment of rail track degradation due to impact loads. Indraratna et al. [22] reported results of field tests on an instrumented track at Bulli, New South Wales, Australia. A typical plot of vertical cyclic stress transmitted to the ballast under an axle load of about 25 tonnes and a train speed of about 60 km/hour is shown in Figure 2.9.

It could be observed that while most of the maximum vertical cyclic stress range is up to 230 kPa, one peak reached to a value as high as 415 kPa. This high magnitude of stress was subsequently found to correspond with the arrival of wheel-flat proving that large dynamic impact stresses are generated in the ballast by wheel imperfections,

Table 2.4 Limiting P_2 forces relating to track and vehicle characteristics [31, 32, 33].

Track Class	Max P_2 force Locomotives (kN)	Max P_2 force Other Rolling Stock (kN)	K_t MN/m	C_t kNs/m	M_t kg
IXC	295	230	117	56	338
IX	295	230	117	56	151
IC	295	230	110	52.5	310
I	295	230	110	52.5	135
2	230	230	100	48	117
3	200	200	95.8	45.9	106
4	180	180	90.3	43.2	95
5	130	130	83.6	40	85

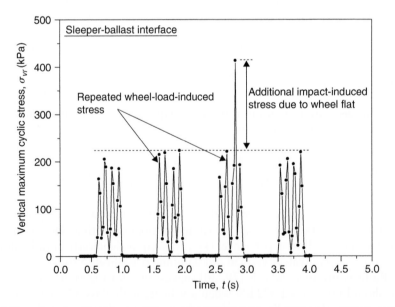

Figure 2.9 Typical measured vertical cyclic stresses transmitted to the ballast by coal train with wagons (100 tons) having wheel irregularity (modified after Indraratna et al., [22]).

and should be carefully assessed and accounted for in the design and maintenance of ballasted tracks.

2.4 LOAD TRANSFER MECHANISM

Typical distribution of wheel load to the rails, sleepers, ballast, subballast and subgrade, is shown in Figure 2.10. Shenton [34] studied the distribution of sleeper/ballast contact pressure in real tracks. This study indicated that as the typical ballast size was in the range of 25–50 mm and the typical width of a sleeper was 250 mm, the number of particles involved in directly supporting the sleeper was relatively small. Shenton [34] estimated that a sleeper which has been placed in track for a while may only be supported by 100–200 contact points. This implies that the measurement of actual sleeper/ballast contact stress is extremely difficult. However, British Railways attempted to measure the sleeper/ballast contact pressure in real track and those measurements are shown in Figure 2.11. The distribution of contact pressure is very erratic

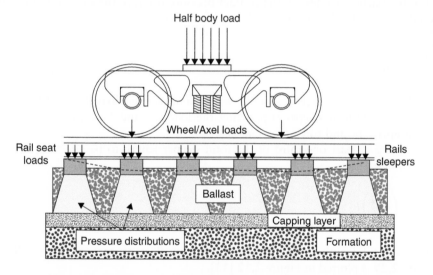

Figure 2.10 Typical wheel load distribution in track (Courtesy RailCorp).

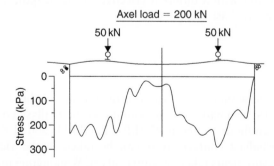

Figure 2.11 Measurements of sleeper/ballast contact pressure (modified after Shenton, [34]).

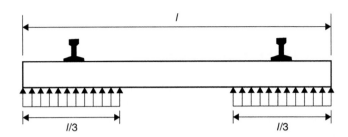

Figure 2.12 Simplified sleeper/ballast contact pressure (modified after Jeffs and Tew, [15]).

and varies from test to test. Nevertheless, these field measurements (Fig. 2.11) provide a sound indication of the maximum pressure exerted by the sleeper to the underlying ballast for a given axle load.

For the purpose of design, the contact pressure between the sleeper and ballast is generally assumed to be uniform and simplified by the following expression [15]:

$$P_a = \left(\frac{q_r}{BL}\right)F_2 \tag{2.28}$$

where, P_a = average contact pressure, q_r = maximum rail seat load, B = width of sleeper, L = effective length of sleeper supporting the load q_r, and F_2 = a factor depending on the sleeper type and track maintenance.

Assuming at least one third of the total sleeper length to be effective, Equation 2.28 becomes:

$$P_a = \left(\frac{3q_r}{Bl}\right)F_2 \tag{2.29}$$

where, l = total length of sleeper. The sleeper/ballast contact pressure following Equation 2.29 is plotted in Figure 2.12 [15].

In the Japanese track Standards, a similar distribution of sleeper/ballast contact pressure is assumed but with a different effective sleeper length, as shown in Figure 2.13 [18] and is expressed as:

$$P_a = \left(\frac{q_r}{2aB}\right)F_2 \tag{2.30}$$

where, a = distance between the rail head centre and edge of the sleeper.

Atalar et al. [18] estimated the maximum sleeper/ballast contact stress for a train speed of 385 km/hour to about 479 kPa. Esveld [1] stated that the maximum permissible sleeper/ballast contact stress can be taken in the vicinity of 500 kPa. The laboratory measurements taken by University of Wollongong underneath sleepers in the laboratory and in real tracks give values in the order of 350–400 kPa.

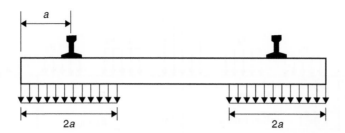

Figure 2.13 Load transfer to ballast assumed by Japanese Standards (modified after Atalar et al., [18]).

2.5 STRESS DETERMINATION

In order to calculate the maximum vertical stress on the subgrade, various methods have been developed based on a two dimensional stress distribution for a plane strain situation. The ballast, subballast, and subgrade create a three layer, linearly elastic system that has to be transformed into an equivalent single layer.

2.5.1 Odemark method

In 1949, Odemark [35] proposed an empirical method to convert a multi-layered system into a single layer system. The maximum vertical stress on the subgrade in the actual three layer system then correlates with the maximum vertical stress in the equivalent half space at a distance from the surface. The equivalent for $N-1$ layers is given by the expression:

$$\tilde{h} = \left\{ \begin{aligned} h_1 &\left(\frac{E_1}{E_{N_L}} \cdot \frac{1-\mu_{N_L}^2}{1-\mu_1^2} \right)^{1/3} + h_2 \left(\frac{E_2}{E_{N_L}} \cdot \frac{1-\mu_{N_L}^2}{1-\mu_2^2} \right)^{1/3} + \cdots \\ &+ h_{N-1} \left(\frac{E_{N_L-1}}{E_{N_L}} \cdot \frac{1-\mu_{N_L}^2}{1-\mu_{N_L-1}^2} \right)^{1/3} \end{aligned} \right\} \tag{2.31}$$

where, $\tilde{h}$ = equivalent depth; h_i = thickness of the ith layer; E_i = Young's modulus of elasticity at the ith layer; and μ_i = Poisson's ratio at the ith layer.

In this method, once a multi-layer has been transformed, calculations are only valid within the lowest layer considered during the transformation (layer N_L). If layers exist beneath layer N_L, it is implicitly assumed that they have elastic properties equal to those found in layer N_L [36]. This method can only approximate the multi-layer theory of elasticity when the elastic moduli decrease with depth ($E_i/E_{i+1} > 2$), and where layers are relatively thick and the transformed thickness of each layer is larger than the radius of the loaded area [36].

2.5.2 Zimmermann method

Figure 2.14 shows the stress pattern on the ballast bed along the length of the track. The stress for each sleeper is assumed to be evenly distributed over its surface area.

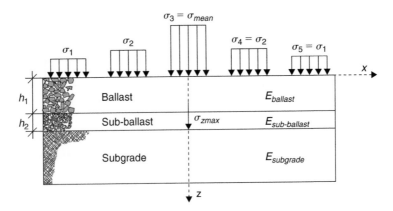

Figure 2.14 Stress pattern on the ballast bed along the length of the rail track (modified after Esveld, [1]).

An equivalent strip load then replaces the even distribution of stresses per sleeper across the width of the sleeper. By superimposing the individual load contribution of each sleeper, and by factoring in the thickness and elasticity of the upper ballast and subballast layers, the maximum vertical stress on the subgrade is then evaluated [1]. The dynamic amplitude is incorporated by using the amplification factor or impact factor given in Equation 2.20. The magnitude of this stress beneath the various sleepers caused by the effective wheel load Q is:

$$\sigma_i = \sigma_{max} \cdot \eta(x_i) \tag{2.32}$$

In the above equation:

$$\sigma_{max} = DAF \cdot \frac{Qa}{2LA_{sb}} \tag{2.33}$$

$$\eta(x_i) = e^{\frac{-x_i}{L}}\left[\cos\frac{x_i}{L} + \sin\frac{x_i}{L}\right] \quad x_i \geq 0 \tag{2.34}$$

$$L = \sqrt[4]{\frac{4EI}{k}} \tag{2.35}$$

where, DAF is solved using Equation (2.20) with $t = 1$, $Q =$ effective wheel load (kN), $a =$ sleeper spacing (m), $A_{sb} =$ contact area between sleeper and ballast bed for a third of the sleeper (m²), $L =$ characteristic length (m), $EI =$ bending stiffness of the rail (kN-m²), $k =$ foundation coefficient of continuous support (kN/m²), and $x_i =$ lateral distance from the point of interest to the centre of the ith sleeper.

In this method of longitudinal beam calculation, the rail is defined as an infinite beam on a continuous elastic support [37]. This assumption holds for a beam of finite length if the length is greater than $2\pi L$.

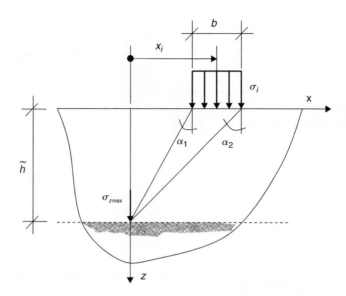

Figure 2.15 Stress due to strip load on half space (modified after Esveld, [1]).

The vertical stress in an elastic half space loaded by an evenly distributed strip load shown in Figure 2.15 can be determined using the two dimensional theory of elasticity. Thus, the compressive stress is given by:

$$\sigma_{zi} = \sigma_i \cdot f(x_i) \tag{2.36}$$

in which:

$$f(x_i) = \frac{1}{\pi}\left[\alpha_1 - \alpha_2 + \frac{1}{2}(\sin 2\alpha_1 - \sin 2\alpha_2)\right] \tag{2.37}$$

$$\alpha_1 = \arctan \frac{x_i + \frac{b}{2}}{\tilde{h}} \tag{2.38}$$

$$\alpha_2 = \arctan \frac{x_i - \frac{b}{2}}{\tilde{h}} \tag{2.39}$$

where, $b = $ sleeper width (m).

In this method, the contributions to the maximum vertical stress on the formation can be determined for each strip load according to:

$$\sigma_{z\max} = \sum_i \sigma_{zi} \tag{2.40}$$

Only a few of the strip loads in the vicinity of the maximum load need to be considered because of the decrease in strip load according to Equation (2.34) and the load spreading under a strip load according to Equation (2.37).

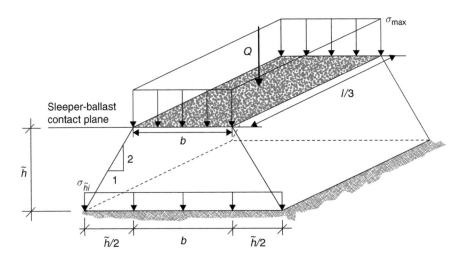

Figure 2.16 Approximation by 2:1 method for the calculation of the induced vertical stress at depth due to an applied load Q.

2.5.3 Trapezoidal approximation (2:1 method)

The 2:1 approximation is a simple method for determining the change in vertical stress with depth. This method assumes that the stress dissipates with depth in the form of a trapezoid that has 2:1 (vertical:horizontal) inclined sides, as illustrated in Figure 2.16. Jeffs and Tew [15] indicated that the load spread method gives an average value of vertical stress at any given horizontal plane within the loaded area below the sleeper. With a rectangular sleeper, the average sleeper-ballast contact pressure σ_{max} for a third of its total length is first converted into a total concentric vertical load Q on the sleeper. The vertical stress at the equivalent depth beneath the sleeper would then be:

$$\sigma_{z\,max} = \frac{Q}{(b + \tilde{h})\left(\frac{l}{3} + \tilde{h}\right)} \tag{2.41}$$

In the above,

$$Q = \sigma_{max} \cdot A_{sb} \tag{2.42}$$

where, $l =$ sleeper length (m).

2.5.4 Arema recommendations

In the design practice for North American railroads, the AREMA Engineering Manual [38] recommends four equations for determining the pressure applied to the subgrade

Table 2.5 AREMA engineering manual equations [38].

Method	Equation
Talbot equation	$\sigma_{z\,max} = \dfrac{16.8\sigma_{max}}{\tilde{h}^{1.25}}$
Japanese National Railways equation	$\sigma_{z\,max} = \dfrac{50\sigma_{max}}{10 + \tilde{h}^{1.35}}$
Boussinesq equation	$\sigma_{z\,max} = \dfrac{6P_{stat}}{2\pi\tilde{h}^2}$
Love equation	$\sigma_{z\,max} = \sigma_{max}\left[1 - \left(\dfrac{1}{1 + (r/\tilde{h})^2}\right)^{3/2}\right]$
	$\sigma_{max} = \dfrac{2P_{stat}}{A_{sb}}(FS)$

by the ballast. It should be emphasized however, that these equations disregard the effect of the subballast layer on the load transfer mechanism to the subgrade surface. The AREMA manual specifies a minimum ballast and subballast thickness of 305 mm and 150 mm respectively. The recommended equations are listed in Table 2.5. In these equations,

$\tilde{h}$ = equivalent thickness in inches except for the Japanese National Railways which is in centimeters,

$\sigma_{z\,max}$ = subgrade stress,

σ_{max} = sleeper-ballast contact stress,

P_{stat} = static rail seat load (lb),

FS = factor of safety, and

r = radius of a circle whose area equals the sleeper bearing area A_{sb} (in).

The static rail seat load is different from the static wheel load. Atalar et al. [18] reported that part of the wheel load is transmitted to the adjacent sleepers and 40% to 60% of the wheel load is resisted directly beneath the wheel. An assumption of 50% resisted wheel load is believed to be reasonable.

The Talbot empirical formula was developed from a number of full scale laboratory tests performed at the University of Illinois [39]. Several different types of ballast were tested, including sand, slag, crushed stone, and gravel, with stresses from applied static loads measured at various depths and locations under several sleepers. The Japanese National Railways equation, on the other hand, was developed for narrow gauge tracks.

The Boussinesq and Love equations were both based on the theory of elasticity. The Boussinesq solution assumed that the rail seat load is a point load on the surface of the substructure that forms a semi-infinite, elastic, and homogeneous mass [40].

The Love formula, meanwhile, was an extension of the Boussinesq results in which the load supplied by the sleeper to the ballast was represented as a uniform pressure over a circular area equal to the sleeper bearing area.

Li and Selig [16] identified the following limitation of the methods described above:

- oversimplification of the actual situation for tracks under heavier axle loads and higher train speeds;
- not reflecting the effect of repeated dynamic loads on subgrade conditions;
- not considering the granular layer properties, and
- assumption of a homogeneous half space that represents ballast, subballast and subgrade layers without considering the properties of individual layers.

Yet these methods provide simple, easy to use solutions instead of the complex, tedious, multilayer theories or finite element techniques. The vertical stress distribution in the subgrade becomes practically uniform when the thickness of construction is greater than 600 mm. Sleepers spaced from 630 to 790 mm apart had a negligible influence on the vertical stress level in the subgrade for a unit load applied to the sleeper.

A simplified example of calculating subgrade stress using different methods is given for following loading and track data:

Velocity of train $= V = 110\,\text{km/h}$
Effective wheel load $= Q = 175\,\text{kN}$
Diameter of wheel $= D_w = 0.97\,\text{m}$
Sleeper spacing $= a = 0.495\,\text{m}$
Sleeper length $= l = 2.5\,\text{m}$
Sleeper width $= b = 0.26\,\text{m}$
Depth of ballast $= h_b = 0.38\,\text{m}$
Depth of subballast $= h_{sb} = 0.15\,\text{m}$
Young's modulus of elasticity of ballast $= E_b = 310\,\text{MPa}$
Young's modulus of elasticity of subballast $= E_{sb} = 125\,\text{MPa}$
Young's modulus of elasticity of subgrade $= E_{su} = 55\,\text{MPa}$
Poisson's ratio of ballast $= \mu_b = 0.3$
Poisson's ratio of subballast $= \mu_{sb} = 0.35$
Poisson's ratio of subgrade $= \mu_{su} = 0.45$

The track can be assumed to be good with usual levelling defects without depressions.

(a) Equivalent depth calculation:
Odemark method (Equation 2.31) $= \tilde{h} = 0.84\,\text{m}$.

(b) Impact factor calculation:
AREA Method (Equation 2.8) $= \phi = 1.059$
ORE Method (Figure 2.6) $= \phi = 1.32$
Atalar Method (Equation 2.14) $= \phi = 2.73$
Eisenmann Method (Equation 2.20) $= \phi = 1.27$

Table 2.6 Calculation of $\sigma_{z\,max}$.

i	x_i [m]	$\eta(x_i)$ Eq. 2.34	σ [kPa] Eq. 2.32	α_1 Eq. 2.38	α_2 Eq. 2.39	$f(x_i)$ Eq. 2.37	σ_{zi} [kPa] Eq. 2.36
1	−1.0	0.66	101.5	0.928	0.798	0.035	3.6
2	−0.5	0.89	136.6	0.641	0.411	0.109	14.9
3	0.0	1.00	153.8	0.154	−0.154	0.194	29.9
4	0.5	0.89	136.6	0.641	0.411	0.109	14.9
5	1.0	0.66	101.5	0.928	0.798	0.035	3.6
					$\Sigma\sigma_{zi} =$	$\sigma_{z\,max} =$	66.8 kPa

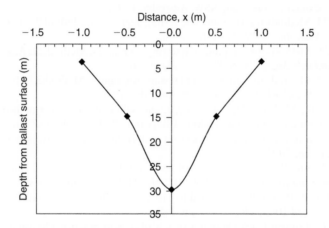

Figure 2.17 Variation of stress with depth from ballast surface.

Table 2.7 Summary of Induced subgrade stress obtained using different methods.

Stress calculation method	Impact factor calculation method	Induced subgrade stress (kPa)
Zimmermann	Eisenmann	85.0
Zimmermann	ORE	88.2
2:1 Approximation	Eisenmann	97.0
2:1 Approximation	ORE	100.7
Talbot's	–	85.8
Japanese	–	49.9
Boussinesq's	–	237.9
Love's ($r = 0.263$ m, $FS = 2$)	–	423.4

(c)   Stress calculation (Zimmermann Method):
Contact area between sleeper and ballast bed for a third of the sleeper $=$ $A_{sb} = lb/3 = 0.22$ m². By substituting the required values in Equation (2.33), $\sigma_{max} = 154$ kPa is obtained. Further, $\sigma_{z\,max}$ can be calculated as shown in Table 2.6 and the results obtained are also plotted in Figure 2.17.

Summary of subgrade stress obtained using different methods is given in Table 2.7.

REFERENCES

1. Esveld, C.: Modern Railway Track. MRT-Productions, *The Netherlands*, 2001.
2. Selig, E.T. and Waters, J.M.: Track Technology and *Substructure Management. Thomas Telford, London*, 1994.
3. Key, A.J.: Behaviour of Two Layer Railway Track Ballast under Cyclic and Monotonic Loading. PhD Thesis, University of Shefield, UK, 1998.
4. Jeffs, T.: Towards ballast life cycle costing. *Proc. 4th International Heavy Haul Railway Conference*, Brisbane, 1989, pp. 439–445.
5. AS 2758.7: Aggregates and rock for engineering purposes, Part 7: *Railway ballast. Standards Australia*, NSW, Australia, 1996.
6. AS 1141.22: Methods for sampling and testing aggregates, Method 22: Wet/dry strength variation. *Standards Australia*, NSW, Australia, 1996.
7. AS 1141.23: Methods for sampling and testing aggregates, Method 23: Los Angeles value. *Standards Australia*, NSW, Australia, 1996.
8. AS 1141.21: Methods for sampling and testing aggregates, Method 21: Aggregate crushing value. *Standards Australia*, NSW, Australia, 1996.
9. AS 1141.27: Methods for sampling and testing aggregates, Method 27: Resistance to wear by attrition. *Standards Australia*, NSW, Australia, 1996.
10. Indraratna, B., Khabbaz, H., Lackenby, J. and Salim, W.: Engineering behaviour of railway ballast – a critical review, Technical Report 1, Rail-CRC Project No. 6, Cooperative Research Centre for Railway Engineering and Technologies, University of Wollongong, NSW, Australia, 2002.
11. Gaskin, P.N. and Raymond, G.: Contribution to selection of railroad ballast. *Transportation Engineering Journal*, ASCE, Vol. 102, No. TE2, 1976, pp. 377–394.
12. Raymond, G.P.: Research on railroad ballast specification and evaluation. *Transportation Research Record* 1006, TRB, 1985, pp. 1–8.
13. Doyle, N.F.: Railway Track Design: A review of current practice. Occasional paper no. 35, *Bureau of Transport Economics*, Commonwealth of Australia, Canberra, 1980.
14. Fender, K.: TGV: High Speed Hero. *Trains Magazine*, August 2010, Kalmbach, Vol. 70, No. 8.
15. Jeffs, T. and Tew, G.P.: A review of track design procedures, Vol. 2, Sleepers and Ballast, *Railways of Australia*, 1991.
16. Li, D. and Selig, E.T.: Method for railroad track foundation design, I: Development. *Journal of Geotechnical and Geoenvironmental Engineering*, ASCE, Vol. 124. No. 4, 1998, pp. 316–322.
17. Office of Research and Experiments (ORE): Stresses in Rails, Question D71, Stresses in the rails, the ballast and the formation resulting from traffic loads. Report No. D71/RP1/E, Int. Union of Railways, Utrecht, Netherlands, 1965.
18. Atalar, C., Das, B.M., Shin, E.C. and Kim, D.H.: Settlement of geogrid-reinforced railroad bed due to cyclic load. *Proc. 15th Int. Conf. on Soil Mech. and Geotech. Engg.*, Istanbul, Vol. 3, 2001, pp. 2045–2048.
19. Eisenmann, J.: Germans gain a better understanding of track structure, *Railway Gazette International*, Vol. 128, No. 8, 1972, pp. 305.
20. Office of Research and Experiments (ORE): Summary Report, Question D71, Stresses in the rails, the ballast and the formation resulting from traffic loads. Report No. D71/RP1/E, Int. Union of Railways, Utrecht, Netherlands, 1970.
21. Jenkins, H.M., Stephenson, J.E., Clayton, G.A., Morland, J.W. and Lyon, D.: The effect of track and vehicle parameters on wheel/rail vertical dynamic forces. *Railway Engineering Journal*, Vol. 3, 1974, pp. 2–16.

22. Indraratna, B., Nimbalkar, S., Christie, D, Rujikiatkamjorn, C. and Vinod, J.S.: Field assessment of the performance of a ballasted rail track with and without geosynthetics. Journal of *Geotechnical and Geoenvironmental Engineering*, ASCE, Vol. 136, No. 7, 2010, pp. 907–917.
23. Anastasopoulos, I., Alfi, S., Gazetas, G., Bruni, S. and Leuven, A.V.: Numerical and experimental assessment of advanced concepts to reduce noise and vibration on urban railway turnouts. *Journal of Transportation Engineering*, ASCE, Vol. 135, No. 5, 2009, pp. 279–287.
24. Nielsen, J.C.O. and Johansson, A.: Out of round railway wheels – a literature survey. *Proc. Instn. Mech. Engrs.*, Part F: J. Rail and Rapid Transit, Vol. 214, No. F2, 2000, pp. 79–91.
25. Andersson, C. and Dahlberg, T.: Wheel/rail impacts at a railway turnout crossing. *Proceedings of Institution of Mechanical Engineers*, Vol. 212, Part F, 1998, pp. 123–134.
26. Li, D. and Davis, D.: Transition of railroad bridge approaches. *Journal of Geotechnical and Geoenvironmental Engineering*, ASCE, Vol. 131, No. 11, 2005, pp. 1392–1398.
27. Dukkipati, R.V. and Dong, R.: Impact loads due to wheel flats and shells" *Vehicle System Dynamics*, Vol. 31, 1999, pp. 1–22.
28. Frederick, C.O. and D.J. Round. Vertical Track Loading, *Track Technology*, Thomas Telford Ltd, 1985, London.
29. British Rail Safety and Standards Board (1993), GM/TT0088 Permissible track forces for railway vehicles. Issue 1, Revision A, *Rail Safety and Standards Board*, London.
30. British Rail Safety and Standards Board (1995), GM/RC2513 Commentary on Permissible Track Forces for Railway Vehicles. Issue 1, *Rail Safety and Standards Board*, London:
31. Australasian Railway Association (2003), Volume 4: Track, Civil and Electrical Infrastructure, Part 1: Identification and Classification of Wheel defects, Code of Practice for the Defined Interstate Rail Network, *Australasian Railway Association*.
32. QR (2001), STD/0026/TEC Rollingstock Dynamic Performance, *Safety Management System*, Version: 2, QR.
33. Rail Infrastructure Corporation (2002), RSU120 General Interface Requirements, Version: 2.0, *Rail Infrastructure Corporation* (Rail Corp).
34. Shenton, M.J.: Deformation of railway ballast under repeated loading conditions. In: Kerr (ed.): Railroad Track Mechanics and Technology. *Proc. of a symposium held at Princeton Univ.*, 1975, pp. 387–404.
35. Odemark, N.: Undersokning av elasticitetegenskaperna hos olika jordarter samt teori for berakning av belagningar eligt elasticitesteorin', Statens Vaginstitute, meddelande, Vol. 77, 1949, Stockholm, Sweden.
36. Ullidtz, P.: Modelling Flexible Pavement Response and Performance, Narayana Press, Odder, Denmark, 1998.
37. Ebersohn, W. and Selig, E.T.: *Introduction to Multi Disciplinary Concepts in Railway Engineering*, lecture, Chair in Railway Engineering, University of Pretoria, 1994.
38. AREMA, Practical Guide to Railway Engineering, American Railway Engineering and Maintenance-of-way Association, Simmons-Boardman Publishing Corporation, Maryland, 2003.
39. AREA, First Progress Report, American Railway Engineering Association Bulletin, AREA-ASCE Special Committee on Stresses in Railroad Track, 1918, Vol. 19, No. 205.
40. Poulos, H.G. and Davis, E.H.: *Elastic Solutions for Soil and Rock Mechanics*, John Wiley and Sons, New York, 1974.

Chapter 3

Factors Governing Ballast Behaviour

In general, the mechanical response of ballast is governed by four main factors: (a) characteristics of constituting particles, (e.g. size, shape, surface roughness, particle crushing strength, resistance to attrition etc.), (b) bulk properties of the granular assembly including particle size distribution, void ratio or density and degree of saturation, (c) loading characteristics including current state of stress, previous stress history and applied stress path, and (d) particle degradation, which is a combined effect of grain properties, aggregate characteristics and loading. These factors are discussed in the following Sections.

3.1 PARTICLE CHARACTERISTICS

The physical and mechanical characteristics of individual particles significantly influence the behaviour of ballast under both static and cyclic loading. In the following Sections, various characteristics of individual ballast grains and their influence on the mechanical behaviour of ballast are discussed.

3.1.1 Particle size

Typically, the size of ballast grains varies in the range of 10–60 mm. Due to transportation, handling, placement and compaction of ballast, as well as movement of heavy construction machines over the ballast layer, inevitable changes occur in their asperities. While sharp angular projections are the first to break, some particles may split into halves or even crush into several small pieces. With an increase in the number of train cycles, the ballast particles are further degraded and gradually decrease in size, but even after these changes, more than 90% of ballast grains still remain in the original range of 10–60 mm even after several million loading cycles.

Several researchers have studied the effects of particle size on the mechanical behaviour of ballast and other coarse aggregates, but there are some contradictions amongst their findings. Kolbuszewski and Frederick [1] indicated that the angle of shearing resistance increases with larger particle sizes. They concluded that increasing particle size increases the dilatancy component of the angle of shearing resistance. In contrast, Marachi et al. [2] presented experimental evidence to show and prove that the angle of internal friction decreases with an increase in maximum particle size (Fig. 3.1). Indraratna et al. [3] observed similar findings in their studies and indicated that the

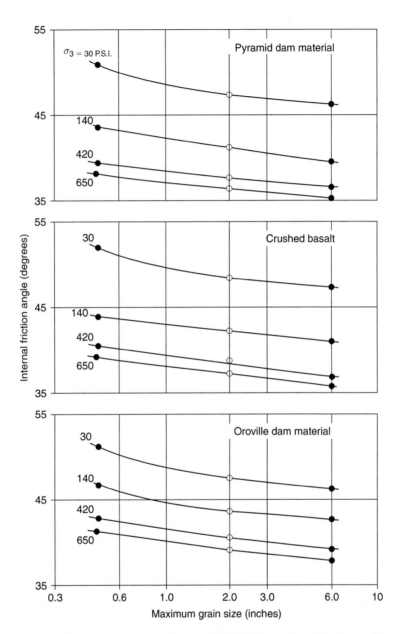

Figure 3.1 Effect of particle size on friction angle (modified after Marachi et al., [2]).

peak friction angle decreased slightly with an increase in grain size at low confining pressure (<300 kPa). They concluded that at high stress levels (>400 kPa), the effect of particle size on friction angle is negligible.

Raymond and Diyaljee [4] observed that larger size ballast with a uniform grading generated higher plastic strains than small-sized uniform ballast. Although smaller

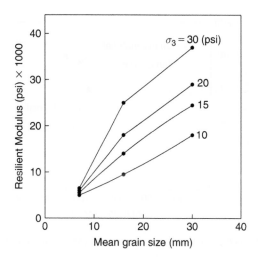

Figure 3.2 Effect of grain size on Resilient Modulus of Ballast (data from Janardhanam and Desai, [5]).

aggregates showed less deformation (i.e. higher resistance) under smaller cyclic loads (amplitudes), those specimens failed immediately after increasing the load amplitude from 140 kPa to 210 kPa. In contrast, larger ballast continued to resist cyclic loading without any sign of failure even after increasing the load amplitude from 140 kPa to 210 kPa. Raymond and Diyaljee concluded that smaller ballast deforms less if the stress level does not exceed a critical value. However, smaller ballast has a lower final compacted strength than larger ballast.

In an attempt to investigate the influence of particle size on ballast behaviour, Janardhanam and Desai [5] conducted a series of true triaxial tests under cyclic loading. They indicated that particle size does not appear to significantly influence ballast strains at various stress levels. They also concluded that volumetric strain is not affected by particle size, but grain size has a significant effect on the resilient modulus of ballast. The modulus increases with the mean grain size at all levels of confinement, and at low confining pressure the relationship is almost linear with the mean grain size (Fig. 3.2). In contrast, Indraratna et al. [3] presented experimental evidence based on monotonic triaxial tests that larger ballast has a smaller deformation modulus and Poisson's ratio compared to smaller aggregates.

Considering the advantages and disadvantages of varying particle size, Selig [6] recommended that ideal ballast should be in the range of 10–50 mm with only a few particles beyond this range. The larger particles stabilise the track and the smaller particles reduce the contact forces between particles and minimise breakage.

3.1.2 Particle shape

Unlike particle size, there is some consensus amongst researchers regarding the effects of grain shape on the mechanical response of ballast and other coarse aggregates. In general, angularity increases frictional interlock between grains which increases

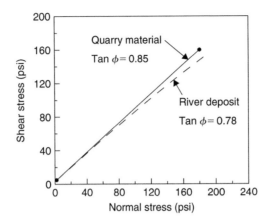

Figure 3.3 Influence of particle shape on strength (data from Holz and Gibbs, [7]).

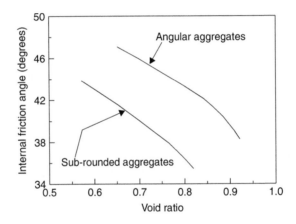

Figure 3.4 Effect of particle shape on friction angle (data from Vallerga et al., [9]).

the shear strength [3, 7, 8]. Holz and Gibbs [7] concluded that the shear strength of highly angular quarried materials is higher than that of relatively sub-angular, or sub-rounded river gravels (Fig. 3.3). Vallerga et al. [9] provided clear evidence that the angle of internal friction is remarkably high for angular aggregates compared to sub-rounded aggregates (Fig. 3.4), while others concluded that the angle of internal friction depends mainly on grain angularity [1, 8]. Jeffs and Marich [10] and Jeffs [11] demonstrated that angular aggregates give less settlement than rounded aggregates. Chrismer [12] indicated that as grain angularity increases, further dilation is required for particle movement which increases the shearing resistance.

Jeffs and Tew [13] reported that the shape of ballast grains depends on the production process and the nature of deposits. Raymond [14] indicated that most specifications restricted the percentage of flaky particles whose aspect ratio exceeds 3, and excluded particles exceeding an aspect ratio of 10. It is thought that because these long but very thin particles can align and form planes of weakness in both vertical and lateral directions, they cannot be used as ballast. The disadvantages of increased flakiness appear to be increased abrasion and breakage, increased permanent strain accumulation under repeated load and decreased stiffness [15]. Most specifications also limit the percentage of misshapen particles, where the term 'misshapen particles' means flat or elongated grains. However, there is uncertainty regarding the allowable percentage of misshapen particles [13]. Raymond [14] stated that cuboidal is the best shape for high quality ballast, an opinion also supported by Jeffs and Tew [13].

3.1.3 Surface roughness

Surface roughness or texture is considered to be one of the key factors that govern the angle of internal friction, hence the strength and stability of ballast. Each grain has the same "roughness" on its surface. The phenomenon 'friction and frictional force' is based on the roughness of the loaded surface, while the shear resistance of ballast and other aggregates depends on the ability of these frictional forces to develop. Raymond [14] concluded that particle shape and surface roughness are of utmost importance and have long been recognised as the major factors influencing track stability. Canadian Pacific Rail preferred surface roughness over particle shape as the key parameter for track stability, and had stringent controls on grain surface rather than direct restrictions on particle shape [14]. Thom and Brown [16, 17] reported an increase in resilient modulus with increasing surface friction of grains, and concluded that the resistance to plastic strain accumulation increases with increasing apparent surface roughness.

Most ballast specifications stipulate crushed or fractured particles, which are defined as grains having a minimum of three crushed faces (i.e. freshly exposed surfaces with a minimum of one third of the maximum particle dimension). These specifications ensure minimum surface roughness of ballast particles, and assume that freshly exposed surfaces have a higher roughness compared to previously exposed surfaces which have been smoothened by mechanical attrition and weathering.

Due to internal attrition of grains under cyclic loading, surface roughness of ballast deteriorates with time (i.e. an increasing number of train passages). Internal attrition also produces fines and is a source of ballast fouling. This reduction in surface roughness by internal attrition and breakage of sharp corners after several million load cycles causes the angle of internal friction and the shear strength of recycled ballast to decrease considerably. Therefore, it is conceivable that the surface roughness of individual particles significantly affects the mechanical behaviour of ballast and ultimately, track stability.

3.1.4 Parent rock strength

The strength of parent rock is probably the most important factor directly governing ballast degradation, and indirectly, settlement and lateral deformation of the track. Parent rock contributes to both compressive and tensile strength. Under the same loading and boundary conditions, weak particles produce more grain breakage and plastic settlement than stronger particles. Although the strength of the parent rock is not usually tested nor required by most ballast specifications (e.g., TS 3402 of

Rail Infrastructure Corporation, NSW), a higher parent rock strength is implied by the selection criteria, which includes petrological examination. High rock strength is also indirectly reflected by other tests such as 'Aggregate crushing value', 'Los Angeles Abrasion value' and 'Wet attrition value'. These test results collectively indicate the durability of ballast and the strength of the parent rock. However, to enhance the quality of ballast during selection, the parent rock strength may also be included in the specifications.

3.1.5 Particle crushing strength

Individual particle crushing strength is an important factor governing particle degradation, including grain splitting and breakage of sharp corners under loading. Particle fracture plays a vital role in the behaviour of crushable aggregates [19]. Particle crushing strength primarily depends upon the strength of the parent rock, grain geometry, the loading point and loading direction. Fracture in rock grains is initiated by tensile failure. The fracture strength can be measured indirectly by diametral compression between flat platens [20]. For a particle of diameter d under diametral compressive force F, the characteristic tensile stress (σ) is given by Jaeger [20] by Equation 3.1.

$$\sigma = \frac{F}{d^2} \tag{3.1}$$

It is relevant to mention here that Equation 3.1 is consistent with the definition for the tensile strength of concrete in the Brazilian test, where a concrete cylinder is compressed diametrically and then split by induced tensile stress. Following Equation 3.1, Mcdowell and Bolton [19] and Nakata et al. [21] described the characteristic particle tensile strength (σ_f), as given by:

$$\sigma_f = \frac{F_f}{d^2} \tag{3.2}$$

where, the subscript f denotes failure.

Festag and Katzenbach [22] categorised grain crushing into particle breakage (fracture) and grain abrasion. Particle breakage is the dissection of grains into parts with nearly the same dimension, a feature that generally occurs under high stress levels. On the other hand, abrasion is a phenomenon where very small particles disintegrate from the grain surface, and this is independent of the stress level. Abrasion takes place in granular materials when particles slip or roll over each other during shear deformation which can occur even at low stress levels. Grain breakage may be absent if the stress level is low compared to particle strength, however, grain abrasion will continue at any stress level. Although the crushing strength of particles is not required by most ballast specifications, it is reflected in the 'Aggregate crushing value' and other standard durability tests.

3.1.6 Resistance to attrition and weathering

The properties of individual grains also govern ballast degradation under traffic loading and environmental changes. Usually, ballast particles are not individually assessed for their capacity to resist attrition and weathering, rather, their resistance is collectively assessed for the aggregate mass. Several standard test methods for quantifying the resistance of ballast against attrition and weathering are available and are used by different railway organisations. These tests include Los Angeles Abrasion (LAA) test, mill abrasion (MA), the Deval test and Sulphate Soundness test etc. [15]. The Los Angeles Abrasion test, the mill abrasion (MA) and Deval tests are commonly used in North America and Europe to measure the attrition resistance of ballast. The Sulphate Soundness test is primarily used to examine the resistance to chemical action of Sodium Sulphate and Magnesium Sulphate (salt). High resistance to attrition and weathering is ensured by specifying certain values in ballast standards and specifications, as shown in Table 2.3 earlier for durability.

3.2 AGGREGATE CHARACTERISTICS

The overall characteristics of the granular mass that govern ballast behaviour include particle size distribution (PSD), void ratio (or density) and the degree of saturation. These characteristics are discussed in the following Sections.

3.2.1 Particle size distribution

The distribution of particle sizes (i.e. gradation) has an obvious and significant influence on track deformation behaviour [13]. Several researchers have studied the effects of particle gradation on the strength and deformation aspects of aggregates. Thom and Brown [16] conducted a series of repeated load triaxial tests on crushed dolomite with similar maximum particle sizes, but varying the gradation from well-graded to uniform. Each grading curve was characterised by an exponent 'n' shown in Figure 3.5(a) where higher values of 'n' represent greater uniformity of particle sizes. According to their results (Figs. 3.5b–e) elastic shear stiffness (modulus) and permeability increase as the grading parameter 'n' increases. As expected, the density and friction angle decrease with the value of 'n'.

Thom and Brown [16] mentioned that optimum dry density was achieved at about $n = 0.3$ for all types of compaction efforts (i.e. heavily compacted, lightly compacted and uncompacted). They also noted that particle size distribution did not significantly influence the angle of internal friction for uncompacted specimens. One significant finding of their research was that uniform gradation provided a higher stiffness compared to well-graded aggregates. In contrast, Raymond and Diyaljee [4] demonstrated that well-graded ballast gives lower settlement compared to single sized ballast (Fig. 3.6). This is not surprising given the higher internal friction associated with well-graded aggregates.

It has been argued that single sized (uniform) ballast has larger void volume than broadly graded ballast [14]. As expected, well-graded or broadly-graded ballast is stronger due to its void ratio being smaller than uniform ballast [13, 14, 23]. However, ballast specifications generally demand uniformly graded aggregates to fulfil its drainage requirements. Since ballast is expected to be a coarse, free draining medium, the optimum gradation should ideally be between uniformly graded coarse aggregates that give almost instantaneous drainage and broadly graded

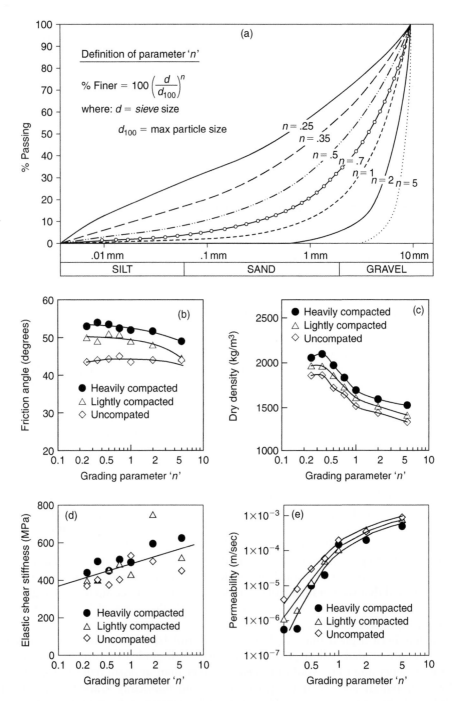

Figure 3.5 (a) Gradation of particles, and its effects on (b) friction angle, (c) density, (d) shear modulus and (e) permeability (inspired by Thom and Brown, [16]).

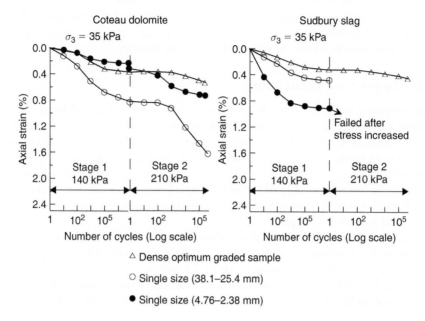

Figure 3.6 Effects of gradation on vertical strains of ballast under cyclic loading (data from Raymond and Diyaljee, [4]).

aggregates that provide higher strength and less settlement at the expense of reduced drainage. Nevertheless, optimum gradation should provide sufficient drainage capacity (hydraulic conductivity) along with sufficient initial density, shear strength, and resilient modulus.

3.2.2 Void ratio (or density)

Researchers have long recognised that the volume of voids in a porous medium (e.g. soil and rock aggregates) compared to the volume of solids (i.e. void ratio) significantly affects its mechanical behaviour [24–28]. It has been well established that aggregates having a lower initial void ratio (i.e. higher initial density) are stronger in shear and generate a smaller settlement than aggregates with a higher initial void ratio (i.e. lower initial density). In widely accepted Critical State Soil Mechanics (CSSM), the significance of void ratio (e) in the mechanical behaviour of soil has been recognised by considering it as a governing state variable along with two other stress invariants, namely, mean effective normal stress p', and deviatoric stress q [26, 27].

All researchers investigating track stability have concluded that an increase in ballast density (i.e. lower void ratio) enhances its strength and stability [29–31]. Selig and Waters [15] concluded that low-density ballast leads to high plastic strains. Indraratna et al. [3] indicated that the critical stage of ballast life is immediately after track construction or maintenance when ballast is in its loosest state (i.e. highest void ratio).

Track stability can be significantly improved by increasing the bulk density of the ballast bed by further compaction or by using broadly-graded aggregates. However, a higher compaction effort also increases the risk of particle breakage and a well-graded ballast contributes to a reduction in drainage characteristics.

3.2.3 Degree of saturation

Ballast response to external mechanical forces is adversely affected by an increased degree of saturation. Water influences track settlement and particle breakage and also leads to trafficability problems. In saturated conditions, subgrade soils soften and mix with water to form a slurry, which under cyclic traffic loading can be pumped up to the ballast layer, as mentioned earlier. Clay pumping is one of the major causes of ballast contamination [15, 32]. Sowers et al. [33] explained that water entering micro-fissures at the contact points between particles increases local stress and leads to increased particle breakage.

Indaratna et al. [28] conducted one-dimensional compression tests to investigate the effects of saturation on the deformation and degradation of ballast. They observed a sudden increase in ballast settlement by about 2.6 mm due to sudden flooding (Fig. 3.7), and reported a further increase in settlement with time (creep) under saturated conditions. They concluded that saturation increased settlement by about 40% of that of dry ballast.

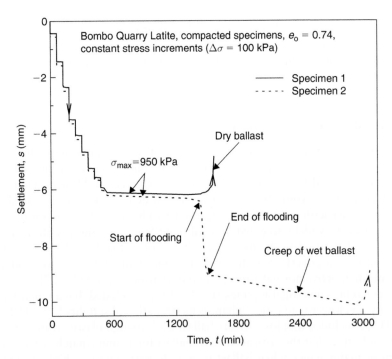

Figure 3.7 Effect of saturation on ballast settlement (modified after Indraratna et al., [28]).

3.3 LOADING CHARACTERISTICS

The deformation and degradation behaviour of ballast is profoundly dependent on the external loading characteristics. The magnitude of confining pressure, previous load history, current state of stress, number of load cycles, loading frequency and amplitudes are among the key parameters that govern track deformation. The effects of these loading variables are discussed in the following Sections.

3.3.1 Confining pressure

Researchers and engineers have recognised the significant effects of confining pressure on the strength and deformation behaviour of soils and granular materials from the earliest days of soil mechanics [24, 25, 34, 35]. Marsal [23] was one of the pioneers who closely studied the effect of confining pressure on the deformation behaviour and particle breakage of rockfills. He tested basalt and granitic gneiss aggregates under high confining pressures (5–25 kg/cm^2), and observed that the shear strength is not a linear function of acting normal pressure. Charles and Watts [36] and Indraratna et al. [37] also reported a pronounced non-linearity of failure envelope for coarse granular aggregates at low confining pressure (Fig. 3.8). Vesic and Clough [35] studied the shear behaviour of sand under low to high pressures and concluded that a mean normal stress exists beyond which the curvature of the strength envelope vanishes and the shear strength is not affected by the initial void ratio. They called it 'breakdown stress' (σ_B), because it represents the stress level at which all dilatancy effects disappear and beyond which particle breakage becomes the only mechanism, in addition to simple slip, by which shear deformation takes place.

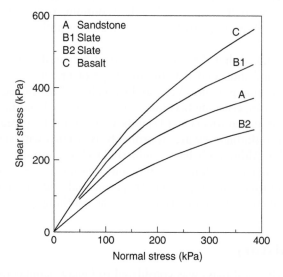

Figure 3.8 Non-linear strength envelop at low confining pressures (data from Charles and Watts, [36]).

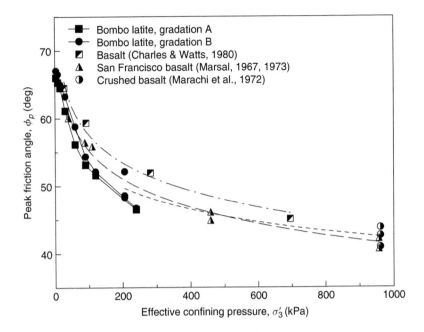

Figure 3.9 Influence of confining pressure on friction angle (modified after Indraratna et al., [3]).

Well documented studies indicate that the angle of internal friction of granular mass decreases with increasing confining pressure [2, 8, 36, 37]. Indraratna et al. [3] presented laboratory experimental results of railway ballast (latite basalt), which revealed that as confining pressure increases from 1 kPa to 240 kPa, the drained friction angle of ballast decreases from about 67° to about 46° (Fig. 3.9). They concluded that the high values of apparent friction angle at low confining pressures are related to low contact forces well below grain crushing strength and the ability of aggregates to dilate at low stress levels.

Marsal [23] noticed that the shearing of rockfill caused a significant amount of particle breakage and indicated that the breakage of granitic gneiss increased with the increase in confining pressure. Vesic and Clough [35] concluded that as the mean normal stress increases, crushing becomes more pronounced and the dilatancy effects gradually disappear. Indraratna et al. [37] indicated that the large reduction in the friction angle at high confining pressures is probably associated with significant crushing of angular particles. Although ballast is subjected to low confinement in track, it also suffers particle breakage, crushing, attrition and wearing under cyclic traffic loading [11, 15, 31]. Indraratna et al. [3] presented experimental evidence that the breakage of latite ballast may increase by about 10 times as the confining pressure increases from 1 kPa to 240 kPa.

3.3.2 Load history

Until the late 1950's, soil mass was considered to behave similar to perfectly plastic solids. Drucker et al. [34] were probably the first, among a few others, who considered soils as work-hardening plastic materials. With their work-hardening theories, they explained the volume change behaviour of clays during loading, unloading and reloading in a consolidation test, and proposed possible yield surfaces for consolidation [34]. Since publishing their concepts and explanations, soil was considered to be a work-hardening plastic material and the researchers have acknowledged the influence of previous load history on the deformation behaviour of soils.

Diyaljee [38] conducted a series of laboratory cyclic tests to investigate the effects of stress history on ballast behaviour. In each test, he applied various cyclic deviatoric stresses (70–315 kPa) in several stages (10,000 cycles each) on identical ballast specimens (same gradation, density and confinement). He found that 2 specimens (T3 and T4, Fig. 3.10a) in stage 2 loading (140 kPa) deformed almost the same as the specimens T5 and T6 in stage 1 with the same load (140 kPa) without any previous stress history, where specimens T3 and T4 had a previous stress history of 70 kPa cyclic loading in stage 1. Stage 1 loading is 50% of stage 2 loading and has an almost negligible influence on the accumulated plastic deformation occurring during stage 2 loading. In contrast, specimens T4 and T9 (Fig. 3.10b) with a maximum load history of 210 kPa, showed a very small increase in plastic strain at 245 kPa cyclic stress compared to specimen T13 at the same loading without any previous load history.

Diyaljee [38] concluded that a previous stress history of more than 50% of the currently applied cyclic deviator stress, significantly decreases the plastic strain accumulation in ballast. However, a previous stress history of less than 50% of the currently applied cyclic deviator stress does not contribute to plastic strain accumulation. His findings agree with the research previously carried out by the Office of Research and Experiments of the International Union of Railways [39].

3.3.3 Current stress state

The current state of stress also influences the deformation and degradation behaviour of ballast. The state of stress is defined by all nine components of stress tensor, σ_{ij}, where, $i = 1, 2, 3$; $j = 1, 2, 3$ [40]. However, due to the difficulties and complexities arising from dealing with these stress elements and their dependencies on axis rotation, invariants of the stress tensor are conventionally employed to describe the state of stress [40]. In soil mechanics, the state of stress and the failure criteria are usually defined by two stress invariants: the mean effective normal stress p', and the deviator stress q [25, 26].

Roscoe and co-researchers developed the first comprehensive stress-strain constitutive model for clay based on the plasticity theory and the critical states, i.e. Cam-clay [25, 26, 41]. They showed that the plastic strain increment depends on the state of stress and other factors. As the state of stress and another state variable (void ratio) of a soil element moves towards the critical state, the rate of plastic shear strain corresponding to any load increment becomes higher. At the critical state, the shear strain continues to increase at a constant stress and constant volume, according to the above theories.

Poorooshasb et al. [42] studied the yielding of sand under triaxial compression and showed that the slope of the plastic strain increment increases from a small value (or zero) to a very high value as the state of stress moves towards the failure envelope (Fig. 3.11). At a stress state close to the failure line, the high slope of the plastic strain

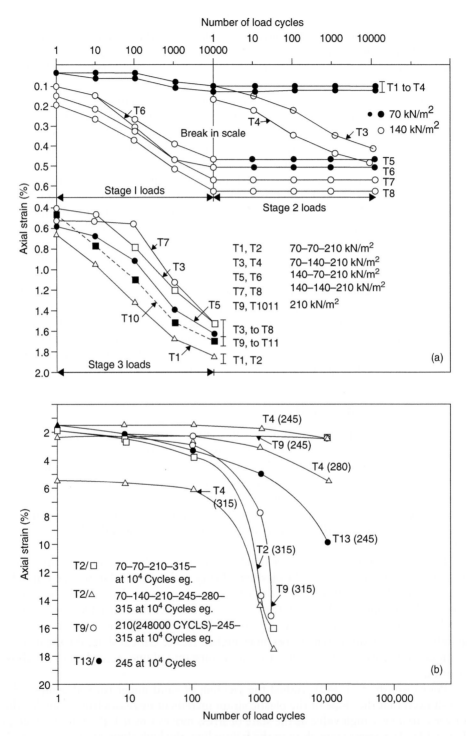

Figure 3.10 Effects of stress history on deformation of ballast under cyclic loading, (a) deviator stress up to 210 kPa, (b) cyclic stress above 210 kPa (inspired by Diyaljee, [38]).

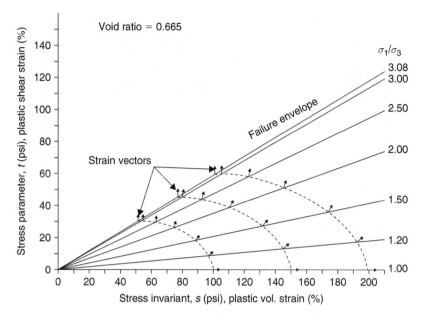

Figure 3.11 Effect of stress state on plastic strains (modified after Poorooshasb et al., [42]).

increment indicates that the plastic shear strain increment is much higher than the plastic volumetric strain increment. Other researchers also reported similar effects of stress state on the plastic deformation of soils and granular aggregates [43–45].

3.3.4 Number of load cycles

Railway engineers have recognised the influence of the number of load cycles on the accumulation of plastic deformation of ballast and other granular media. An increase in the number of load cycles generally increases the settlement and lateral deformation of granular particles, including ballast. However, the degree and rate of deformation at various load cycles are the salient aspects that have been studied by various researchers.

Shenton [46] reported that the track settlement immediately after tamping increased at a decreasing rate with the number of axles (Fig. 3.12a). He also indicated that the track settlement may be approximated by a linear relationship with the logarithm of load cycles (Fig. 3.12b). Raymond et al. [47] also demonstrated that both axial and volumetric strains of dolomitic ballast increased linearly with the logarithm of load cycles, irrespective of the loading amplitude (Fig. 3.13). Similar observations were also reported by others [15, 48]. In contrast, Raymond and Diyaljee [4] presented evidence, as shown in Figure 3.6 earlier, that the accumulated plastic strains of ballast may not be linearly related to the logarithm of load cycles for all ballast types, grading, and load magnitudes. Diyaljee [38] reported that the plastic strain of ballast also increased non-linearly with an increase in logarithm of load cycles at a higher cyclic deviator stress (see Fig. 3.10).

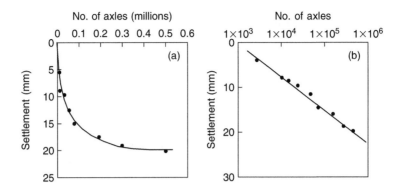

Figure 3.12 Settlement of track after tamping, (a) in plain scale, (b) in semi-logarithmic scale (data from Shenton, [46]).

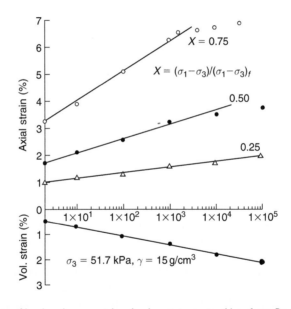

Figure 3.13 Effects of load cycles on axial and volumetric strains (data from Raymond et al., [47]).

Shenton [49] examined a wide range of track settlement data collected from different parts of the world and concluded that the linear relationship of track settlement with the logarithm of load cycles or total tonnage might be a reasonable approximation over a short period of time. However, this approximation can lead to a significant underestimation for a large number of axles (Fig. 3.14).

Jeffs and Marich [10] conducted a series of cyclic load tests on ballast and indicated a rapid increase in settlement initially, followed by a stabilised zone with a linear increase in settlement (Fig. 3.15). They also noticed a sudden increase in the rate

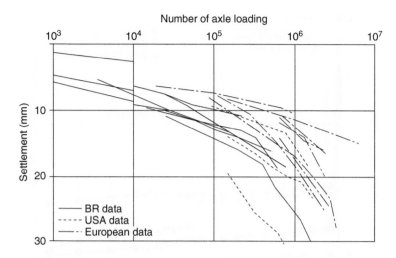

Figure 3.14 Settlement of Track at different parts of the world (modified after Shenton, [49]).

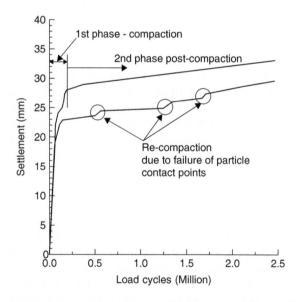

Figure 3.15 Settlement of ballast under cyclic load (data from Jeffs and Marich, [10]).

of settlement in the stabilised (post-compaction) zone, which they attributed to 're-compaction' of ballast. Jeffs and Marich attributed this to the failure of particle contact points within the ballast bed causing a sudden increase in settlement rate. The effect of re-compaction was noticed for about 100,000 load cycles after which the rate of settlement became almost constant.

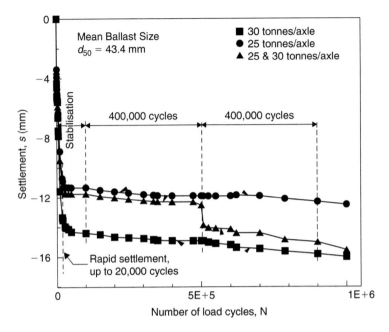

Figure 3.16 Settlement of ballast under cyclic loading (modified after Ionescu et al., [50]).

Ionescu et al. [50] conducted a series of true triaxial tests on latite ballast and concluded that the behaviour of ballast is highly non-linear under cyclic loading (Fig. 3.16). They also reported a rapid increase in initial settlement (similar to Jeffs and Marich, [10]) during the first 20,000 load cycles, followed by a consolidation stage up to about 100,000 cycles. Ionescu et al. [50] indicated that the ballast bed stabilised during this first 100,000 load cycles, after which settlement increased at a decreasing rate.

3.3.5 Frequency of loading

Because train speeds vary from place to place, it is important to study the influence of loading frequency on ballast behaviour. Shenton [46] carried out a series of cyclic loading tests, varying the frequency from 0.1 to 30 Hz, while maintaining other variables such as confining pressures and load amplitude constant. Based on the test results (Fig. 3.17), Shenton concluded that the frequency of loading does not significantly influence deformation behaviour of ballast. However, it was pointed out that these test findings should not be confused with track behaviour, where an elevated train speed increases the dynamic forces and imparts greater stresses on the ballast bed.

Kempfert and Hu [51] reported in-situ measurements of dynamic forces in track resulting from speeds up to 400 km/hour. They found that a speed of up to about 150 km/hour has an insignificant influence on the dynamic vertical stress (Fig. 3.18). These field measurements appear to be consistent with Shenton's laboratory findings described earlier. However, the measured data shows a linear increase in dynamic stress as the speed increases from 150 to about 300 km/hour. Beyond 300 km/hour and up

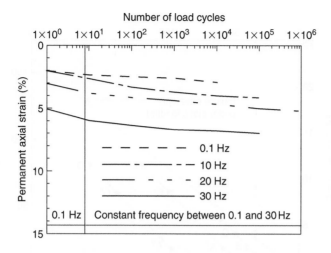

Figure 3.17 Effect of loading frequency on ballast strains (data from Shenton, [46]).

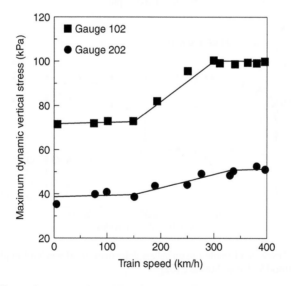

Figure 3.18 Effects of train speed on dynamic stresses (data from Kempfert and Hu, [51]).

to the maximum measured speed (400 km/hour), the effect of speed on dynamic stress becomes insignificant again.

3.3.6 Amplitude of loading

The amplitude of cyclic loading also plays a major role in ballast deformation. Stewart [52] carried out a series of cyclic triaxial tests varying the load amplitudes at every 1,000

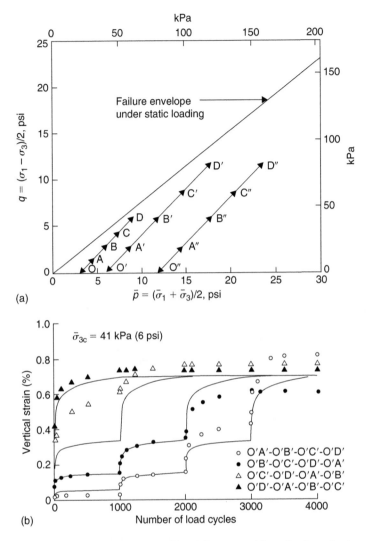

Figure 3.19 Effect of cyclic load amplitude on ballast deformation, (a) test load amplitude, and (b) ballast strain (modified after Stewart, [52]).

cycles to examine the role of load amplitude on ballast deformation. Figure 3.19(a) shows the test load amplitude, and Figure 3.19(b) shows the vertical strain of ballast against the number of load cycles. Stewart explained that the permanent strain in the first cycle increased significantly when the load amplitude was increased. It was noted that an increase in load amplitude beyond the maximum past stress level increased settlement immediately, apart from increasing the final (long term) cumulative strain. Diyaljee [38] and Ionescu et al. [50] reported similar findings in their laboratory investigations. In contrast, decreasing the load amplitude does not seem to

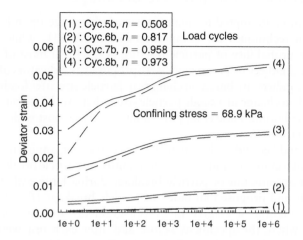

Figure 3.20 Effect of cyclic stress level on ballast strain (modified after Suiker, [53]).

contribute to the accumulated plastic strain [38, 52]. Stewart [52] further verified that the final cumulative strains obtained at the end of various staged, variable-amplitude loading tests (after 4,000 cycles), were independent of the order of applied stresses.

Recently, Suiker [53] studied the effects of load amplitude on ballast behaviour. He referred to the cyclic load amplitude in terms of the ratio between the cyclic stress ratio and the maximum static stress ratio [$n = (q/p)_{cyc}/(q/p)_{stat, max}$]. Suiker concluded that at low cyclic stress level ($n < 0.82$), the rate of plastic deformation of ballast is negligible (Fig. 3.20). In other words, the response of ballast below this cyclic stress level becomes almost elastic. This phenomenon was termed 'shakedown' and will be discussed later in more detail.

3.4 PARTICLE DEGRADATION

The most important geotechnical characteristics of granular materials such as the stress-strain behaviour and strength, volume change and pore pressure development, and variation in permeability depend on the integrity of the particles or the amount of particle crushing that occurs from stress change [54]. All granular aggregates subjected to stresses above normal geotechnical ranges exhibit considerable particle breakage [23, 24, 35, 55–61]. Some researchers indicate that particle breakage can even occur at low confining pressure [54, 60, 62]. The significance of particle degradation on the mechanical behaviour of granular aggregates has been recognised by various researchers [3, 23, 35, 59–63]. In the following Sections, the various methods for quantifying particle breakage, factors affecting particle breakage and the influence of particle breakage on the deformation behaviour of ballast and other granular aggregates are discussed.

3.4.1 Quantification of particle breakage

Several investigators attempted to quantify particle breakage upon loading and proposed their own techniques for computation [23, 57, 61], while others focused primarily on the probability of particle fracture [19, 64]. In most of these methods, different empirical indices or parameters were proposed as indicators of particle breakage. All breakage indices are based on changes in particle size after loading. While some indices are based on change in a single particle size, others are based on changes in overall grain-size distribution. Lade et al. [54] summarised the most widely used breakage indices for comparison.

Marsal [23] and Lee and Farhoomand [57] were the first, among others, who developed independent techniques and indices for quantifying particle breakage. Marsal [23] noticed a significant amount of particle breakage during large-scale triaxial tests on rockfill materials and proposed an index of particle breakage (B_g). Marsal's method involved the evaluation of change in overall grain-size distribution of aggregates after breakage, where the specimens before and after each test were sieved. From the recorded changes in particle gradation, the difference in percentage retained on each sieve size ($\Delta W_k = W_{ki} - W_{kf}$) is computed, where, W_{ki} represents the percentage retained on sieve size k before the test and W_{kf} is the percentage retained on the same sieve size after the test. He noticed that some of these differences were positive and some negative. Theoretically, the sum of all positive values of ΔW_k must be equal to the sum of all negative values. Marsal defined the breakage index B_g, as the sum of the positive values of ΔW_k, expressed as a percentage. The breakage index B_g, has a lower limit of zero indicating no particle breakage, and has a theoretical upper limit of unity (100%) representing all particles broken to sizes below the smallest sieve size used. This method implies that B_g could change if a different set of sieves was used. Therefore, the same set of sieves must be used for all ballast materials if comparisons are to be made with regard to breakage.

Lee and Farhoomand [57] measured the extent of particle breakage while investigating earth dam filter materials. They primarily investigated the effects of particle crushing on the plugging of dam filters and proposed a breakage indicator expressing the change in a single particle size (D_{15}), which is a key parameter in filter design. Later on, Hardin [61] defined two different quantities: the breakage potential B_p, and total breakage B_t, based on changes in grain-size distribution, and introduced the relative breakage index B_r ($=B_t/B_p$), as an indicator of particle degradation. Hardin's relative breakage B_r, has a lower limit of zero and an upper limit of unity. It is relevant to mention here that Hardin's method requires a planimeter or numerical integration technique for computing B_t and B_p. Lade et al. [54] compares the above 3 methods of particle breakage measurements in a graphical form, as shown in Figure 3.21.

Miura and O-hara [60] used the changes in grain surface area (ΔS) as an indicator of particle breakage. Their concept was based on the idea that new surfaces could be generated as the particles were broken, and therefore, the changes in surface area could be used as a measure of particle breakage. With their method, the specific surface area of each particle size (i.e. sieve size) is computed assuming that all grains are perfectly spherical. The sieving data before and after the test, along with the specific surface area are then used to calculate the change in surface area, ΔS. The parameter ΔS

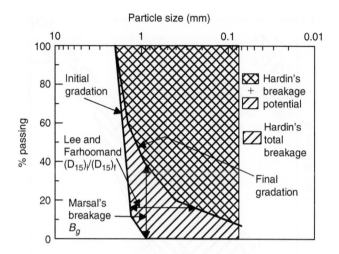

Figure 3.21 Various definitions of particle breakage (inspired by Lade et al., [54]).

has a lower limit of zero and has no theoretical upper limit, which often leads to criticism.

After considering the various methods of particle breakage quantification, Marsal's breakage index B_g, has been adopted in this study due to its simplicity in computation and ability to provide a perception about the degree of particle degradation as a numerical value.

Indraratna et al. [66] and Lackenby et al. [67] introduced a new Ballast Breakage Index (BBI) specifically for railway ballast to quantify the extent of degradation. The evaluation of BBI quantifies the change in the particle size distribution before and after testing (Fig. 3.22). By adopting a linear particle size axis, BBI can be determined from Equation 3.3, where the parameters A and B are defined in Figure 3.22.

$$BBI = \frac{A}{A+B} \tag{3.3}$$

3.4.2 Factors affecting particle breakage

Ballast degradation in general depends on many factors, including load amplitude, frequency, number of cycles, aggregate density, grain angularity, confining pressure and degree of saturation. However, the most significant factor governing ballast breakage is the fracture strength of its constituent particles [65]. Lee and Farhoomand [57] indicated that particle size, angularity, particle size distribution and magnitude of confining pressure influence particle degradation. They concluded that larger particle size, higher grain angularity and uniformity in gradation can increase the extent of particle crushing. Marsal [23] agreed with Lee and Farhoomand [57] with respect to breakage, and pointed out additional fundamental factors such as the average value

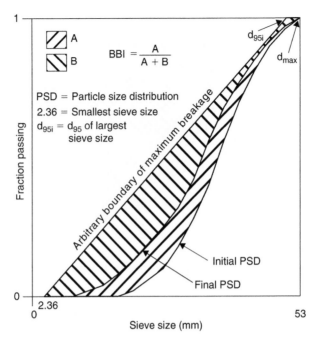

Figure 3.22 Ballast breakage index (BBI) determination (after Indraratna et al., [66], Lackenby et al., [67]).

of contact forces (stresses), strength of particles at contact points, and the number of contacts per particle. The presence of micro-fissures in crushed rocks from the blasting and crushing process is another reason for particle breakage.

Bishop [56] indicated that at high stress levels, particle breakage during shearing is considerably higher than during static consolidation. Lade et al. [54] pointed out that larger grains can contain more flaws or defects, thereby have a higher probability of disintegration. They also indicated that increasing mineral hardness decreases particle crushing. Smaller particles are generally created after fracturing along these defects. As fracturing continues, the subdivided particles contain fewer defects and are therefore, less prone to crushing. McDowell and Bolton [19] reported that the tensile strength of a single particle decreases as the particle size increases.

3.4.3 Effects of principal stress ratio on particle breakage

Particle degradation affects the behaviour of ballast as well as other granular aggregates in rockfill dams and filters. As mentioned earlier in Section 2.4.4, various investigators observed the change in particle sizes (particle degradation) due to change of stress. Some researchers only reported the amount of degradation in terms of breakage indices or factors. A number of others attempted to correlate the computed breakage indices with the strength, dilatancy, and friction angle. However, there is still significant

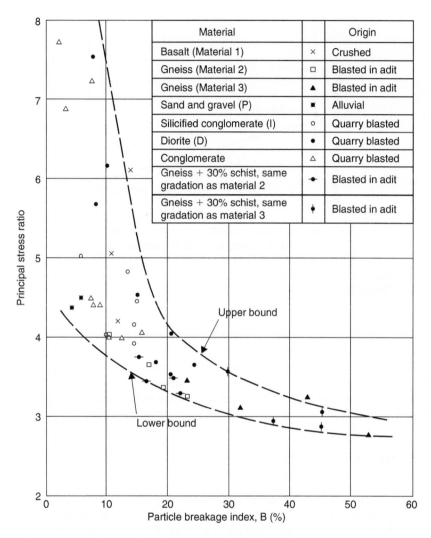

Material		Origin
Basalt (Material 1)	×	Crushed
Gneiss (Material 2)	□	Blasted in adit
Gneiss (Material 3)	▲	Blasted in adit
Sand and gravel (P)	✳	Alluvial
Silicified conglomerate (I)	○	Quarry blasted
Diorite (D)	●	Quarry blasted
Conglomerate	△	Quarry blasted
Gneiss + 30% schist, same gradation as material 2	✦	Blasted in adit
Gneiss + 30% schist, same gradation as material 3	♦	Blasted in adit

Figure 3.23 Effect of particle breakage on principal stress ratio at failure (modified after Marsal, [23]).

research conducted on the specific effects of particle breakage on the mechanical behaviour of ballast and other granular materials.

In an attempt to correlate the strength of aggregates with particle breakage, Marsal [23] plotted the peak principal stress ratio (σ_1/σ_3) against the breakage index B_g (Fig. 3.23). It was concluded that the shear strength decreases with the increasing particle breakage. Although no distinct correlation could be established between the principal stress ratio at failure and smaller values of particle breakage (<15%), Marsal's test data defined a lower bound of σ_1/σ_3 against breakage (Fig. 3.23). In contrast, Miura and O-hara [60] defined the ratio of surface area increment to the plastic work increment (dS/dW) as the particle crushing rate. They reported that the principal

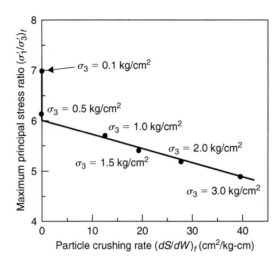

Figure 3.24 Effect of particle crushing rate on principal stress ratio at failure (data from Miura and O-hara, [60]).

stress ratio at failure decreases linearly with increasing particle crushing rate at failure $(dS/dW)_f$, as shown in Figure 3.24.

Indraratna et al. [3] presented a correlation between the particle breakage index, principal stress ratio and peak friction angle of railway ballast, as shown in Figure 3.25. They indicated that both the peak principal stress ratio and peak friction angle of ballast decreased as the breakage index increased at higher confining pressure.

3.4.4 Effects of confining pressure on particle breakage

Indraratna et al. [66] and Lackenby et al. [67] proposed that ballast degradation behaviour under cyclic loading can be distinctly categorised into three zones, namely: The Dilatant Unstable Degradation Zone (DUDZ), Optimum Degradation Zone (ODZ), and Compressive Stable Degradation Zone (CSDZ). These zones are defined by the magnitude of confining pressure (σ'_3) applied to the specimen (i.e. DUDZ: $\sigma'_3 < 30$ kPa, ODZ: 30 kPa $< \sigma'_3 < 75$ kPa, CSDZ: $\sigma'_3 > 75$ kPa). However, the maximum deviator stress magnitude $(q_{max,cyc} = \sigma'_{1\,max} - \sigma'_3)$ and maximum static peak deviator stress $(q_{peak,sta})$ also play an important role in characterising these degradation zones, as explained below.

Dilatant unstable degradation zone (DUDZ)

Specimens subjected to low σ'_3 and increased overall volumetric dilation due to rapid and considerable axial and expansive radial strains are characterised in the DUDZ (Fig. 3.26). Degradation in the DUDZ is the most prominent, with extensive breakage occurring at the onset of loading associated with the maximum axial strain and dilation

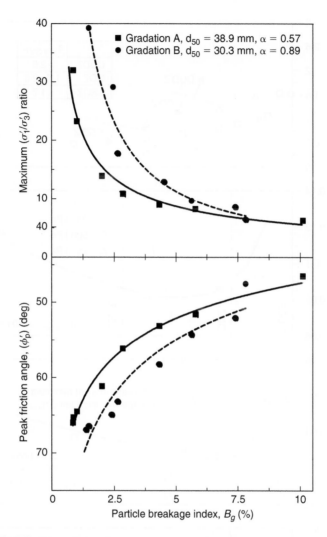

Figure 3.25 Influence of particle breakage on principal stress ratio and friction angle (modified after Indraratna et al., [3]).

rates. The micromechanical processes of degradation in the DUDZ have been discussed by Indraratna et al. [66]. Oda [68] and Cundall et al. [69] have also confirmed that the deviatoric force in a granular material is transmitted mainly through column like structures aligned in the direction of the major principal stress. Consider, for example, a DUDZ ballast specimen (300 mm diameter) subjected to a major principal stress σ'_1 of 780 kPa and σ'_3 of 30 kPa ($q_{max,cyc} = 750$ kPa). This translates to an axial force F of 55 kN, which might be distributed over, at least, 4 ballast columns, thus the induced characteristic stress (defined as F/d^2) on a ballast particle of diameter $d = 40$ mm would be about 8.5 MPa. This stress may not be high enough to cause particle splitting, based on the particle strengths given by Lim et al. [70]. However, if the characteristic stress

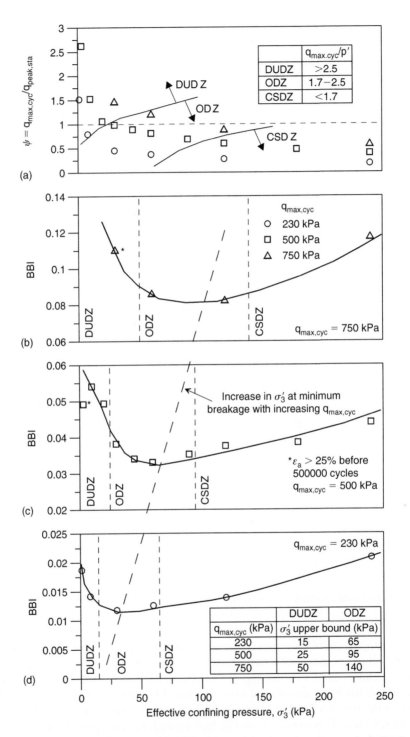

Figure 3.26 Effect of confining pressure σ_3' and maximum deviator stress $q_{max,cyc}$ on the ballast breakage index BBI, and the effect of $q_{max,cyc}$ on the DUDZ, ODZ and CSDZ breakage zones (after Lackenby et al., [67]).

F/a^2 induced in a small size of a in a deforming ballast column is considered, particle fracture is a real possibility. The majority of the degradation in this zone is due to the breakage of angular corners or projections, and very little particle splitting is observed due to ineffectual particle contacts.

For relatively small $q_{max,cyc}$ such as 230 kPa (Fig. 3.26d), the DUDZ σ'_3 range is limited, because, the magnitude of $q_{max,cyc}$ is insufficient to induce significant dilation. As the deviator stress $q_{max,cyc}$ increases (Figs. 3.26c and b) the tendency for dilation is much greater, thus the σ'_3 range of the DUDZ increases. The corresponding upper σ'_3 boundaries for each respective value of $q_{max,cyc}$ are included in Figure 3.26d and are obtained at zero volumetric strain (ε_v). DUDZ degradation can be avoided for latite basalt if value of σ'_3 exceeding 15, 25 or 50 kPa are applied for $q_{max,cyc} = 230$, 500 and 750 kPa, respectively. Undoubtedly, the DUDZ conditions should be avoided as much as possible for optimum stability of rail tracks.

Optimum degradation zone (ODZ)

The range of σ'_3 defining the ODZ is affected by the applied magnitude of $q_{max,cyc}$, (Fig. 3.26). Indraratna et al. [66] argued that a minor increase in σ'_3 would cause an optimum internal contact stress distribution, resulting in reduced stress concentration and tensile stresses, thereby, minimising breakage. Increase in σ'_3 would also lead to lower axial strains, and overall specimen compression (i.e., increase coordination number as also discussed by Oda [68].

Figure 3.26a implies that ODZ specimens generally have ψ ($q_{max,cyc}/q_{peak,sta}$) values ranging from about 0.4 up to 1.2. Increasing the magnitude of $q_{max,cyc}$ also results in a larger ODZ zone (Fig. 3.26d), i.e., 15–65 kPa, 25–95 kPa and 50–140 kPa for $q_{max,cyc} = 230$, 500 and 750 kPa, respectively.

Compressive stable degradation zone (CSDZ)

In the CSDZ, particle movement and dilation is significantly suppressed due to the considerably high confining stress as explained by Indraratna et al. [66]. The σ'_3 boundary between the ODZ and CSDZ can be identified by a 'flattening out' of ε_v (Fig. 3.26d). The reduced mobility of particles and the highly stressed but relatively secure contact points are the most significant differences between the ODZ and CSDZ. While corner degradation is still predominant, particle splitting also occurs through weak planes (microcracks and other flaws). Moreover, the fatigue of particles becomes more noticeable in the CSDZ (Indraratna et al., [66]). Within highly confined granular assembly, the vertical force chains are more isotropic due to lateral resistance from surrounding particles. Irrespective of the lower ψ ratios in the CSDZ, breakage is more pronounced in this zone compared to the ODZ. CSDZ is encountered when $\sigma'_3 > 65$, 95 and 140 kPa for $q_{max,cyc} = 230$, 500 and 750 kPa, respectively. Figure 3.26 illustrates that the confining pressure directly controls breakage influences. If railway organizations were to increase train axle loads, an increased ballast confinement system would be crucial to minimize ballast degradation. Possible methods of ballast confinement have been discussed by Lackenby et al. [67] and as illustrated in Figure 3.27.

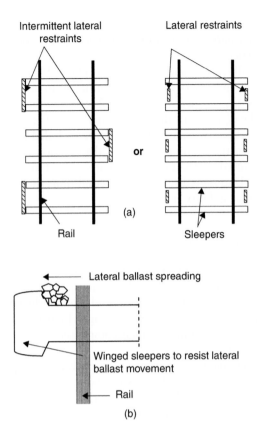

Figure 3.27 Potential methods of increasing confining pressure using: (a) Intermittent lateral restraints (after Indraratna et al., 2004), and (b) Winged sleepers (Lackenby et al., [67]).

REFERENCES

1. Kolbuszewski, J. and Frederick, M.R.: The significance of particle shape and size on the mechanical behaviour of granular materials. *Proc. European Conference on the Soil Mechanics and Foundation Engineering*, 1963, pp. 253–263.
2. Marachi, N.D., Chan, C.K. and Seed, H.B.: Evaluation of properties of rockfill materials. *J. of the Soil Mech. and Found. Div.*, ASCE, Vol. 98, No. SM1, 1972, pp. 95–114.
3. Indraratna, B., Ionescu, D. and Christie, H.D.: Shear behaviour of railway ballast based on large-scale triaxial tests. *J. of Geotechnical and Geoenvironmental Engineering*, ASCE, Vol. 124. No. 5, 1998, pp. 439–449.
4. Raymond, G.P. and Diyaljee, V.A.: Railroad ballast sizing and grading. *J. of the Geotechnical Engineering Division*, ASCE, Vol. 105. No. GT5, 1979, pp. 676–681.
5. Janardhanam, R., and Desai, C.S.: Three-dimensional testing and modeling of ballast. *J. of Geotechnical Engineering*, ASCE, Vol. 109, No. 6, 1983, pp. 783–796.
6. Selig, E.T.: Ballast for heavy duty track. In: *Track Technology*, Proc. of a Conf. organized by the Inst. of Civil Engrs and held at Univ. of Nottingham, 1984, pp. 245–252.

7. Holtz, W.G. and Gibbs, H.J.: Triaxial shear tests on pervious gravelly soils. *J. of the Soil Mech. and Found. Div., ASCE*, Vol. 82, No. SM1, 1956, pp. 867.1–867.22.

8. Leps, T.M.: Review of shearing strength of rockfill. *J. of the Soil Mech. and Found. Div., ASCE*, Vol. 96, No. SM4, 1970, pp. 1159–1170.

9. Vallerga, B.A., Seed, H.B., Monismith, C.L. and Cooper, R.S.: Effect of shape, size and surface roughness of aggregate particles on the strength of granular materials, *ASTM STP 212*, 1957, pp. 63–76.

10. Jeffs, T. and Marich, S.: Ballast characterictics in the laboratory. *Conference on Railway Engineering*, Perth, 1987, pp. 141–147.

11. Jeffs, T.: Towards ballast life cycle costing. *Proc. 4th International Heavy Haul Railway Conference*, Brisbane, 1989, pp. 439–445.

12. Chrismer, S.M.: Considerations of factors affecting ballast performance. *AREA Bulletin AAR Research and Test Dept. Report No. WP-110*, 1985, pp. 118–150.

13. Jeffs, T. and Tew, G.P.: *A review of track design procedures, Vol. 2, Sleepers and Ballast*, Railways of Australia, 1991.

14. Raymond, G.P.: Research on railroad ballast specification and evaluation. *Transportation Research Record 1006*, TRB, 1985, pp. 1–8.

15. Selig, E.T. and Waters, J.M.: *Track Technology and Substructure Management*. Thomas Telford, London, 1994.

16. Thom, N.H. and Brown, S.F.: The effect of grading and density on the mechanical properties of a crushed dolomitic limestone. *Proc. of the 14th Australian Road Research Board Conference*, Vol. 14, Part. 7, 1988, pp. 94–100.

17. Thom, N.H. and Brown, S.F.: The mechanical properties of unbound aggregates from various sources. In: Jones and Dawson (eds.): *Unbound Aggregates in Roads.*, Butterworth, London, 1989, pp. 130–142.

18. T.S. 3402: *Specification for supply of aggregate for ballast*. Rail Infrastructure Corporation of NSW, Sydney, Australia, 2001.

19. McDowell, G.R. and Bolton, M.D.: On the micromechanics of crushable aggregates. *Geotechnique*, Vol. 48, No. 5, 1998, pp. 667–679.

20. Jaeger, J.C.: Failure of rocks under tensile conditions. *Int. J. of Rock. Min. Sci.*, Vol. 4, 1967, pp. 219–227.

21. Nakata, Y., Kato, Y. and Murata, H.: Properties of compression and single particle crushing of crushable soil. *Proc. 15th Int. Conf. on Soil Mech. and Geotech. Engg.*, Istanbul, Vol. 1, 2001, pp. 215–218.

22. Festag, G. and Katzenbach, R.: Material behaviour of dry sand under cyclic loading. *Proc. 15th Int. Conf. on Soil Mech. and Geotech. Engg.*, Istanbul, Vol. 1, 2001, pp. 87–90.

23. Marsal, R.J.: Large scale testing of rockfill materials. *J. of the Soil Mech. and Found. Div., ASCE*, Vol. 93, No. SM2, 1967, pp. 27–43.

24. Terzaghi, K. and Peck, R.B.: *Soil mechanics in engineering practice*, John Wiley & Sons, Inc., New York, 1948.

25. Roscoe, K.H., Schofield, A.N. and Wroth, C.P.: On yielding of soils. *Geotechnique*, Vol. 8, No. 1, 1958, pp. 22–53.

26. Roscoe, K.H., Schofield, A.N. and Thurairajah, A.: Yielding of clays in states wetter than critical. *Geotechnique*, Vol. 13, No. 3, 1963, pp. 211–240.

27. Schofield, A.N. and Wroth, C.P.: *Critical state soil mechanics*. McGraw Hill, 1968.

28. Indraratna, B., Ionescu, D., Christie, H.D. and Chowdhury, R.N.: Compression and degradation of railway ballast under one-dimensional loading. *Australian Geomechanics*, December, 1997, pp. 48–61.

29. Gaskin, P.N., Raymond, G. and Powell, A.G.: Response of railroad ballast to vertical vibration. *Transportation Engineering Journal*, ASCE, Vol. 104, 1978, pp. 75–87.

30. Profillidis, V.A.: *Railway Engineering*, Avebury Technical, Aldershot, 1995.

31. Indraratna, B., Ionescu, D., and Christie, H.D.: State-of-the-Art Large Scale Testing of Ballast. *Conference on Railway Engineering (CORE 2000)*, Adelaide, 2000, pp. 24.1–24.13.

32. Indraratna, B., Salim, W. and Christie, D.: Improvement of recycled ballast using geosynthetics. *Proc. 7th International Conference on Geosynthetics*, Nice, France, 2002, pp. 1177–1182.

33. Sowers, G.F., Williams, R.C. and Wallace, T.S.: Compressibility of broken rock and the settlement of rockfills. *Proc. of the 6th International Conf. on Soil Mechanics and Foundation Engineering*, Vol. 2, 1965, pp. 561–565.

34. Drucker, D.C., Gibson, R.E. and Henkel, D.J.: Soil mechanics and work-hardening theories of plasticity. *Transactions, ASCE*, Vol. 122, 1957, pp. 338–346.

35. Vesic, A.S. and Clough, G.W.: Behavior of granular materials under high stresses. *J. of the Soil Mech. and Found. Div., ASCE*, Vol. 94, No. SM3, 1968, pp. 661–688.

36. Charles, J.A. and Watts, K.S.: The influence of confining pressure on the shear strength of compacted rockfill. *Geotechnique*, Vol. 30, No. 4, 1980, pp. 353–367.

37. Indraratna, B., Wijewardena, L.S.S. and Balasubramaniam, A.S.: Large-scale triaxial testing of greywacke rockfill. *Geotechnique*, Vol. 43, No. 1, 1993, pp. 37–51.

38. Diyaljee, V.A.: Effects of stress history on ballast deformation. *J. of Geotechnical Engineering, ASCE*, Vol. 113, No. 8, 1987, pp. 909–914.

39. Office of Research aand Experiments (ORE).: Optimum adaptation of the conventional track to future traffic, Question D117. *Report No. 5, Int. Union of Railways*, Utrecht, Netherlands, 1974.

40. Chen, W.F. and Saleeb, A.F.: *Constitutive Equations for Engineering Materials*. John Wiley and Sons, 1982.

41. Roscoe, K.H. and Burland, J.B.: On the generalized stress-strain behaviour of wet clay. In: *Engineering Plasticity*, 1968, pp. 535–609.

42. Poorooshasb, H.B., Holubec, I. and Sherbourne, A.N.: Yielding and flow of sand in triaxial compression: Part I. *Canadian Geotechnical Journal*, Vol. 3, No. 4. 1966, pp. 179–190.

43. Dafalias, Y.F. and Herrmann, L.R.: Bounding surface formulation of soil plasticity. In: Pande and Zienkiewicz (eds.): *Soil Mechanics – Transient and Cyclic Loads*. 1982, pp. 253–282.

44. Mroz, Z. and Norris, V.A.: Elastoplastic and viscoplastic constitutive models for soils with application to cyclic loading. In: Pande and Zienkiewicz (eds.): *Soil Mechanics – Transient and Cyclic Loads*. 1982, pp. 173–217.

45. Lade, P.V.: Elasto-plastic stress-strain theory for cohesionless soil with curved yield surfaces. *International Journal of Solids and Structures*, Vol. 13, 1977, pp. 1019–1035.

46. Shenton, M.J.: Deformation of railway ballast under repeated loading conditions. In: Kerr (ed.): *Railroad Track Mechanics and Technology*. Proc. of a symposium held at Princeton Univ., 1975, pp. 387–404.

47. Raymond, G.P., Gaskin, P.N. and Svec, O.: Selection and performance of railroad ballast. In: Kerr (ed.): *Railroad Track Mechanics and Technology*. Proc. of a symposium held at Princeton Univ., 1975, pp. 369–385.

48. Raymond, G.P. and Bathurst, R.J.: Repeated-load response of aggregates in relation to track quality index. *Canadian Geotech. Journal*, Vol. 31, 1994, pp. 547–554.

49. Shenton, M.J.: Ballast deformation and track deterioration. In: *Track Technology*. Proc. of a Conf. organized by the Inst. of Civil Engineers and held at the Univ. of Nottingham, 1984, pp. 242–252.

50. Ionescu, D., Indraratna, B. and Christie, H.D.: Behaviour of railway ballast under dynamic loads. *Proc. 13th Southeast Asian Geotechnical Conference*, Taipei, 1998, pp. 69–74.

51. Kempfert, H.G. and Hu, Y.: Measured dynamic loading of railway underground. *Proc. 11th Pan-American Conf. on Soil Mech. and Geotech. Engg.*, Brazil, 1999, pp. 843–847.

52. Stewart, H.E.: Permanent strains from cyclic variable amplitude loadings. *J. of Geotechnical Engineering, ASCE*, Vol. 112, No. 6, 1986, pp. 646–660.
53. Suiker, A.S.J.: *The mechanical behaviour of ballasted railway tracks*. PhD Thesis, Delft University of Technology, The Netherlands, 2002.
54. Lade, P.V., Yamamuro, J.A. and Bopp, P.A.: Significance of particle crushing in granular materials. *J. of Geotech. Engg., ASCE*, Vol. 122, No. 4, 1996, pp. 309–316.
55. Hirschfield, R.C. and Poulos, S.J.: High pressure triaxial tests on a compacted sand and an undisturbed silt. *Lab Shear Testing of Soils*, ASTM STP 361, 1963, pp 329–339.
56. Bishop, A.W.: The strength of soils as engineering materials, *Geotechnique*, Vol. 16, No. 2, 1966, pp. 91–128.
57. Lee, K.L. and Farhoomand, I.: Compressibility and crushing of granular soil in anisotropic triaxial compression. *Canadian Geotechnical Journal*, Vo. 4, No. 1, 1967, pp. 69–86.
58. Lee, K.L. and Seed, H.B.: Drained strength characteristics of sands. *J. of the Soil Mech. and Found. Div., ASCE*, Vol. 93, No. SM6, 1967, pp. 117–141.
59. Bilam, J.: Some aspects of the behaviour of granular materials at high pressures. In: *Stress-strain behaviour of soils*. Proc. Roscoe Mem. Symp., 1971, pp. 69–80.
60. Miura, N. and O-hara, S.: Particle crushing of decomposed granite soil under shear stresses. *Soils and Foundations*, Vol. 19, No. 3, 1979, pp. 1–14.
61. Hardin, B.O.: Crushing of soil particles. *J. of Geotechnical Engineering, ASCE*, Vol. 111, No. 10, 1985, pp. 1177–1192.
62. Indraratna, B. and Salim, W.: Modelling of particle breakage of coarse aggregates incorporating strength and dilatancy. *Geotechnical Engineering*, Proc. Institution of Civil Engineers, London, Vol. 155, Issue 4, 2002, pp. 243–252.
63. Ueng, T.S. and Chen, T.J.: Energy aspects of particle breakage in drained shear of sands, *Geotechnique*, Vol. 50, No. 1, 2000, pp. 65–72.
64. McDowell, G.R., Bolton, M.D. and Robertson, D.: The fractal crushing of granular materials. *J. Mech. Phys. Solids*, Vol. 44, No. 12, 1996, pp. 2079–2102.
65. Indraratna, B. and Salim, W.: Deformation and degradation mechanics of recycled ballast stabilised with geosynthetics. *Soils and Foundations*, Vol. 43, No. 4, 2003, pp. 35–46.
66. Indraratna, B., Lackenby, J., and Christie, D.: Effect of Confining Pressure on the Degradation of Ballast under Cyclic Loading. *Geotechnique*, Institution of Civil Engineers, UK, Vol. 55, No. 4, 2005, pp. 325–328.
67. Lackenby, J., Indraratna, B and McDowel, G.: The Role of Confining Pressure on Cyclic Triaxial Behaviour of Ballast. *Geotechnique*, Institution of Civil Engineers, UK Vol. 57, No. 6. 2007, pp. 527–536.
68. Oda, M.: Deformation Mechanism of Sand in Triaxial Compression Tests. *Soils and Foundations*, Vol. 12. No. 4, 1972, pp. 45–63.
69. Cundall, P.A., Drescher, A., and Strack, O.D.L.: Numerical Experiments on Granular Assemblies; Measurements and Observations. *IUTAM Conference on Deformation and Failure of Granular Materials*, Delft, 1982, pp. 355–370.
70. Lim, W.L., McDowell, G.R., and Collop, A.C.: Quantifying the Relative Strengths of Railway Ballasts. *Geotechnical Engineering*, Vol. 158. No. GE2, 2005, pp. 107–111.

Chapter 4

State-of-the-art Laboratory Testing and Degradation Assessment of Ballast

In this Chapter, the authors describe the laboratory investigation of ballast response under monotonic, cyclic and impact loadings, using state-of-the-art large-scale cylindrical, prismoidal triaxial and drop-weight impact equipment. The entire testing equipment and experimental procedure have been developed at the University of Wollongong. In order to study the strength, deformation and degradation characteristics of both fresh and recycled ballast, a series of monotonic triaxial tests was conducted in the laboratory using the large cylindrical triaxial apparatus. The crushing strengths of fresh and recycled ballast grains were then studied in a separate series of single particle crushing tests. In order to investigate the deformation and degradation behaviour of fresh and recycled ballast under cyclic loading, a small section of track was simulated in the prismoidal triaxial chamber. Representative field lateral stresses were applied to the ballast specimens and a cyclic vertical load equivalent to a typical 25 tonne/axle train load was applied to the specimens. To enhance the engineering performance of recycled ballast in track, an attempt was made to stabilise recycled ballast in the laboratory model using various types of geosynthetics. In order to investigate progressive degradation of fresh ballast subjected to impact loading, a series of impact tests was conducted using the high-capacity drop-weight impact machine. The performance of shock mats in the attenuation of dynamic impact loads and subsequent mitigation of ballast degradation was studied. The details of these new items of equipment, test materials, specimen preparation and test procedures are described in the following Sections.

4.1 MONOTONIC TRIAXIAL TESTING

The strength, deformation, and degradation behaviour of ballast under monotonic loading was investigated using the large-scale cylindrical triaxial apparatus. Consolidated Drained (CD) triaxial shearing tests were conducted on ballast specimens at various effective confining pressures. The conventional triaxial apparatus is one of the most versatile and widely used laboratory methods for obtaining the deformation and strength characteristics of geomaterials [1]. Despite its wide acceptance as the principal geotechnical testing apparatus, it is impractical and almost impossible to conduct a shear test on a ballast specimen in the conventional triaxial apparatus, because of large grain sizes. According to Australian Standards, ballast grains can be 63.0 mm maximum [2], while the diameters of the conventional triaxial specimens are 37–50 mm. Therefore, to conduct a shear test on a ballast specimen, one needs to either scale down

Figure 4.1 Large-scale triaxial apparatus built at the University of Wollongong, (a) triaxial cell and loading frame, and (b) control panel board.

the ballast grains to fit within a conventional triaxial apparatus or fabricate a larger testing rig.

Many researchers have indicated that the strength and deformation characteristics of aggregates are influenced by particle size [3, 4, 5]. Because of the inevitable size-dependent dilation and particle crushing mechanism, the disparity between the actual size of ballast in track and scaled down aggregates for testing in a conventional triaxial apparatus may give misleading or inaccurate results of strength and deformation parameters [6]. To overcome this problem, it is imperative to conduct large-scale triaxial testing of field-size ballast so that realistic strength-deformation and degradation characteristics are obtained. This is the primary reason why a large-scale triaxial facility was designed and built at the University of Wollongong [7].

4.1.1 Large-scale triaxial apparatus

The large-scale triaxial apparatus (Fig. 4.1) can accommodate specimens 300 mm in diameter and 600 mm high. The main components of the apparatus are: (a) cylindrical triaxial chamber, (b) axial loading unit, (c) cell pressure control unit in combination of air and water pressure, (d) cell pressure and pore pressure measurement system, (e) axial deformation measuring device, and (f) volumetric change measurement unit. The change in volume of a specimen during consolidation and drained shearing is measured by a coaxial piston located within a small cylindrical chamber connected to the main cell, in which the piston moves up or down depending on the increase or decrease in volume.

A combination of air and water is used to apply confining pressure to the test specimens. Any change in specimen volume during shearing will affect cell water pressure which is minimised by compressed air in the pressure control chamber. Cell pressure

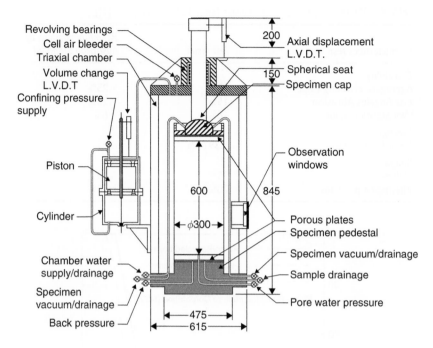

Figure 4.2 Schematic illustration of large-scale cylindrical triaxial apparatus (after Indraratna et al., [1]).

can be decreased by opening an exhaust valve and increased by a control valve, which allows compressed air into the pressure control chamber.

A vertical load is applied via a pump connected to the hydraulic loading unit (Fig. 4.2), and measured by a pressure transducer connected to the loading unit. Cell and pore water pressures are measured by two transducers. Vertical deformation of the specimen and movement of the co-axial piston of the volumetric measurement device are measured by two linear variable differential transducers (LVDT). The details of the triaxial apparatus are shown in Figure 4.2.

4.1.2   Characteristics of test ballast

4.1.2.1   Source of ballast

Fresh and recycled ballast specimens were sheared under monotonic drained loading using the large triaxial apparatus. Fresh ballast was collected from Bombo quarry (NSW), a major source for the Rail Infrastructure Corporation (RIC) of NSW. Recycled ballast was collected from Chullora stockpiles (Sydney), where discarded waste ballast was screened and the fine particles separated from coarse grains in a recycling plant.

4.1.2.2   Properties of fresh ballast

As fresh ballast was part of the ballast delivered to the track site, its particle size, gradation, and other index properties were as specified by Technical Specification TS

Table 4.1 Characteristics of fresh ballast (after Indraratna et al., [1]).

Characteristics test	Test result	Recommendations by Australian Standard
Durability		
Aggregate crushing value	12%	<25%
Los Angeles Abrasion	15%	<25%
Wet attrition value	8%	<6%
Strength		
Point load index	5.39 MPa	–
Shape		
Flakiness	25%	<30%
Misshapen particles	20%	<30%

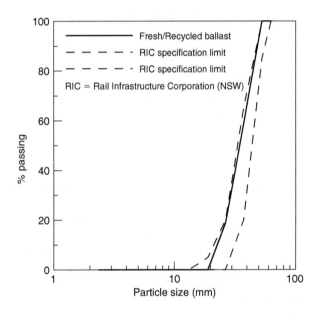

Figure 4.3 Particle size distribution of ballast tested (adapted from TS 3402, [8]).

3402 of RIC [8], and represents sharp angular coarse aggregates of crushed volcanic basalt (latite). The basalt is a fine-grained, dense-looking black aggregate, with the essential minerals being plagioclase (feldspar) and augite (pyroxenes).

Although a variety of parent rocks are used as the source of ballast in different parts of the world, igneous and sedimentary rocks are most widely used because they generally have high hardness and compressive strength, and are resistant to weathering. The common mineral groups are pyroxenes, quartz and feldspar. The specific minerals constituting parent rock govern the physical and mechanical properties of ballast. The durability, shape and strength of fresh ballast used in the laboratory study are summarised in Table 4.1. The grain size distribution (both fresh and recycled) including the RIC specification [8] is shown in Figure 4.3. The selected grain size distribution used in the laboratory testing (Fig. 4.3) is typical of ballast gradations used by the railway organisations (e.g. RIC).

To avoid the influence of particle size and gradation on experimental results, a single particle size distribution (Fig. 4.3) was selected within the given range of ballast specification [8]. The same gradation curve was followed when preparing the test specimens, both fresh and recycled. The sample size ratio is defined by the ratio between the diameter of triaxial specimen and maximum particle size. Many researchers have argued that as the sample size ratio approaches 6, the size effects become negligible [1, 4, 5]. A maximum ballast size of 53 mm was used in the monotonic triaxial shearing, the corresponding sample size ratio becoming 5.7, which was considered to be large enough to minimise the effect of sample size.

4.1.2.3   Properties of recycled ballast

Discarded ballast removed from the track during the renewal operation had been stock-piled in the specified yard. With the volume of waste ballast increasing daily, various railway organisations considered recycling some ballast partly to road construction and other projects, and some back to the track. With this objective in mind, Rail Infrastructure Corporation (RIC) commissioned a recycling plant at their Chullora yard near Sydney. Recycled ballast used in the laboratory investigation was collected from Chullora after screening off the fine particles in the recycling plant.

A physical examination indicated that about 90% of the recycled ballast was semi-angular crushed rock fragments, while the remaining 10% consisted of semi-rounded river gravels and other impurities (cemented materials, sleeper fragments, nuts, bolts, fine particles etc.) [9]. Most of the semi-angular rock particles were almost the same size and shape as fresh ballast, while the obvious difference was that these were less angular, had less asperities, and were dirtier. Fine particles were clearly visible around recycled ballast grains even after passing through the screening operation. It is anticipated that its strength, bearing capacity and resiliency will be less due to reduced angularity, greater heterogeneity and containing more impurities than fresh ballast.

4.1.3   Preparation of ballast specimens

All load cells, pressure transducers and LVDTs should be calibrated before preparing the test specimens. To prepare the specimen for triaxial testing, a 5 mm thick cylindrical rubber membrane was placed around the pedestal of the base plate and clamped with 2 steel bands. The membrane was stiff enough to stand by itself. The membrane was then temporarily supported by a steel cylindrical split mould clamped together with nuts and bolts. The ballast was carefully sieved using standard sieves, and different proportions of particle size were mixed together as per the selected gradation curve shown in Figure 4.3. The mixed ballast was then placed inside the rubber membrane and then compacted in four layers, each approximately 150 mm thick, using a hand-held vibratory hammer. The bulk unit weights of the specimens were 15.4–15.6 kN/m^3, which represent typical ballast density in the field. To minimise particle breakage during vibration, a 5 mm thick rubber pad was placed underneath the vibrator. After compaction, a steel cap was placed on top of the specimen, the membrane was clamped securely to the top cap with 2 steel bands and the split mould was

Figure 4.4 Triaxial chamber, split mould and a ballast specimen.

then removed (Fig. 4.4). The triaxial cylinder was then placed around the specimen and connected to the base plate. A rubber o-ring was placed between the cylindrical chamber and the base plate, and high vacuum silicon grease was applied along the edges to make the triaxial cell watertight.

4.1.4 Test procedure

After preparing the specimen, the triaxial cell was placed inside the loading frame and the specimen was filled with water through the base plate. The triaxial chamber was also filled with water and left overnight to saturate the aggregates. The preselected test confining pressure was applied to the specimen after achieving the Skempton's pore pressure parameter $B > 0.97$ [10]. The test specimens were isotropically consolidated to preselected confining pressures of 10 to 300 kPa before shearing, to investigate the influence of confining pressure on the strength, deformation and degradation of ballast. Raymond and Davies [11] indicated that lateral stress in rail track is unlikely to exceed 140 kPa, as mentioned earlier in Chapter 2. Nevertheless, the behaviour of ballast was investigated over a wider range of confining pressures. The range of confining pressures (10–300 kPa) applied in ballast testing is expected to cover all possible lateral stresses in track and is consistent with previous research (e.g. [1, 11]). Table 4.2 shows the details of confining pressures applied in the monotonic triaxial tests.

The confining pressure was increased in several steps and the corresponding change in volume of the specimen was recorded. After consolidating the specimen to its pre-selected pressure (see Table 4.2), the vertical load was increased using a hydraulic pump to commence shearing. Fully drained compression tests were conducted at an axial strain rate of 0.25% per minute, which allowed excess pore pressure to dissipate completely. The pressure transducers and LVDTs were connected to the digital panel board and a datalogger (DT800), supported by a host computer. All load, pressure and displacement measurements were recorded by the datalogger. Shearing was continued

Table 4.2 Confining pressures applied in monotonic triaxial testing.

Type of Ballast Tested	Effective Confining Pressure (kPa)
Fresh Ballast	10
Fresh Ballast	50
Fresh Ballast	100
Fresh Ballast	200
Fresh Ballast	300
Recycled Ballast	10
Recycled Ballast	50
Recycled Ballast	100
Recycled Ballast	200
Recycled Ballast	300

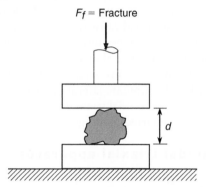

Figure 4.5 Schematic of ballast grain fracture test (modified after Indraratna and Salim, [13]).

until the vertical strain of ballast reached about 20%. Additional triaxial tests were conducted on fresh ballast terminating shearing at 0%, 5% and 10% axial strains to study the variation of ballast breakage with increasing strains. The ballast specimens were recovered at the end of each test, then dried and sieved, and changes in particle size were recorded. All vertical and lateral stress measurements were corrected for membrane effect as per Duncan and Seed's [12] procedure.

4.2 SINGLE GRAIN CRUSHING TESTS

As mentioned in Chapter 3, the crushing strength of individual particles is a key parameter governing ballast degradation. To assess crushing strength characteristics, single grain crushing tests were conducted on various sizes of fresh and recycled ballast. The schematic illustration of the grain crushing test is shown in Figure 4.5, where a single grain was placed between the top and bottom platens of a compression machine. The initial particle diameter (d) was measured before applying compression. The maximum load at which a particle fractured (F_f) was recorded and the corresponding tensile strength was calculated using Equation 3.2 (Chapter 3).

4.3 CYCLIC TRIAXIAL TESTING

Ideally, ballast should be tested in a real track under actual loading conditions. However, these tests are costly, time consuming, and disrupt traffic schedules. Moreover, many variables which affect the proper formulation of definitive ballast relationship, are often difficult to control in the field [14]. Therefore, laboratory experiments simulating field load and boundary conditions are usually carried out on ballast specimens. With the assistance of Rail Services Australia (now incorporated within Rail Infrastructure Corporation, NSW), a state-of-the-art prismoidal triaxial apparatus was designed and built at the University of Wollongong to investigate the response of a ballasted track under cyclic loading.

Several investigators have used large testing chambers with rigid and fully restrained walls to study ballast behaviour under cyclic loading (e.g. [15, 16, 17]). The lateral movement of ballast in real railway tracks is not fully restrained, particularly in the direction perpendicular to the rails [18]. The confinement offered by fully restrained cell walls is therefore, a major shortcoming in physical modelling of ballast in the laboratory. Consequently, some investigators developed semi-confined devices for ballast modelling [14, 19]. To simulate lateral deformation of ballast in actual real tracks, the vertical walls of the prismoidal triaxial rig were designed and built to allow free lateral movements under imparted loadings.

4.3.1 Large prismoidal triaxial apparatus

The large prismoidal triaxial rig used in this study can accommodate specimens 800 mm long, 600 mm wide, and 600 mm high. Figure 4.6(a) shows the prismoidal triaxial chamber and Figure 4.6(b) is a schematic of the triaxial apparatus including specimen set-up. This is a true triaxial apparatus where three independent principal stresses can be applied in the three mutually orthogonal directions.

A system of hinge and ball bearings enables the vertical walls to move laterally. Since each wall of the rig can move independently in the lateral directions, the ballast specimen is free to deform laterally under cyclic vertical load and lateral pressures. The lateral confinement offered by the shoulder and crib ballast in the actual track is not sufficient to restrain lateral movement of ballast. Therefore, the prismoidal triaxial rig with unrestrained sides provides an ideal facility for physical modelling of ballast under cyclic loading. This particular design of the chamber correctly simulates realistic track conditions, which permit lateral strains during loading.

The cyclic vertical load (σ_1) is provided by a servo-hydraulic actuator and the load is transmitted to the ballast through a 100 mm diameter steel ram and a rail/sleeper arrangement (Fig. 4.7a). Intermediate and minor principal stresses (σ_2 and σ_3, respectively) are applied via hydraulic jacks, and are measured by an assembly of load cells (Fig. 4.7b).

Sleeper settlement and lateral deformations of the vertical walls could be measured by 18 electronic potentiometers. Two pressure cells, one beneath the sleeper and the other at the ballast/capping interface, could be placed inside the chamber to monitor

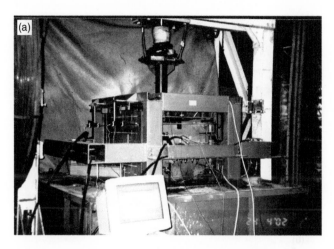

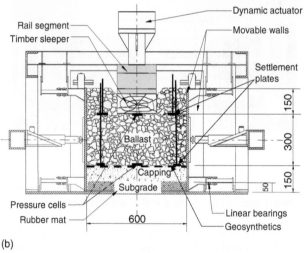

Figure 4.6 (a) Prismoidal triaxial chamber, (b) schematic illustration of cyclic triaxial rig (after Indraratna and Salim, [13]).

ballast stresses. Eight settlement plates were installed at each of the sleeper/ballast and ballast/capping interfaces to measure vertical strain. To get high quality real time data, all load cells, pressure cells and electronic potentiometers need to be connected to a data logger and supported by a host computer. This fully instrumented equipment can precisely measure all vertical and lateral loads and associated deformations.

4.3.2 Materials tested

4.3.2.1 Ballast, capping and clay characteristics

As mentioned earlier, fresh and recycled ballast specimens were tested under representative cyclic loading. The properties of fresh and recycled ballast were discussed earlier

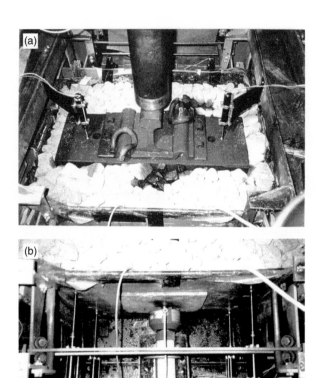

Figure 4.7 (a) Top of triaxial chamber showing sleeper and rail, and (b) load-cell, hydraulic jack and potentiometers attached to a vertical wall of the chamber.

in Section 4.2.2. A thin layer of compacted clay was used in the laboratory model to simulate the subgrade of a real track. A capping layer comprising sand-gravel mixture was used between the ballast and the clay layers. As mentioned earlier in Chapter 2, the capping layer (subballast) also acts as a filter preventing the 'pumping' of clay subgrade to the ballast. The particle size distribution of ballast (both fresh and recycled) and the capping materials, including the specification [8], are shown in Figure 4.8. Table 4.3 shows the grain size characteristics of fresh ballast, recycled ballast and the capping materials used by the authors in cyclic triaxial tests.

Remoulded alluvial soft clay from the South Coast of Sydney was used to represent track subgrade in the laboratory model. Table 4.4 shows the index properties of the clay used in the specimens. The clay has been classified as CH (high plasticity clay) based on the Casagrande Plasticity Chart.

4.3.2.2 Characteristics of geosynthetics

Three types of geosynthetics were used to stabilise recycled ballast in the laboratory model. These included: (a) geogrid, (b) woven-geotextile, and (c) geocomposite, a

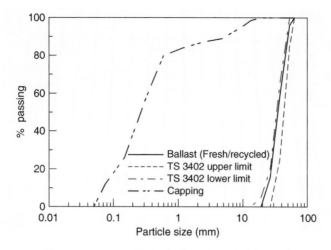

Figure 4.8 Particle size distribution of ballast and capping materials (modified after Indraratna and Salim, [13]).

Table 4.3 Grain size characteristics of ballast and capping materials.

Material	Particle shape	d_{max} (mm)	d_{min} (mm)	d_{50} (mm)	C_u	C_c
Fresh ballast	Highly angular	63.0	19.0	35.0	1.6	1.0
Recycled ballast	Semi-angular	63.0	19.0	35.0	1.6	1.0
Capping	Angular to rounded	19.0	0.05	0.26	5.0	1.2

Table 4.4 Soil properties of clay used in cyclic test specimens (data from Redana, [20]).

Soil Properties	Values
Clay content (%)	40–50
Silt content (%)	45–60
Water content, w (%)	40
Liquid limit, w_L (%)	70
Plastic limit, w_P (%)	30
Plasticity Index, PI (%)	40
Unit weight, γ (t/m^3)	1.7
Specific Gravity, G_s	2.6

combination of geogrid and non-woven geotextile bonded together. The physical, structural and geotechnical characteristics of these geosynthetics are described below.

Geogrid

The geogrid used to stabilise recycled ballast in the laboratory model was a bi-oriented geogrid (Fig. 4.9) supplied by Polyfabrics Australia Pty Ltd. It was made

Figure 4.9 Typical bi-oriented polypropylene geogrid.

Table 4.5 Physical properties of a typical geogrid (courtesy, Polyfabrics Australia Pty Ltd).

Physical Characteristics	Data			
Structure	Bi-oriented geogrid			
Mesh Type	Rectangular apertures			
Standard Colour	Black			
Polymer Type	Polypropylene			
Carbon Black Content	2%			

Dimensional Characteristics	Unit	Data		Notes
Aperture size MD	mm	40		b,d
Aperture size TD	mm	27		b,d
Mass per unit area	g/m^2	420		b

		Data		
Technical Characteristics	Unit	MD	TD	Notes
Tensile strength at 2% strain	kN/m	10.5	10.5	b,c,d
Tensile strength at 5% strain	kN/m	21	21	b,c,d
Peak tensile strength	kN/m	30	30	a,c,d
Yield point elongation	%	11	10	b,c,d

Notes:
95% lower confidence limit values, ISO 2602
Typical values
Tests performed using extensometers
MD: machine direction (longitudinal to the roll)
TD: transverse direction (across roll width).

of polypropylene, and manufactured by extrusion and biaxial orientation to enhance its tensile properties. These geogrids have high tensile strength, high elastic modulus, and strong resistance to construction damage and environmental exposure and are generally used for soil stabilisation and reinforcing embankments. Having large

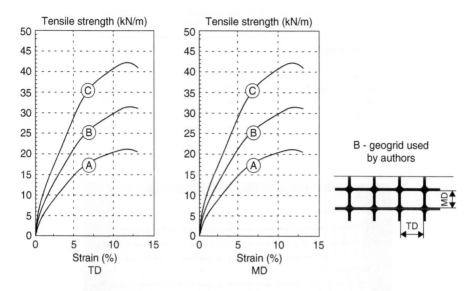

Figure 4.10 Typical load-deformation response of 3 different types of geogrids (courtesy, Polyfabrics Australia Pty Ltd).

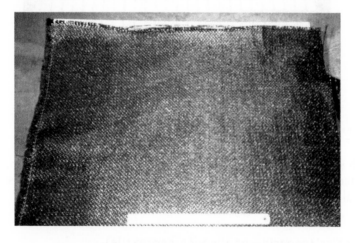

Figure 4.11 Typical polypropylene woven-geotextile.

apertures (>25 mm), geogrids provide strong mechanical interlock with coarse ballast grains. The physical size, strength and technical characteristics of the geogrid used by the authors are given in Table 4.5 and typical load-deformation behaviour is shown in Figure 4.10.

Woven-geotextile

A typical polypropylene woven-geotextile (Fig. 4.11) supplied by Amoco Chemicals Pty Ltd, Australia, was also used to stabilise recycled ballast in the laboratory. It

Table 4.6 Properties of polypropylene woven-geotextile (courtesy, Amoco Chemicals Pty Ltd, Australia).

Characteristics	Unit	Data
Mass	g/m^2	>450
Tensile strength	kN/m	>80
Pore size	mm	<0.30
Flow rate	litres/m^2/sec	>30

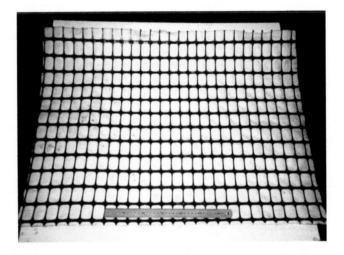

Figure 4.12 A typical bonded geocomposite.

was a high strength material having a tensile strength of over 80 kN/m, with good particle retention characteristics and high flow capacity. The physical, strength and geotechnical properties are summarised in Table 4.6.

Geocomposite (geogrid+non-woven geotextile)

A geogrid-geotextile geocomposite supplied by Polyfabrics Australia Pty Ltd, was also used to stabilise recycled ballast. These geocomposites are manufactured by bonding a geogrid and non-woven polypropylene geotextile together. Adding a non-woven geotextile to geogrid enables this composite to provide filtering and separating functions. Due to having large apertures (>25 mm, see Table 4.5), geogrid alone cannot provide these functions effectively. In a geocomposite, the geogrid component makes a strong mechanical interlock with the ballast grains and provides reinforcement, while the non-woven geotextile filter separates and allows partial in-plane drainage. Figure 4.12 shows a typical geocomposite used by the authors in the laboratory model study. The physical and mechanical characteristics of the geocomposite are given in Table 4.7.

Table 4.7 Characteristics of bonded geocomposite (courtesy, Polyfabrics Australia Pty Ltd).

Geogrid Physical Characteristics	Data			
Structure	Bi-oriented geogrid			
Mesh Type	Rectangular apertures			
Standard Colour	Black			
Polymer Type	Polypropylene			
Carbon Black Content	2%			

Geotextile physical Characteristics	Unit	Data		
Mass per unit area	g/m^2	140		
Polymer type	–	Polypropylene		

Dimensional Characteristics	Unit	Data		Notes
Geogrid Aperture size MD	mm	40		b,d
Geogrid Aperture size TD	mm	27		b,d
Mass per unit area	g/m^2	560		b

Technical Characteristics	Unit	Data		Notes
		MD	TD	
Peak tensile strength	kN/m	30	30	a,c,d
Yield point elongation	%	11	11	b,c,d

Notes:
95% lower confidence limit values, ISO 2602
Typical values
Tests performed using extensometers
MD: machine direction (longitudinal to the roll)
TD: transverse direction (across roll width).

4.3.3 Preparation of test specimens

A small track section including subgrade, capping, ballast, sleeper and rail, was simulated inside the triaxial chamber (see Fig. 4.6b) to represent a real track in the laboratory. A compacted clay layer (50 mm thick) was placed at the bottom of the triaxial chamber to model the subgrade of a real track. A relatively thin layer of clay was used in the laboratory model due to the limited height of the triaxial chamber. It is expected that a thicker subgrade of a specific thickness will equally affect the deformation and degradation response of various ballast specimens. Moreover, the vertical strains of ballast are computed by excluding the deformation of the capping and subgrade layers. In this respect, the thickness of clay layer used in the laboratory model is expected to have an insignificant influence on the test results, especially when comparing the response of different ballast specimens with and without the geosynthetic inclusion.

A 100 mm thick sand-gravel mixture (capping layer) was used above the clay layer to represent the subballast. Both the load bearing ballast (300 mm thick) and crib ballast (150 mm thick) layers consisted of either fresh or recycled ballast. The load bearing ballast was placed above the compacted capping layer. An assembly of timber sleeper

Table 4.8 Cyclic triaxial testing of ballast.

Type of Ballast	Type of Geosynthetics used	Test condition
Fresh Ballast	–	Dry
Fresh Ballast	–	Wet
Recycled Ballast	–	Dry
Recycled Ballast	–	Wet
Recycled Ballast	Geogrid	Dry
Recycled Ballast	Geogrid	Wet
Recycled Ballast	Woven-geotextile	Dry
Recycled Ballast	Woven-geotextile	Wet
Recycled Ballast	Geocomposite	Dry
Recycled Ballast	Geocomposite	Wet

and rail section was placed above the compacted load bearing ballast, and the space between the sleeper and vertical walls was filled with crib ballast. One layer of geosynthetics (geogrid, woven-geotextile or geocomposite) was placed at the ballast/capping interface (i.e. the weakest interface) to improve the performance of recycled ballast. To completely recover the load bearing ballast after the test, 2 layers of thin, loose, geotextiles were placed above and below the ballast layer for the purpose of separation only.

A vibratory hammer was used to compact the ballast and capping layers. To achieve representative field density, compaction was carried out in several layers, each about 75 mm thick. A 5 mm thick rubber pad was used beneath the vibrator to minimise particle breakage. Each test specimen was compacted to nearly the same initial density. The bulk unit weights of the compacted ballast and capping layers were in the order of $15.3 \, kN/m^3$ and $21.3 \, kN/m^3$, respectively. The initial void ratio (e_o) of the ballast layer was approximately 0.74.

4.3.4 Cyclic triaxial testing

A total of 10 cyclic triaxial tests were carried out on fresh and recycled ballast, with and without geosynthetic inclusion. To study the effect of saturation, 5 specimens were tested dry and the remaining ones were tested wet, with all specimens having identical loading and boundary conditions. Table 4.8 gives the details of the cyclic triaxial testing of ballast.

In addition to the cyclic tests, one slow repeated load test was conducted on recycled dry ballast without any geosynthetics. The repeated load test was carried out at various pre-selected load cycles (i.e., before applying any cyclic load, after 100,000 load cycles, and after 500,000 load cycles). This test was carried out to study the stress-strain response of ballast for a number of load cycles, and also to examine how the stress-strain response evolves during the course of cyclic loading.

4.3.4.1 Magnitude of cyclic load

The maximum sleeper/ballast contact stress must be ascertained before commencing any cyclic load test. As discussed earlier in Chapter 2, the maximum sleeper/ballast contact stress depends on many factors, including wheel static load and train speed. The static axle load depends on the type of vehicle, and may vary from 70–350 kN [21].

For establishing the maximum sleeper/ballast contact stress to apply in the laboratory cyclic load tests, a nominal axle load of 250 kN was assumed, which corresponds to a static wheel load of about 125 kN.

Following the design method proposed by Li and Selig [22], the design wheel load for a train speed of 100 km/hour with a wheel diameter of 0.97 m, was computed to be 192 kN (Equation 2.5). Atalar et al. [23] reported that part of this wheel load is transmitted to the adjacent sleepers, and 40–60% of the wheel load is resisted by the sleeper directly beneath the wheel. Assuming 50% of the design wheel load as the rail seat load and $F_2 = 1$, $l = 2.5$ m, Equation 2.12 gives an average contact pressure of 440 kPa. Assuming $a = 0.5$ m and $B = 0.26$ m, the stress distribution shown in Figure 2.13 and computed using Equation (2.30), gives an average sleeper/ballast contact stress of about 370 kPa.

Based on the above estimations, the maximum cyclic vertical stress for the laboratory investigations carried out by the authors was selected to be 460 kPa. The corresponding maximum vertical load for the laboratory model translated to about 73 kN, which was consistent with a previous study by Ionescu et al. [24].

4.3.4.2 Test procedure

Small lateral pressures ($\sigma_2 = 10$ kPa and $\sigma_3 = 7$ kPa) were applied to the triaxial specimens through the hydraulic jacks to simulate field confinement. In a real track, the confinement is generally developed by the weight of crib and shoulder ballast, along with particle frictional interlock (i.e. lateral earth pressure or K_o–effect). An initial vertical load of 10 kN was applied to the specimens to stabilise the sleeper and ballast, and to serve as a reference for all settlement and lateral movement measurements. In this state, initial readings of all load cells, pressure cells, potentiometers, and settlement plates were taken.

The cyclic vertical load was applied by a dynamic actuator with a maximum load of 73 kN, at a frequency of 15 Hz. The total number of load cycles applied in each test was half a million. The cyclic load was halted at selected load cycles to record the settlement, lateral displacement and load magnitude readings. For wet tests (see Table 4.8), the ballast specimens were gradually flooded with water before applying the cyclic load, and water was supplied during cyclic loading to maintain 100% saturation. At the end of each test, the ballast specimens were recovered, sieved, and any change in the particle size distribution was recorded for breakage assessment.

The repeated load test was carried out using the prismoidal triaxial rig at selected interval of load cycles, including the start of cyclic loading. In this test, the vertical load was slowly increased from the initial value to the maximum 73 kN, and then decreased to its initial value. This loading-unloading procedure was repeated for several cycles. During the repeated load test, all load and deformation measurements were continuously recorded using the datalogger (DT800).

4.4 IMPACT TESTING

Two types of impact testing apparatus are prevalent in practice for more than two decades viz. drop weight hammer and pendulum machine. The drop weight hammer is the most commonly adopted technique worldwide, as it can simulate repeated impact loading resembling actual track conditions [28, 29]. Therefore, a high-capacity drop-weight impact testing equipment was constructed at the University of Wollongong, which is currently the largest in Australia.

Installing resilient mats such as rubber pads (shock mats) in rail tracks can attenuate the dynamic impact force substantially. A shock mat when provided at the bottom of the ballast layer is often called a ballast mat or a sub-ballast mat, and when provided at the top (i.e. at the interface between sleeper and ballast) is usually called an under-sleeper pad or soffit pad [21]. The effectiveness of ballast mats in reducing noise along stiff tracks (e.g. concrete bridges and tunnels) and controlling vibration along open tracks has been studied by previous researchers [26, 30]. However, to the knowledge of the authors, no study has yet been reported on quantifying the role of shock mats in reducing ballast degradation. In view of this, a series of laboratory tests has been carried out to evaluate the effectiveness of shock mats in mitigating ballast breakage.

4.4.1 Drop weight impact testing equipment

A high capacity drop-weight impact testing equipment was used for the present tests. The impact testing equipment consists of a free-fall hammer of 5.81 kN weight that can be dropped from a maximum height of 6 m with an equivalent maximum drop velocity of 10 m/s. The drop hammer is attached to rollers and is guided through runners on the vertical columns which provides very low friction during a free fall. To eliminate surrounding noise and ground motion, a strong isolated floor is used. The strong isolated floor is made of reinforced concrete using high strength concrete and owns a significantly higher fundamental frequency than the testing rig. The large concrete foundation ($5.0 \times 3.0 \times 2.5$ m) is built over a compacted sand bed and surrounded by 50 mm thick shock absorber material. The impact test rig can house test specimen within a working area of 1800×1500 mm. Figure 4.13(a) shows the large scale drop weight impact testing equipment and Figure 4.13(b) is a schematic of the impact test set-up.

4.4.2 Test instrumentation

The test measurements include the impact load-time history, the transient acceleration-time history, the vertical and horizontal deformations and the particle breakage. The device used for measuring the transient impact loads is the dynamic load cell. The accelerometer is used to capture records of transient accelerations, and the sample deformations are obtained after each blow. The details of test instruments are shown in Figure 4.14.

The impact load is the contact force between the impactor and the test specimen as the drop-weight hammer strikes the shaft. It is monitored and recorded by a dynamic load cell (capacity of 1200 kN), mounted on the drop-weight hammer and connected to a computer controlled data acquisition system.

The acceleration is measured by using a piezoelectric accelerometer (capacity of 10,000g, where g is the gravitational acceleration). In order to eliminate the ground loop noise interfering with the measurements, the inner body of the accelerometer is electrically isolated from the mounting surface. The spectral analysis features in LabView8 could facilitate transfer of the test data in time domain into the frequency

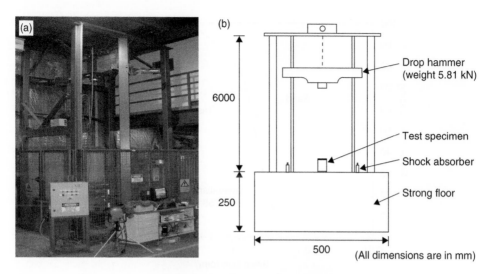

Figure 4.13 (a) Drop weight impact testing equipment, (b) schematic illustration of impact testing rig.

Figure 4.14 Instrumentation details (load-cell and accelerometer).

domain. The vertical and lateral deformations of the test specimen were measured by manual measurements after each blow.

4.4.3 Materials tested

4.4.3.1 Ballast and sand characteristics

The properties of fresh ballast were discussed earlier in Section 4.2.2. A thin layer of compacted sand was used in the laboratory physical model to simulate a typical

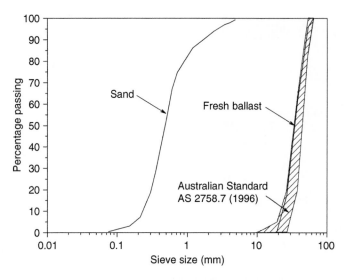

Figure 4.15 Particle size distribution of ballast and subgrade materials.

Table 4.9 Grain size characteristics of ballast and sand materials.

Material	Particle shape	d_{max} (mm)	d_{min} (mm)	d_{50} (mm)	C_u	C_c
Fresh ballast	Highly angular	63.0	19.0	35.0	1.6	1.0
Sand	Well graded	4.75	0.075	0.48	2.3	1.0

'weak' subgrade. The particle size distribution of fresh ballast and the sand materials, including the specification [8], are shown in Figure 4.15. Table 4.9 shows the grain size characteristics of fresh ballast, and the sand materials in impact tests.

4.4.3.2 Characteristics of shock mat

The elastic pad or rubber mat (shock mat) used in the current study was supplied by Phoenix AG (Australia) Pty. Ltd (Fig. 4.16). These shock mats have high compressive and high impact strength and are typically installed as protective layer over several bridges in Australia. In the current study, shock mats are used to study its effectiveness in the attenuation of high frequency impact loads and subsequent mitigation of ballast deformations and degradation. The physical, structural and geotechnical characteristics of these shock mats are described in Table 4.10.

4.4.4 Preparation of test specimens

The ballast was thoroughly cleaned, dried, sieved through a set of standard sieves (aperture size 63: 2.36 mm). A rigid steel plate ($D = 300$ mm, $t = 50$ mm) was used to represent a hard base condition such as bridge deck or rock structure etc, where the

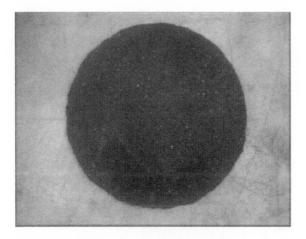

Figure 4.16 Typical shock mat.

Table 4.10 Physical properties of a typical shock mat (courtesy, Phoenix AG Australia Pty Ltd).

Physical Characteristics	Data		
Structure	Recycled rubber granulates		
Particle Size	1–3 mm		
Standard Colour	Black		
Binding Type	Polyurethane elastometer		
Surface	Fine granulate structure		

Dimensional Characteristics	Unit	Data	Notes
Length (rolls)	mm	6000 mm	a
Width (rolls)	mm	1250 mm	a
Thickness	mm	10 mm	a
Density	kg/m^3	920	a

Technical Characteristics	Unit	Data	Notes
Tensile Strength	kN/m^2	600	a,b
Elongation at Break	%	80	a,b
Thermal Resistance	°C	−30 to 80	
Flaming Rating		B2	c

Notes:
a Typical values
b Test method based on DIN 53571
c Test method based on AS 1530, part 3 and DIN 4102.

breakage due to impact loads becomes pronounced in the field. In order to simulate relatively weak subgrade conditions, a vibrocompacted well graded sand cushion of 100 mm thickness was provided below the ballast bed. In engineering practice, a shock mat layer of thickness in the range of 10–60 mm is used in rail tracks either as under-sleeper mats or as ballast mats. In view of this, three layers of shock mat accounting to

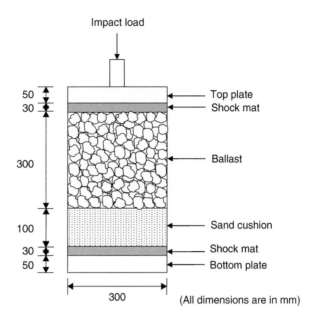

Figure 4.17 Schematic diagram of a typical test specimen.

a total thickness of 30 mm were used. The details of the typical test sample are shown in Figure 4.17.

A vibratory hammer was used to compact the ballast and sand materials. The ballast specimens ($H = 300$ mm, $D = 300$ mm) were compacted in several layers, each about 75 mm thick to simulate field densities for heavy haul tracks. A rubber pad (4 mm thick) was used to minimise the risk of breaking sharp corners and edges of ballast during compaction produced by a vibrator. In order to resemble low track confining pressure in the field, test specimens were confined in a rubber membrane thick enough ($t = 7$ mm) to prevent piercing by sharp particles during testing. The bulk unit weights of the compacted ballast and sand layers were in the order of 15.3 kN/m^3 and 15.9 kN/m^3, respectively.

4.4.5 Impact testing programme

A total of 8 impact load tests were carried out on fresh ballast, with and without shock mat. The efficiency of shock mats was further investigated by varying its position of placement. To study the effect of base conditions, 4 specimens were tested on steel base and the remaining ones were tested on relatively weaker sand base. Table 4.11 gives the details of the impact testing of ballast.

4.4.5.1 Magnitude of impact load

The impact load history is the combined effect of the response of inertial forces and sample resistance. The magnitude and frequency of these impact loads are generally

Table 4.11 Impact testing of ballast.

Test no.	Base condition	Shock mat details
1	Steel	Without shock mat
2	Steel	Shock mat at top of ballast
3	Steel	Shock mat at bottom of ballast
4	Steel	Shock mat at top and bottom of ballast
5	Sand	Without shock mat
6	Sand	Shock mat at top of ballast
7	Sand	Shock mat at bottom of ballast
8	Sand	Shock mat at top and bottom of ballast

much higher than the cyclic dynamic loads caused from the repeated passage of wheels. In particular, the greatest and most common dynamic impact loads are caused due to wheel flats. The typical loading duration produced by the wheel flats can vary from 1 msec to 10 msecs, while the magnitude of the impact force could be as high as 600 kN per rail seat [29]. During field studies on an instrumented full scale track at Bulli, New South Wales, Australia, it was found that where trains had wheel flats, pressures as high as 415 kPa were transmitted to the ballast bed as discussed in chapter 2. In the present study, drop height was selected to produce dynamic stresses in the range of 400–700 kPa simulating a typical wheel-flat [27, 31]. The impact load can be simplified as a shock pulse acting after the static wheel load is removed. Hence in the present study, a static preload was not considered.

4.4.5.2 Test procedure

The drop hammer was hoisted mechanically to the required drop height and released by an electronic quick release system. The impact loading was discontinued after 10 impact blows due to attenuation of strains in the ballast layer. Due to friction of the guiding runner, the velocity of the drop hammer decreases to 98% of the theoretical velocity, hence the required drop height (h_a) was adjusted by the coefficient 0.96 according to energy conservation principle [29].

For data recording purpose, an automatic trigering was enabled using the impact loading signal obtained during free fall of drop-weight hammer. The data sampling frequency rate was set to 50,000 Hz. To reduce noise, the raw impact load-time histories were digitally filtered using a low-pass fourth-order Butterworth filter with a cut-off frequency of 2000 Hz.

REFERENCES

1. Indraratna, B., Ionescu, D. and Christie, H.D.: Shear behaviour of railway ballast based on large-scale triaxial tests. *J. of Geotechnical and Geoenvironmental Engineering, ASCE,* Vol. 124. No. 5, 1998, pp. 439–449.
2. AS 2758.7: Aggregates and rock for engineering purposes, Part 7: Railway ballast. *Standards Australia*, NSW, Australia, 1996.

3. Marsal, R.J.: Large scale testing of rockfill materials. *J. of the Soil Mech. and Found. Div.*, ASCE, Vol. 93, No. SM2, 1967, pp. 27–43.

4. Marachi, N.D., Chan, C.K. and Seed, H.B.: Evaluation of properties of rockfill materials. *J. of the Soil Mech. and Found. Div.*, ASCE, Vol. 98, No. SM1, 1972, pp. 95–114.

5. Indraratna, B., Wijewardena, L.S.S. and Balasubramaniam, A.S.: Large-scale triaxial testing of greywacke rockfill. *Geotechnique*, Vol. 43, No. 1, 1993, pp. 37–51.

6. Indraratna, B., Ionescu, D. and Christie, H.D.: State-of-the-Art Large Scale Testing of Ballast. *Conference on Railway Engineering (CORE 2000)*, Adelaide, 2000, pp. 24.1–24.13.

7. Indraratna, B.: Large-scale triaxial facility for testing non-homogeneous materials including rockfill and railway ballast. *Australian Geomechanics*, Vol. 30, 1996, pp. 125–126.

8. T.S. 3402: *Specification for supply of aggregate for ballast*. Rail Infrastructure Corporation of NSW, Sydney, Australia, 2001.

9. Indraratna, B., Salim, W. and Christie, D.: Improvement of recycled ballast using geosynthetics. *Proc. 7th International Conference on Geosynthetics*, Nice, France, 2002, pp. 1177–1182.

10. Skempton, A.W.: The pore pressure coefficients A and B. *Geotechnique*, Vol. 4, No. 4, 1954, pp. 143–147.

11. Raymond, G.P. and Davies, J.R.: Triaxial tests on dolomite railroad ballast. *J. of the Geotech. Engg. Div.*, ASCE, Vol. 104, No. GT6, 1978, pp. 737–751.

12. Duncan, J.M. and Seed, H.B.: Corrections for strength test data. *J. of the Soil Mech. and Found. Div.*, ASCE, Vol. 93, No. 5, 1967, pp. 121–137.

13. Indraratna, B. and Salim, W.: Deformation and degradation mechanics of recycled ballast stabilised with geosynthetics. *Soils and Foundations*, Vol. 43, No. 4, 2003, pp. 35–46.

14. Jeffs, T. and Marich, S.: Ballast characterictics in the laboratory. *Conference on Railway Engineering*, Perth, 1987, pp. 141–147.

15. Atalar, C., Das, B.M., Shin, E.C. and Kim, D.H.: Settlement of geogrid-reinforced railroad bed due to cyclic load. *Proc. 15th Int. Conf. on Soil Mech. Geotech. Engg.*, Istanbul, Vol. 3, 2001, pp. 2045–2048.

16. Raymond, G.P. and Bathurst, R.J.: Repeated-load response of aggregates in relation to track quality index. *Canadian Geotech. Journal*, Vol. 31, 1994, pp. 547–554.

17. Eisenmann, J., Leykauf, G. and Mattner, L.: Deflection and settlement behaviour of ballast. *Proc. 5th International Heavy Haul Railway Conference*, Beijing, 1993, pp. 193–227.

18. Indraratna, B., Salim, W., Ionescu, D. and Christie, D.: Stress-strain and degradation behaviour of railway ballast under static and dynamic loading, based on large-scale triaxial testing. *Proc. 15th Int. Conf. on Soil Mech. and Geotech. Engg*, Istanbul, Vol. 3, 2001, pp. 2093–2096.

19. Norman, G.M. and Selig, E.T.: Ballast performance evaluation with box tests, *AREA Bul.* 692, Vol. 84, 1983, pp. 207–239.

20. Redana, I.W.: Effectiveness of vertical drains in soft clay with special reference to smear effects. PhD Thesis, University of Wollongong, Australia, 1999.

21. Esveld, C.: *Modern Railway Track*. MRT-Productions, The Netherlands, 2001.

22. Li, D. and Selig, E.T.: Method for railroad track foundation design, I: Development. *J. of Geotechnical and Geoenvironmental Engineering*, ASCE, Vol. 124, No. 4, 1998, pp. 316–322.

23. Atalar, C., Das, B.M., Shin, E.C. and Kim, D.H.: Settlement of geogrid-reinforced railroad bed due to cyclic load. *Proc. 15th Int. Conf. on Soil Mech. and Geotech. Engg.*, Istanbul, Vol. 3, 2001, pp. 2045–2048.

24. Ionescu, D., Indraratna, B. and Christie, H.D.: Behaviour of railway ballast under dynamic loads. *Proc. 13th Southeast Asian Geotechnical Conference*, Taipei, 1998, pp. 69–74.

25. Jenkins, H.M., Stephenson, J.E., Clayton, G.A., Morland, J.W. and Lyon, D.: The effect of track and vehicle parameters on wheel/rail vertical dynamic forces. *Railway Engineering Journal*, Vol. 3, 1974, pp. 2–16.

26. Anastasopoulos, I., Alfi, S., Gazetas, G., Bruni, S. and Leuven, A.V.: Numerical and experimental assessment of advanced concepts to reduce noise and vibration on urban railway turnouts. *Journal of Transportation Engineering, ASCE*, Vol. 135, No. 5, 2009, pp. 279–287.

27. Indraratna, B., Nimbalkar, S., Christie, D., Rujikiatkamjorn, C. and Vinod, J.S.: Field assessment of the performance of a ballasted rail track with and without geosynthetics. *Journal of Geotechnical and Geoenvironmental Engineering, ASCE*, Vol. 136, No. 7, 2010, pp. 907–917.

28. Wang, N.: Resistance of concrete railroad ties to impact loading. *PhD thesis*, Department of Civil Engineering, University of British Columbia, Canada, 1996.

29. Kaewunruen, S. and Remennikov, A.: Dynamic crack propagations in prestressed concrete sleepers in railway track systems subjected to severe impact loads. *Journal of Structural Engineering, ASCE*, Vol. 136, No. 6, 2010, pp. 749–754.

30. Auersch, L.: Dynamic axle loads on tracks with and without ballast mats: numerical results of three-dimensional vehicle-track-soil models. *Journal of rail and rapid transit, Proceedings of the Institution of Mechanical Engineers*, Vol. 220, 2006, pp. 169–183.

31. Christie, D.: In Situ Measurement of Ballast Settlement and Dilation and Pressure Distribution through Ballast. *CRC Press*, 2008, pp. 1–11.

Chapter 5

Behaviour of Ballast with and without Geosynthetics and Energy Absorbing Mats

The strength, deformation and degradation behaviour of fresh and recycled ballast has been studied in a series of monotonic triaxial shearing tests using a large-scale triaxial apparatus (Fig. 4.1, Chapter 4). The effects of confining pressure on friction angle, dilatancy, stress-ratio and particle breakage were particularly examined. The stress-strain behaviour (both fresh and recycled ballast) under cyclic loading has also been investigated in a large prismoidal triaxial chamber (Fig. 4.6) simulating a small track section. The stabilisation aspects of recycled ballast using various types of geosynthetics were also studied in these model tests. To quantify ballast degradation, each specimen was sieved before and after testing. The crushing strengths of ballast grains were determined by conducting a series of single particle crushing tests. The behaviour of fresh ballast has been studied in a series of impact loading tests using large-scale drop-weight impact equipment (Fig. 4.13, Chapter 4). This Chapter describes the strength, deformation and degradation behaviour of fresh and recycled ballast under monotonic, cyclic and impact loadings. The effectiveness of various geosynthetics in stabilising recycled ballast is presented and discussed through laboratory model test results. The benefits of shock mats in the effective mitigation of ballast degradation under impact loads are also discussed.

5.1 BALLAST RESPONSE UNDER MONOTONIC LOADING

5.1.1 Stress-strain behaviour

A series of isotropically consolidated drained triaxial tests were conducted by the authors on fresh and recycled ballast using large-scale cylindrical triaxial apparatus, as mentioned earlier in Chapter 4. The variations of deviator stresses $(q = \sigma_1' - \sigma_3')$ and volumetric strains $(\varepsilon_v = \varepsilon_1 + 2\varepsilon_3)$ with the corresponding shear strains $[\varepsilon_s = 2/3(\varepsilon_1 - \varepsilon_3)]$ for the fresh and recycled ballast under monotonic triaxial loading are shown in Figs. 5.1 and 5.2, respectively. The parameters σ_1' and σ_3' represent the major and minor principal effective stresses and the corresponding strains are denoted by ε_1 and ε_3, respectively.

Figures 5.1 and 5.2 clearly show that the shear behaviour of both fresh and recycled ballast, is non-linear. No distinct failure plane was observed in these shear tests even after 20% axial straining. It is evident that an increase in confining pressure increases the deviator stress, as expected. At low confinement (≤ 100 kPa), the volume

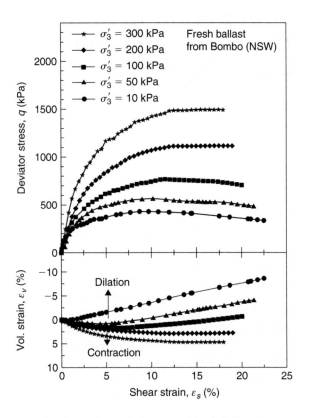

Figure 5.1 Stress-strain and volume change behaviour of fresh ballast in isotropically consolidated drained shearing (modified after Indraratna and Salim, [1]).

of ballast increases (i.e. dilation, represented by negative ε_v) during drained shearing [1–4]. Higher confining pressure tends to shift the overall volumetric strain towards contraction (i.e. ε_v becomes positive). A state of peak deviator stress $(\sigma'_1 - \sigma'_3)_p$, can be regarded as 'failure' for ballast. At low confining pressure, a peak deviator stress (i.e. failure) is evident (Figs. 5.1–5.2), followed by a post–peak strain softening associated with the volume increase [1–4].

Figures 5.1 and 5.2 indicate that the stress-strain behaviour of recycled ballast under monotonic triaxial shearing is generally similar to fresh ballast, except that the shear strength of recycled ballast under the same confinement is considerably lower than that of fresh ballast. To compare the stress-strain and strength characteristics of fresh and recycled ballast under triaxial compression directly, the stress-strain and volume change data of Figures 5.1 and 5.2 are re-plotted together in Figure 5.3. For clarity, only 3 sets of test data at confining pressures of 10, 100 and 300 kPa are plotted in this figure.

Figure 5.3 clearly shows that the recycled ballast tested by the authors has a lower peak deviator stress $(\sigma'_1 - \sigma'_3)_p$, compared to the fresh ballast. Owing to the

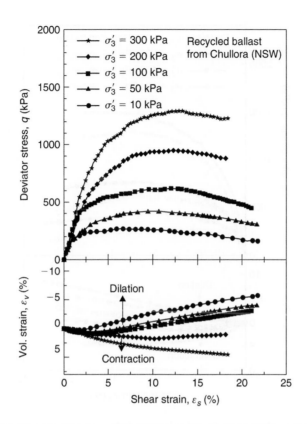

Figure 5.2 Stress-strain and volumetric response of recycled ballast under triaxial shearing (after Salim, [2]).

sharp corners breaking off under previous traffic loading cycles, recycled ballast generally has less angularity than fresh ballast. Less angularity and fine dust accumulated around recycled ballast grains reduce the frictional interlock and the shear strength [2]. Figure 5.3 also indicates that fresh ballast dilates more than recycled ballast at low confinement (e.g. 10 kPa), which is attributed to the higher angularity of fresh ballast. Dilatancy is suppressed at higher confinement (e.g. 300 kPa), and both fresh and recycled ballast continue to contract at a decreasing rate as the shear strain increases [1–4].

Since ballast is a coarse granular medium, its response to loading is expected to be comparable to other granular media (for example, rockfill and coarse sands). Most rockfills tested in the laboratory are almost similar in size, shape and source (i.e. parent rock) of ballast. However, one significant difference between the test conditions of rockfill and the field ballast is the confining pressure. Since rockfill used in dams is usually subjected to medium to high pressure, the mechanical behaviour of rockfill has been studied under high confining pressures (2.5–4.5 MPa) in the past [5, 6]. Subsequently, some researchers concentrated their study on rockfill at low to medium confining pressures (<1 MPa), realising that the normal stress on the critical failure

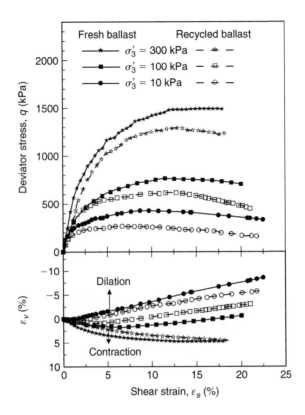

Figure 5.3 Comparison of stress-strain and volumetric behaviour between fresh and recycled ballast under triaxial drained shearing.

surface of a rockfill dam would not be so high [7, 8]. In contrast, ballast on railway tracks is subjected to much less confining pressure. Raymond and Davies [9] indicated that the lateral stress in ballast is unlikely to exceed 140 kPa. Despite this difference in confinement, the stress-strain behaviour of rockfill under monotonic loading may be compared with that of ballast, considering the granular nature, particle size, shape and parent rock of these two coarse media.

Figures 5.4(a) and 5.4(b) show the comparison between the stress-strain and volume change behaviour of ballast and rockfill under monotonic loading. These figures indicate that the stress-strain and volume change response of ballast under monotonic triaxial shearing is closely comparable to that of rockfills. However, one difference is evident that at low confining pressure (<100 kPa), ballast exhibits dilatant behaviour in triaxial compression [3, 9], whereas at higher confinement, both ballast and rockfill show overall contraction at failure [3, 7–9].

Figures 5.5 and 5.6 show the variation of deviator stress ratio $(\eta = q/p')$ with increasing shear strain (ε_s) for the fresh and recycled ballast, respectively, where p' is the mean effective normal stress. These figures reveal that the deviator stress ratio increases rapidly to a peak value at low confinement and then decreases gradually as the

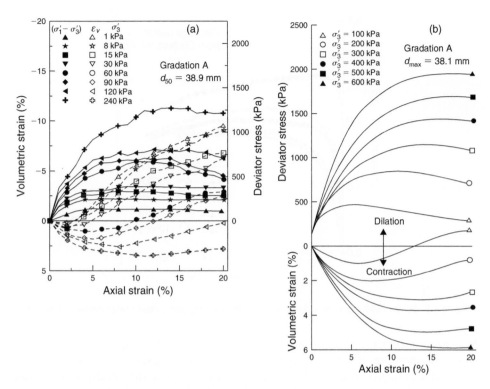

Figure 5.4 Stress-strain-volume change behaviour, (a) ballast (modified after Indraratna et al., [3]), and (b) greywacke rockfills (modified after Indraratna et al., [8]).

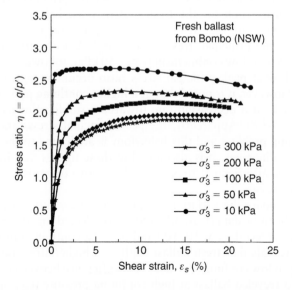

Figure 5.5 Stress ratio (η) versus shear strain plots for fresh ballast under drained shearing.

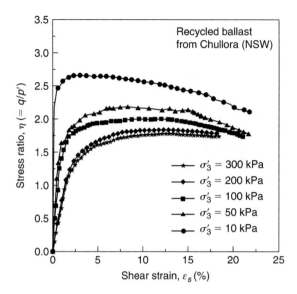

Figure 5.6 Stress ratio (η) versus shear strain plots for recycled ballast under drained shearing.

shear strain increases. At higher confining pressure (≥ 200 kPa), however, the deviator stress ratio increases at a decreasing rate with the increasing shear strain and reaches a stable value at higher strain levels. It is noted in Figs. 5.5–5.6 that all stress ratio-strain data approach a common stress ratio (η) value as the shear strain increases, irrespective of the confining pressures. Apparently, both fresh and recycled ballast exhibit a similar variation in deviator stress ratio with the increasing shear strain [2].

Figures 5.7 and 5.8 show the variation of effective principal stress ratio (σ'_1/σ'_3) with increasing shear strain for the fresh and recycled ballast, respectively. These results clearly demonstrate that at low confining pressure, both fresh and recycled ballast exhibit a higher principal stress ratio (σ'_1/σ'_3). A peak value of principal stress ratio is clearly evident at low confinement (e.g. ≤ 50 kPa), followed by strain softening. In contrast, no distinct peak principal stress ratio occurs at higher confining stress (e.g. ≥ 200 kPa). However, it is noted that at a higher confining pressure, the principal stress ratio increases at a decreasing rate towards a stable value as the shear strain increases (Figs. 5.7–5.8). Apparently, the peak principal stress ratio decreases with increasing confining pressure and this behaviour is attributed to the absence of dilatancy at higher confinement. It is also noted in these figures that irrespective of the confining pressures, all principal stress ratio data move towards a common value as the shear strain increases.

In order to compare the variations of principal stress ratio (σ'_1/σ'_3) of fresh and recycled ballast, the same data (Figs. 5.7–5.8) are re-plotted together in Figure 5.9. For clarity, only 3 sets of test data at 10, 100 and 300 kPa confining pressures are plotted here. Figure 5.9 reveals that fresh ballast exhibits a higher stress ratio than recycled ballast, especially at low confining pressure. The difference between the principal stress ratio of fresh and recycled ballast at high confining pressure (e.g. 300 kPa) becomes insignificant. This behaviour is primarily attributed to the inhibition of dilatancy at higher confining pressure [2].

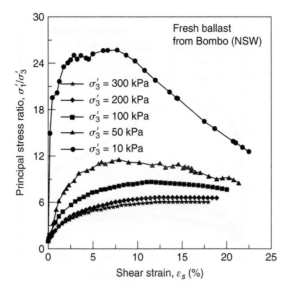

Figure 5.7 Variation of effective principal stress ratio with shear strain for fresh ballast (modified after Indraratna and Salim, [1]).

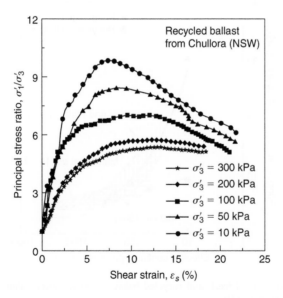

Figure 5.8 Variation of effective principal stress ratio with shear strain for recycled ballast.

5.1.2 Shear strength and stiffness

The shear strengths of fresh and recycled ballast in terms of principal stress ratio at failure $(\sigma'_1/\sigma'_3)_f$, are plotted against the effective confining pressure in Figure 5.10.

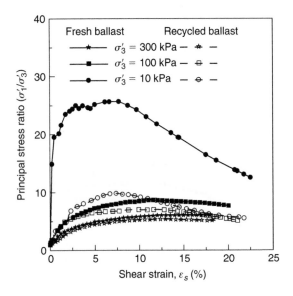

Figure 5.9 Comparison of principal stress ratio between fresh and recycled ballast in drained triaxial shearing.

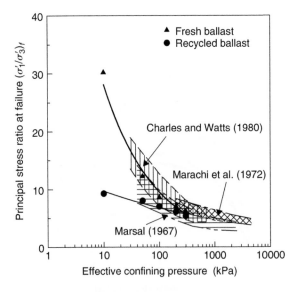

Figure 5.10 Shear strength of fresh and recycled ballast (modified after Salim, [2] and inspired by Indraratna et al., [3]).

Selected rockfill data from the previous studies [5–7] are also plotted for comparison. Since the imparted confining stresses during triaxial testing of ballast were relatively low compared to those in previous studies on rockfill, fresh ballast exhibited a relatively higher stress ratio at failure than rockfill. The test results clearly show that the failure

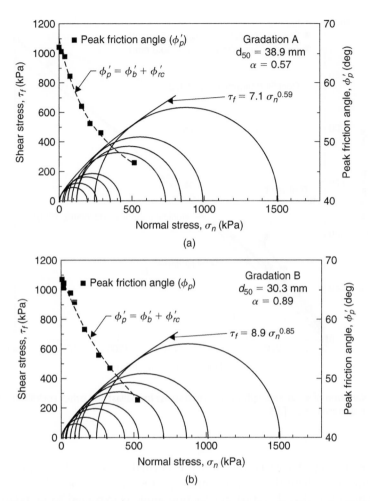

Figure 5.11 Mohr-Coulomb failure envelopes for latite basalt: (a) Gradation A; (b) Gradation B (modified after Indraratna et al., [3]).

stress ratio of recycled ballast is significantly lower than the fresh ballast, especially at low confinement [2]. It is also noted that in general, the principal stress ratio at failure decreases with increasing confining pressure both for ballast and rockfill.

The conventional Mohr-Coulomb strength envelopes provide a convenient approach to link shear strength of basalt with effective confining pressure (σ'_3). Indraratna et al. [3] demonstrated these envelopes for latite basalt (Fig. 5.11). At lower range of stresses, the shear strength envelope is markedly curved and passes through the origin (zero cohesion), as expected for granular materials. In fact, normal stresses below 400 kPa are usually representative of typical ballasted foundations (Indraratna et al., [23]).

The apparent friction angle (ϕ'_p) corresponding to the peak deviator stress of the ballast can be estimated by drawing a tangent from the origin to each Mohr circle

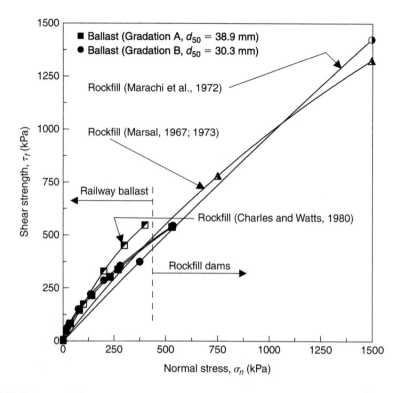

Figure 5.12 Variation of shear strength with normal stress for ballast and other rockfills (modified after Indraratna et al., [3]).

of effective stresses. Its variation with the normal stress is also plotted in Figure 5.11. In practice, where the lateral confining stress (hence, normal stress) is usually small in railway tracks, the apparent friction angle is expected to be relatively high ($\phi'_p > 40°$). However, at large normal stress levels, the apparent friction angle becomes considerably smaller, approaching a value of the order of 35°.

Indraratna et al. [3] summarised the variation of shear strength with the normal stress for ballast and other rockfill aggregates (basalt), as shown in Figure 5.12. They pointed out that the shear strength envelopes of these coarse aggregates were non-linear, especially ballast that was sheared under low confining pressure. At higher confining stress, the shear strength envelopes tend to become linear and the conventional Mohr-Coulomb (linear) analysis may be employed to describe the shear strength of aggregates. Indraratna et al. [3] also stated that the shear strength of ballast could be expressed by extending a normalised shear strength criterion originally proposed by Indraratna et al. [8], as given by:

$$\frac{\tau_f}{\sigma_c} = m \left(\frac{\sigma_n}{\sigma_c}\right)^n \tag{5.1}$$

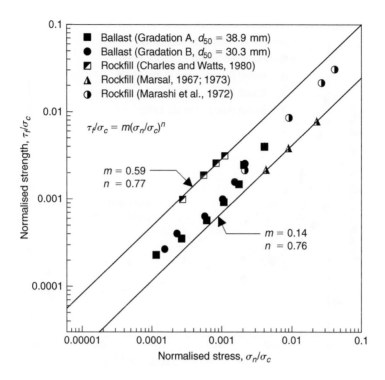

Figure 5.13 Normalised shear strength variation with normalised normal stress for ballast and other rockfills (modified after Indraratna et al., [3]).

Table 5.1 Values of coefficients m and n for the normalised failure criterion (after Indraratna et al., [3]).

Material	σ_c (MPa)	σ_3' range (kPa)	Value of coefficient	
			m	n
Ballast (Gradation A)	130	1–240	0.18	0.69
Ballast (Gradation B)	130	1–240	0.14	0.65
Rockfill (Marachi et al., [6])	175	200–4500	0.55	0.90
Rockfill (Marsal, [5, 10])	175	400–2470	0.14	0.76
Rockfill (Charles & Watts, [7])	360	30–500	0.59	0.77

where, τ_f is the shear stress at failure, σ_n is the normal stress on the failure plane, σ_c is the uniaxial compressive strength of the parent rock, and m and n are dimensionless constants. Indraratna et al. [3] presented their ballast experimental data together with other previous rockfill data [5–7, 10] in a normalised form plotted in log-scales, as shown in Figure 5.13. The values of m and n for their test ballast and the rockfills are given in Table 5.1 [3].

The peak friction angles (ϕ_p) of fresh and recycled ballast determined from the drained triaxial compression tests conducted by the authors are plotted against the

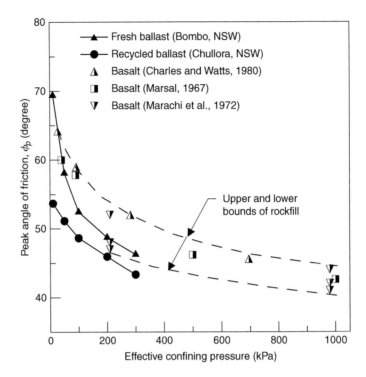

Figure 5.14 Variation of peak friction angle of fresh and recycled ballast with effective confining pressure (after Salim, [2] and inspired by Indraratna et al., [3]).

effective confining pressure, as shown in Figure 5.14. The ϕ_p values of other crushed basalt (rockfill) obtained at relatively higher confining pressures by the previous researchers [5–7] are also plotted here for comparison. Figure 5.14 reveals that the peak friction angle of both fresh and recycled ballast decreases with the increasing confining pressure, and this behaviour is attributed primarily to the decrease in dilation at elevated confining stress. These test results are consistent with the findings of previous research on rockfill [5–7]. The influence of dilatancy and particle breakage on the friction angle of ballast at various confining pressures will be discussed further in Chapter 7. Figure 5.14 also confirms that recycled ballast has a lower frictional strength than fresh ballast. The test results reveal that the peak friction angles of fresh and recycled ballast tested by the authors decrease from 69° to 46° and 54° to 43°, respectively, as the effective confining pressure increases from 10 to 300 kPa.

The variation of initial elastic modulus (E_i) of fresh and recycled ballast with effective confining pressure is illustrated in Figure 5.15. These data points indicate that the initial deformation modulus is linearly related to the effective confining pressure for both fresh and recycled ballast used by the authors. As expected, the initial elastic modulus increases with the increasing confining pressure. Obviously, fresh ballast exhibits a higher elastic modulus than recycled ballast (Figure 5.15) due to the higher angularity and better frictional interlock in fresh aggregates [2].

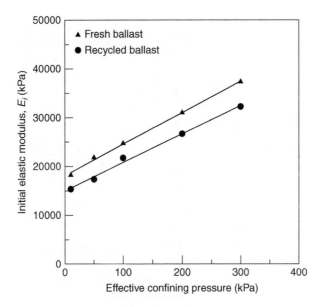

Figure 5.15 Initial deformation modulus of fresh and recycled ballast at various confining pressures.

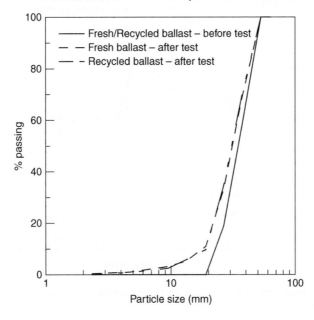

Figure 5.16 Change in particle size of ballast shown in conventional gradation plots.

5.1.3 Particle breakage in triaxial shearing

As mentioned earlier in Chapter 4, changes in grain size resulting from shearing were recorded in each test. Figure 5.16 shows the change in ballast gradation plotted in

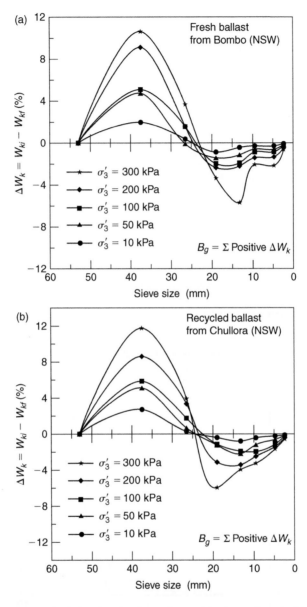

Figure 5.17 Alternative method showing the change in particle size under triaxial shearing, (a) fresh ballast, and (b) recycled ballast.

conventional semi-logarithmic grain size distribution curves. It is obvious that small changes in ballast size cannot be clearly illustrated in the conventional gradation plots. Therefore, an alternative technique extending the method proposed by Marsal [5] was adopted, where the differences in percentage retained before and after testing (ΔW_k) were plotted against the sieve size. Figures 5.17(a) and (b) show the variations of ΔW_k with sieve size for the fresh and recycled ballast tested by the authors.

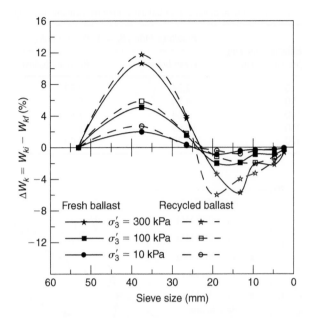

Figure 5.18 Comparison of particle breakage between fresh and recycled ballast.

Figures 5.17(a) and (b) clearly indicate that the breakage of ballast under triaxial compression increases with increasing confining pressure. It is relevant to mention here that a positive ΔW_k for a given sieve size represents a decrease in percentage retained in that sieve due to particle breakage. In contrast, a negative ΔW_k in a smaller sieve indicates an increase in percentage retained in that sieve resulting from the passing of broken particles through the larger sieves. Figures 5.17(a) and (b) reveal that larger particles (>30 mm) are more vulnerable to breakage than smaller grains for both fresh and recycled ballast.

To compare the degradation characteristics between fresh and recycled ballast, the data of Figs. 5.17(a) and (b) are re-plotted together, as shown in Figure 5.18. Three sets of experimental data (at 10, 100 and 300 kPa) are presented here for clarity. Figure 5.18 confirms that recycled ballast suffers higher particle breakage than fresh ballast. A large number of hairline micro-cracks in recycled ballast grains resulting from previous loading cycles is believed to be a major cause of this behaviour. The presence of micro-cracks decreases the crushing strength of recycled ballast which is also confirmed by the single grain crushing test results discussed later in this Chapter. Recycled ballast is, therefore, more vulnerable to degradation and requires external reinforcing agents to strengthen its resistance against breakage in order to compete with fresh ballast as a potential construction material. The values of Marsal's [5] breakage index B_g, for the fresh and recycled ballast tested by the authors under monotonic triaxial loading are given in Table 5.2.

The influence of strain levels on the degree of particle breakage was investigated by the authors by conducting additional triaxial tests and terminating shearing at

Table 5.2 Particle breakage of ballast under monotonic loading.

Effective confining pressure (kPa)	Breakage Index, $B_g = \Sigma$ Positive ΔW_k	
	Fresh ballast	Recycled ballast
10	2.34	2.99
50	4.74	5.77
100	6.64	7.60
200	10.69	11.95
300	14.29	15.68

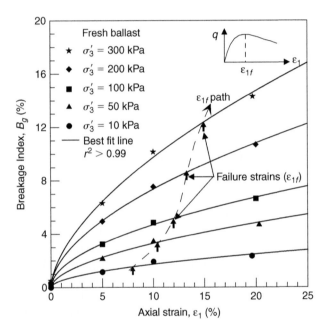

Figure 5.19 Variation of particle breakage of fresh ballast with axial strain (modified after Indraratna and Salim, [11]).

0%, 5% and 10% axial strains, and then computing the breakage indices from the measurements of grain size changes, as mentioned earlier in Chapter 4. Indraratna and Salim [11] presented the variations of B_g values with axial strains of ballast, as shown in Figure 5.19. The failure strains (ε_{1f}) are indicated and the locus of failure strains is also shown in the figure.

Figure 5.19 shows that the degree of particle breakage increases non-linearly with increasing axial strain and that the magnitude of breakage also increases with higher confining pressure. The trend lines of breakage indices are shown as the solid lines in this figure. It is noted that the particle breakage continues to increase even after the peak deviator stress (or failure). These test results also indicate that the rate of particle breakage $dB_g/d\varepsilon_1$, (i.e. slope) is initially high and decreases with increasing axial strain towards a constant.

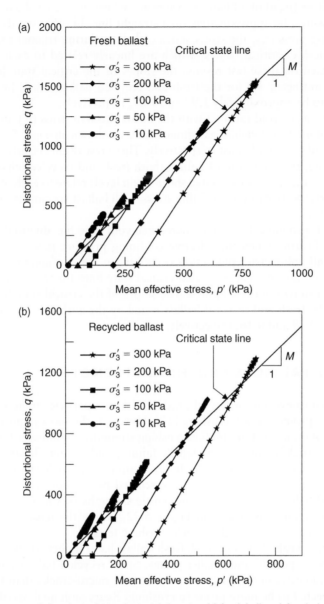

Figure 5.20 Variation of p' and q in drained triaxial shearing, (a) fresh ballast (after Indraratna and Salim, [1]), and (b) recycled ballast.

5.1.4   Critical state of ballast

The variations of deviator stress (q) with the mean effective stress (p') for the fresh and recycled ballast under triaxial drained shearing are shown in Figs. 5.20(a) and (b), respectively. An increase in confining pressure increases the mean effective stress, which leads to a higher deviator stress. These results show that at the end of drained shearing, the states of stress (p', q) of all ballast specimens, which were consolidated to various confining pressures, lie approximately on a straight line. In other words, irrespective of the confining pressures, the stress states of ballast during triaxial shearing move towards unique (i.e. critical) states, which are linearly related to each other in the p'-q plane. Based on these test results, the slopes of the critical state lines (i.e. the critical state parameter, M) for the fresh and recycled ballast tested by the authors are estimated to be approximately 1.90 and 1.67, respectively.

The variations of void ratio (e) with the mean effective stress (p') during drained shearing are plotted in a semi-logarithmic scale, as shown in Figs. 5.21(a) and (b) for the fresh and recycled ballast, respectively. These test results clearly show that in drained shearing, the void ratio of ballast (both fresh and recycled) changes as such that the states of the specimens at large shear strain levels relate to each other in a very specific way. Irrespective of the confining stresses, all ballast specimens move towards the critical states.

Figure 5.21 also indicates that an increase in void ratio (i.e. dilation) is associated with drained shearing when the effective confining pressure is low (≤ 100 kPa). In contrast, overall volumetric contraction occurs when the confining pressure is high (e.g. 200 kPa and above). The estimated critical state lines for the fresh and recycled ballast are also shown in these figures. The slopes of the critical state lines in e-ln p' plane (λ) for the fresh and recycled ballast tested by the authors are estimated to be approximately 0.19 and 0.16, respectively.

5.2 SINGLE PARTICLE CRUSHING STRENGTH

Angularity, coarseness, uniformity of gradation, lower particle strength, stress level and anisotropy promote grain crushing [12]. However, the most important factor is the resistance of grains to fracture (i.e. crushing strength). As indicated in Chapter 2, fracture in a particle is initiated by tensile failure, and the tensile strength of rock grains is represented by Equation 2.16. Interpretations of particle strength and failure of granular materials especially sand are discussed by various researchers [13–15]. Indraratna and Salim [16] presented the tensile strengths of various sized fresh and recycled ballast grains, as shown in Figure 5.22, where the tensile strengths were determined from a series of single particle crushing tests.

Figure 5.22 reveals that in general, fresh ballast has a higher tensile strength than recycled ballast, especially the smaller grains. Since recycled ballast has undergone millions of load cycles in the past, it contains more micro-cracks than fresh ballast, hence it is expected to be more prone to crushing. Regression analysis of the particle strength data indicates that recycled ballast generally has about 35% lower tensile strength than fresh ballast. This lower crushing strength of recycled ballast is directly responsible for its higher particle breakage under triaxial shearing compared to fresh ballast (see Fig. 5.18).

Figure 5.22 also indicates that for both fresh and recycled ballast, the tensile strength decreases linearly with increasing grain size. McDowell and Bolton [13] and Nakata et al. [14] reported a similar trend for particle strength for sand and limestone aggregates. This is because larger particles contain more flaws and have a higher probability of defects [15]. Fracturing larger particles along these defects (cracks) creates

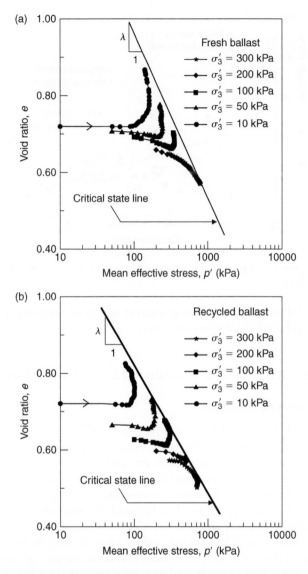

Figure 5.21 Variation of void ratio in drained shearing, (a) fresh ballast (after Indraratna and Salim, [1]), and (b) recycled ballast.

smaller particles. The subdivided particles contain fewer defects and are less likely to fracture. In other words, smaller grains are more resistant to crushing and larger grains are more vulnerable to breakage. The grain crushing test findings (Fig. 5.22) are also consistent with the particle breakage results obtained in monotonic triaxial shearing (see Fig. 5.17) where larger grains exhibited higher particle breakage. Figure 5.22 indicates that the degree of scatter of the strength data from its best-fit line is higher for recycled ballast than for fresh aggregates. This is attributed to the heterogeneity of

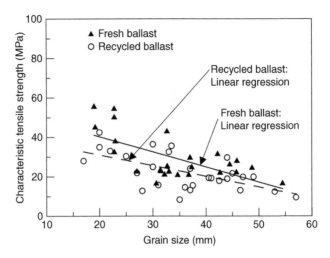

Figure 5.22 Single grain crushing strength of fresh and recycled ballast (modified after Indraratna and Salim, [16]).

recycled ballast (obtained from different sources and mixed together), whereas, fresh ballast contains relatively homogeneous minerals.

5.3 BALLAST RESPONSE UNDER CYCLIC LOADING

5.3.1 Settlement response

The deformation behaviour of fresh and recycled ballast under representative cyclic loading was investigated in the laboratory using the prismoidal triaxial apparatus. These model tests were conducted in both dry and wet states to study the effects of saturation. Figures 5.23(a) and (b) show the settlement of fresh and recycled ballast (dry and wet) with and without inclusion of geosynthetics. As expected, fresh dry ballast gives the least settlement (Fig. 5.23a). It is believed that the higher angularity of fresh ballast contributes to better particle interlock and therefore, causes less settlement. Recycled ballast alone, being less angular, exhibits significantly higher settlement than fresh ballast, especially when wet (saturated). This is not surprising given that reduced angularity of recycled ballast results in lower friction angle (see Fig. 5.14) and lower deformation modulus (Fig. 5.15) compared to fresh ballast. The cyclic test results reveal that wet recycled ballast (without any geosynthetic inclusion) generates the highest settlement (Fig. 5.23b). This is because water acts as a lubricant, thereby reducing frictional resistance and promoting particle slippage.

Figure 5.23 depicts the benefits of using geosynthetics in recycled ballast (both dry and wet). Each of the three types of geosynthetics used by the authors decreases the settlement of recycled ballast considerably. However, the geocomposite (geogrid bonded with non-woven geotextiles) stabilises recycled ballast remarkably well, as revealed in the test results (Fig. 5.23). The combination of reinforcement by the geogrid and the filtration and separation functions provided by the non-woven geotextile component

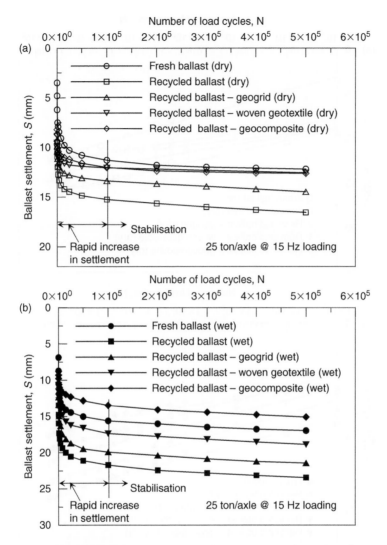

Figure 5.23 Settlement response of fresh and recycled ballast under cyclic loading, (a) in dry condition, and (b) in wet condition.

(of the geocomposite) minimises the lateral spreading and fouling of recycled ballast, especially when wet. The non-woven geotextile also prevents the fines moving up from the capping and subgrade layers, thus, keeps recycled ballast relatively clean.

In contrast, the geogrid can only stabilise recycled ballast marginally, especially in wet conditions, because its large apertures (>25 mm) cannot prevent the fines migrating from the capping and subgrade layers. The woven-geotextile decreases the settlement of recycled ballast effectively when dry (Fig. 5.23a). However, owing to its limited filtration capacity with the aperture size less than 0.30 mm, the woven-geotextile is not as effective as the geocomposite when wet (Fig. 5.23b). Despite these

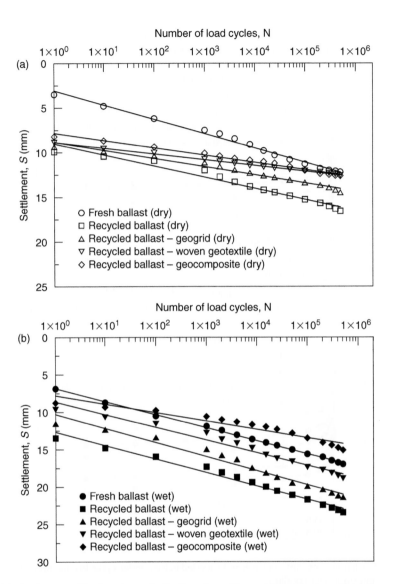

Figure 5.24 Settlement of fresh and recycled ballast plotted in semi-logarithmic scale (a) dry specimens, (b) wet specimens.

differences in the settlement behaviour, Figure 5.23 shows one common feature: initially the settlement increases rapidly in all specimens. It is also noted that all ballast specimens stabilise within about 100,000 load cycles, beyond which the settlement increase is marginal.

The settlement data of Figure 5.23 are re-plotted in a semi-logarithmic scale, as shown in Figure 5.24. The non-linear variation of ballast settlement with increasing load cycles (Fig. 5.23) becomes linear in the semi-logarithmic plot (Fig. 5.24). The linear trend lines of settlement data are shown as solid lines in these figures. Figure 5.24

reveals that the inclusion of geosynthetics in recycled ballast decreases the settlement in the first cycle of loading, and also decreases the overall settlement rate. These model test results also indicate that ballast settlement under cyclic loading may be represented by a semi-logarithmic relationship as given by:

$$S = a + b \cdot \ln N \tag{5.2}$$

where, S is the ballast settlement, N is the number of load cycles, and a and b are two empirical constants, depending on the type of ballast, type of geosynthetics used, initial density and the degree of saturation.

5.3.2 Strain characteristics

The difference between the settlement of ballast at the sleeper/ballast interface and the settlement of ballast/capping interface (measured by the settlement plates) was used to calculate the average vertical strain (major principal strain, ε_1) of the load bearing ballast layer. Figures 5.25(a) and (b) show the average vertical strain of ballast against the number of load cycles plotted in a semi-logarithmic scale for the dry and wet specimens, respectively. These results demonstrate an appreciable reduction in the vertical strain of recycled ballast when geosynthetics are included. In particular, all three types of geosynthetics used by the authors decrease the vertical strain of recycled ballast in dry state (Fig. 5.25a). However, in wet conditions, the geocomposite appears to be the most effective, where the vertical strain of recycled ballast decreases close to that of fresh ballast without geosynthetics (Fig. 5.25b), for the same reasons given in Section 5.3.1. Geogrid alone decreases the vertical strain of recycled ballast marginally when wet, and the woven-geotextile stabilises recycled ballast moderately in saturated condition.

Figure 5.25 indicates that the vertical strain of ballast linearly increases with the logarithm of load cycles, and may be expressed by a function similar to Equation 5.2, as given by:

$$\varepsilon_1 = c + d \cdot \ln N \tag{5.3}$$

where, ε_1 is the major (vertical) principal strain, N is the number of load cycles, and c and d are two empirical constants.

The lateral strains of ballast (intermediate principal strain ε_2, and minor principal strain ε_3) were calculated from the lateral deformation measurements of the vertical walls and the initial lateral dimensions of the test specimens. The lateral strain perpendicular to the sleeper (i.e. parallel to the rails) is the intermediate principal strain (ε_2), which corresponds to the intermediate principal stress (σ_2). The strain parallel to the sleeper is the minor principal strain (ε_3) and it corresponds to the minor principal stress (σ_3).

The intermediate principal strains (ε_2) of ballast are plotted against the logarithm of load cycles, as shown in Figs. 5.26(a) and 5.26(b). These results reveal that initially recycled ballast (without stabilisation) gives higher lateral strain because of its reduced angularity and friction compared to fresh ballast. These laboratory results also indicate that at increased load cycles, the intermediate principal strains of both fresh and

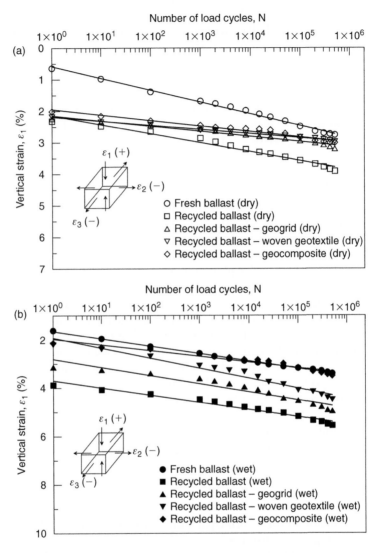

Figure 5.25 Vertical strain of ballast layer under cyclic loading, (a) in dry condition, and (b) in wet condition.

recycled ballast almost converge to one value. Inclusion of geosynthetics in recycled ballast decreases the intermediate principal strain in both dry and wet conditions. The superiority of the geocomposite over the other two types of geosynthetics in terms of minimising lateral strain is convincing, as revealed in Figure 5.26. It is also noted that the use of geosynthetics in recycled ballast decreases ε_2 below that of fresh ballast at higher number of load cycles. The model test results indicate that the intermediate principal strain may also be represented by a semi-logarithmic function, similar to Equation 5.2.

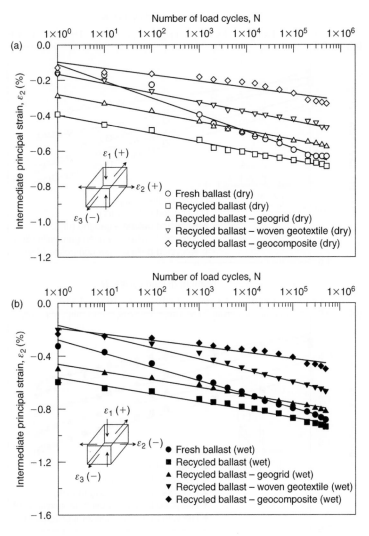

Figure 5.26 Intermediate principal strain of ballast under cyclic loading, (a) in dry condition, and (b) in wet condition.

The variations of minor principal strain (ε_3) of ballast with increasing load cycles are shown in Figures 5.27(a) and (b) in semi-logarithmic scales. These results reveal that both the geocomposite and woven-geotextile decrease the minor principal strain of recycled ballast effectively, whether dry or wet. In contrast, the geogrid decreases the lateral strain of recycled ballast only slightly. Figure 5.27(b) clearly shows that recycled ballast gives significantly higher lateral strain (ε_3) compared to fresh ballast in saturated conditions. The test results also indicate that the minor principal strain of recycled ballast decreases appreciably when stabilised with the geocomposites. A decrease in the rate of lateral strain in recycled ballast (i.e. slope) by the use of woven-geotextiles

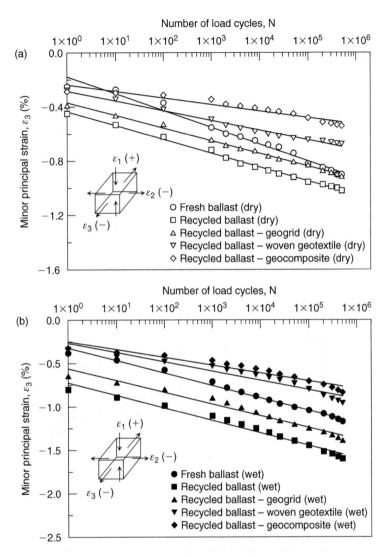

Figure 5.27 Minor principal strain of ballast under cyclic loading, (a) in dry condition, and (b) in wet condition.

or geocomposites is clearly evident in these results (Fig. 5.27a). More significantly, recycled ballast stabilised with geocomposites or woven-geotextiles, exhibits lateral strain (ε_3) less than that of fresh ballast (without any geosynthetics) at higher number of load cycles. This has significant bearing in the maintenance of rail tracks. The reduction in the lateral movement of ballast with the inclusion of geocomposites decreases the need for additional layers of crib and shoulder ballast during maintenance operation.

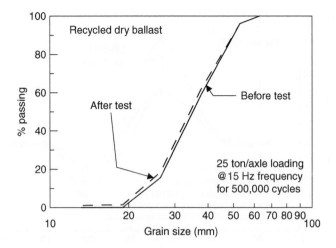

Figure 5.28 Conventional plot of particle size distribution of ballast before and after test (modified after Indraratna and Salim, [16]).

5.3.3 Particle breakage

In order to quantify particle breakage under cyclic loading, the load bearing ballast layer was isolated from the crib ballast and capping layer by placing thin loose geotextiles above and below the load bearing ballast. These loose geotextiles did not resist any lateral movement; they were separators only and useful in recovery of complete ballast specimens at the end of testing. Each specimen was sieved before and after the test, and changes in ballast grading were recorded. Figure 5.28 shows the change in particle size distribution in a conventional gradation plot. Only one specimen data (dry recycled ballast) is shown in this figure for the purpose of clarity.

Since small changes in particle size cannot be clearly illustrated in conventional gradation plots (Fig. 5.28), an alternative method was adopted, as explained earlier in Section 5.1.3. Figures 5.29(a) and (b) show the variations of ΔW_k with various sieve sizes for the dry and wet specimens, respectively.

Figure 5.29 indicates that recycled ballast alone suffers higher particle breakage than fresh ballast, either wet or dry. Use of geosynthetics decreases the breakage of recycled ballast almost to that for fresh ballast without any geosynthetics. It is also clear from this figure that larger particles are more vulnerable to breakage. This observation is in agreement with the lower tensile strength of larger grains found in single particle crushing tests (see Fig. 5.22). The values of Marsal's [5] breakage index B_g, for the fresh and recycled ballast with and without inclusion of geosynthetics are given in Table 5.3.

It may be concluded that particle breakage in recycled ballast is approximately 95–97% higher than fresh ballast. Saturation increases ballast degradation slightly (about 8%). Geosynthetics (either geogrid, woven-geotextile or geocomposite) decrease the breakage of recycled ballast by 40–48%, which means the breakage index (B_g) comes down close to the value of fresh ballast without any geosynthetics.

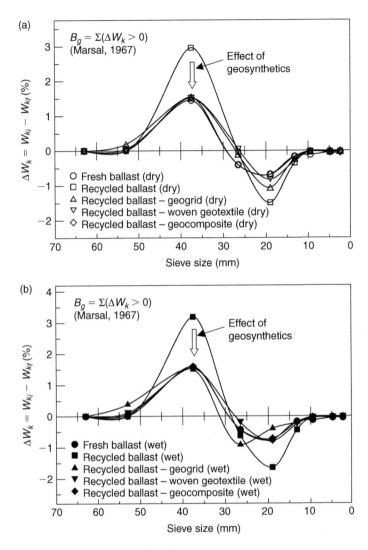

Figure 5.29 Change in particle size of ballast under cyclic loading, (a) in dry conditions, and (b) in wet conditions.

5.4 BALLAST RESPONSE UNDER REPEATED LOADING

The stress-strain response of recycled dry ballast under repeated loading carried out by the authors in the prismoidal triaxial chamber at different intervals of cyclic loading is shown in Figure 5.30. Before the cyclic load was applied, the stiffness of the recycled ballast was relatively low. This is because the ballast was relatively loose (initial bulk unit weight = 15.3 kN/m³) at the beginning of loading. With the increase in vertical load and associated deformation during the first cycle of repeated load, the aggregates

Table 5.3 Particle breakage of ballast under cyclic loading.

Type of Ballast	Type of Geosynthetics used	Test Condition	B_g
Fresh Ballast	–	Dry	1.50
Fresh Ballast	–	Wet	1.63
Recycled Ballast	–	Dry	2.96
Recycled Ballast	–	Wet	3.19
Recycled Ballast	Geogrid	Dry	1.70
Recycled Ballast	Geogrid	Wet	1.88
Recycled Ballast	Woven-Geotextile	Dry	1.56
Recycled Ballast	Woven-Geotextile	Wet	1.64
Recycled Ballast	Geocomposite	Dry	1.54
Recycled Ballast	Geocomposite	Wet	1.60

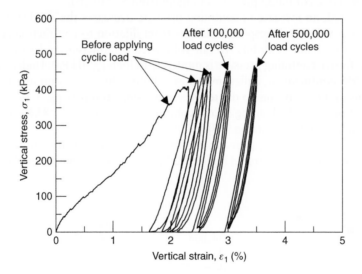

Figure 5.30 Stress-strain plots in repeated load test at various stages of cyclic loading (modified after Indraratna and Salim, [16]).

re-arranged themselves, therefore, the void ratio decreased, which resulted in higher stiffness (Fig. 5.30).

The unloading path (Fig. 5.30) indicates a non-linear resilient behaviour with some strain recovery, while the plastic strain remains significant after unloading was completed. The reloading path apparently becomes almost linear with increasing strain, while the subsequent unloading path remains non-linear. Each loading-unloading path generates a hysteresis loop. The area covered in the loop represents the amount of energy dissipated during that loading-unloading stage. Figure 5.30 also indicates that during the initial stage of cyclic loading (cycles 1–5), the mean slope of the hysteresis loop increases rapidly with the higher number of cycles. This confirms that the resilient modulus of ballast increases with the increase in load repetition. As the load cycle increases, the resilient modulus increases further (Fig. 5.30), as a result of cyclic shakedown and densification [17, 18].

5.5 EFFECT OF CONFINING PRESSURE

Although the effective confining pressure is recognised as a key parameter governing the strength and deformation behaviour of geomaterials, it is often overlooked in the design of railway tracks. The track substructure is essentially self-supporting with minimal lateral constraint provided by the frictional resistance of load bearing ballast and shoulder ballast. The effects of confining pressure (σ_3') on the deformation and breakage of ballast under drained cyclic loading have been studied by Indraratna et al. [19] and Lackenby [20] using the large-scale triaxial apparatus (Fig. 4.1) with a cyclic actuator. Figure 5.31 shows the variations of volumetric strain with increasing axial strain for effective confining pressures ranging from 8–240 kPa.

As expected, an increase in confining pressure decreases the axial strain and the corresponding volumetric response changes from dilation to contraction (Fig. 5.31). It is evident that ballast exhibits overall volumetric contraction under cyclic loading when the effective confining pressure is moderate to high (≥ 30 kPa). Under monotonic loading at low confinement, ballast indicates a slight volumetric contraction at smaller axial strains and then begins to dilate with increasing strains (Figs. 5.1–5.2). In contrast, ballast under cyclic loading at low confining pressures (e.g. ≤ 8 kPa) dilates with no appreciable initial contraction [19].

The effects of confining pressure on the axial and volumetric strains of ballast under cyclic loading are illustrated Figure 5.32 [20]. These results clearly show that the

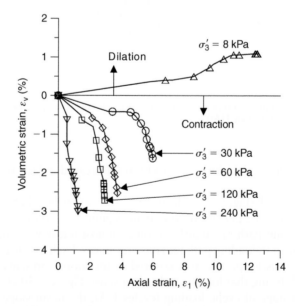

Figure 5.31 Volumetric response of ballast under cyclic loading at various confining pressures (modified after Indraratna et al., [19]).

axial strain decreases with the increasing confining pressure. Ballast exhibits volume increase (i.e. dilation, represented by negative volumetric strain) at small confining pressure ($\sigma'_3 < 30$). As the confining pressure increases ($\geq 30\,$kPa), ballast becomes progressively more contractive (Fig. 5.32).

Under cyclic loads, the effects of confining pressure on the breakage of ballast are shown in Figure 5.33. The breakage intensity is divided into two regions, namely 'stable' and 'unstable' zones [19]. At low confining pressure (1–8 kPa), ballast deforms rapidly and axial strain becomes large (Fig. 5.32) because of large lateral expansion and volumetric dilation. This higher axial strain causes increased particle breakage (Fig. 5.33) and this behaviour is consistent with the laboratory findings presented earlier (Fig. 5.19). Due to small confining pressure, specimens in the unstable degradation zone are characterised by a limited co-ordination number and small particle-to-particle contact areas [19].

As the confining pressure increases (45–60 kPa), axial strain decreases significantly due to the change in volumetric behaviour from dilation to contraction. At this pressure range, small axial strain in combination with moderate interparticle contact stress decreases the extent of particle breakage. As the confining pressure increases further

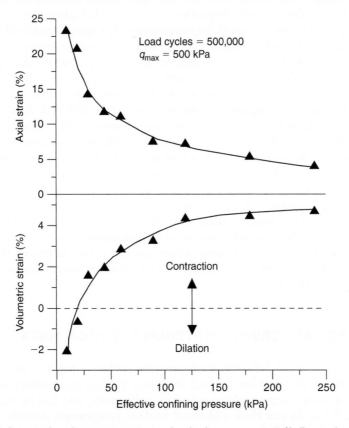

Figure 5.32 Influence of confining pressure on axial and volumetric strains of ballast under cyclic loading (modified after Lackenby, [20]).

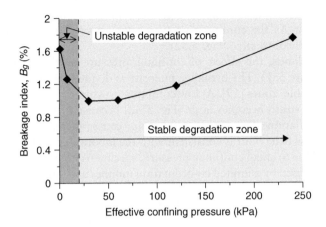

Figure 5.33 Effect of confining pressure on particle breakage (modified after Indraratna et al., [19]).

(>60 kPa), interparticle contact stress becomes increasingly high and the scope of particles to slide and roll over each other is thereby reduced. At higher confining pressures, although the axial strains are small, higher interparticle contact stress leads to increasing grain breakage. Particles in this stress domain fail not only at the beginning of loading when the axial strain rates are the greatest, but also by fatigue at higher load cycles. Based on the available test results, Indraratna et al. [19] concluded that the optimum confining pressure where the breakage would be minimal appeared to be at the start of the stable degradation zone, (approximately 45–60 kPa). Although these preliminary results indicate the significant role of confining pressure, further laboratory experiments at higher q_{max} will need to be conducted to verify optimum confining stress.

Even though the direct measurements of in-situ track confining pressures are not available, it is estimated that the induced lateral stresses in track are in the order of 10–20 kPa [19], which clearly fall in the unstable degradation zone (Fig. 5.33). An increase in track confinement is expected to reduce particle breakage, increase the bearing capacity and dynamic resilience, thereby improve the track performance. The following simple techniques were proposed by Indraratna et al. [19] for increasing ballast confinement: (1) by decreasing sleeper spacing, (2) by increasing the height of shoulder ballast, and (3) by using intermittent lateral restraints at various parts of the track (Chapter 3, Fig. 3.27).

5.6 ENERGY ABSORBING MATERIALS: SHOCK MATS

The ballast in a typical ballasted track provides resiliency for low frequency loading (secondary suspension) but for high frequency loading (i.e. primary suspension), other resilient components such as rail pads, soffit mats, and ballast mats are necessary (Fig. 5.34). In fact these additional resilient components actually restore the elasticity to the ballast. Ballast mats below the ballast layer are mostly suitable to help mitigate ground vibration on viaducts (bridge) and protect the concrete deck, and for

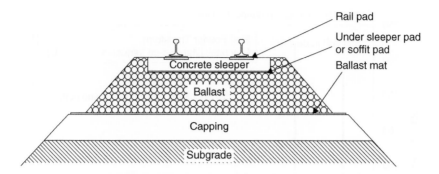

Figure 5.34 Overview of various elastic elements in a ballasted track.

mitigating structural noise. They also prevent particles of ballast from being crushed which improves the durability of the track [21].

In situations where it is highly imperative to use a reduced thickness of ballast such as on a bridge deck, ballast mats are preferred because they protect against degradation [22]. Soffit pads are usually used below concrete sleepers so they are also called under sleeper pads (USPs). Soffit pads are quite effective in reducing the vertical transfer of dynamic stresses because they increase the contact area which subsequently reduces the contact stresses between the sleeper and particles of ballast. The use of USPs has increased in recent years, mainly in the newly built high speed railway tracks in Central Europe [21]. A stiff track structure can create severe dynamic loading under operating conditions, leading to significant failure of track components and a subsequent increase in maintenance. Installing resilient mats such as rubber pads (ballast mat, soffit pad) in rail tracks can attenuate the dynamic forces and improve overall performance. The ability of ballast mats to reduce structural noise and vibration under the rail tracks has been studied extensively, but there is lack of proper studies which deal with the benefits of ballast mat and soffit pad in reducing ballast degradation. In view of this, a series of laboratory tests has been carried out to evaluate the effectiveness of shock mats in mitigating ballast breakage.

The impact load-time histories under a single impact load are shown in Figs. 5.35 and 5.36 for steel base and sand base conditions respectively. Two types of distinct peak forces viz. P_1; an instantaneous sharp peak with very high frequency and P_2; a gradual peak of smaller magnitude with relatively lesser frequency are observed during impact loading. It is also evident that, often multiple P_1 type peaks occur followed by the distinct P_2 type peak. The multiple P_1 peaks are caused when the drop hammer is unrestrained vertically, so that after the first impact it is rebounded to impact the specimen again.

The observed benefits of a shock mat are twofold: (a) it attenuates the impact force and (b) it reduces the impulse frequencies thereby extending the time duration of impact. Even without a shock mat, a ballast bed on a weak subgrade leads to a decreased magnitude and increased duration of impact force as compared to a stiffer subgrade (Figs. 5.35 and 5.36). Not surprisingly, the benefits of shock mats in softer subgrade will be less pronounced, as a weak subgrade itself serves as a flexible cushion, hence the beneficial role of the ballast mat remains under-utilised. Naturally the shock

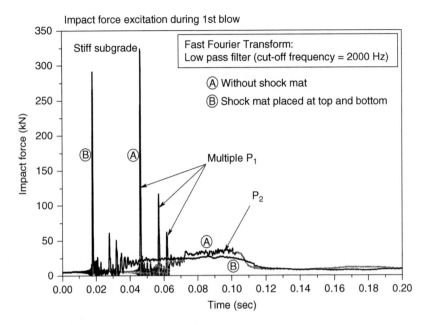

Figure 5.35 Typical impact force responses observed for steel base.

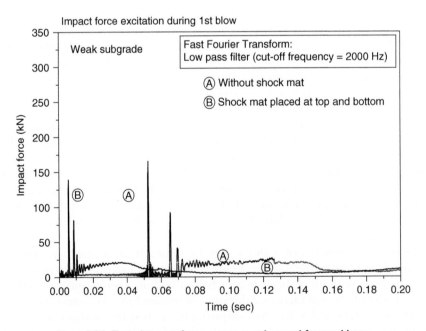

Figure 5.36 Typical impact force responses observed for sand base.

Table 5.4 Particle breakage of ballast under impact loading.

Test No.	Base condition	Shock mat details	BBI
1	Steel	Without shock mat	0.170
2	Steel	Shock mat at top of ballast	0.145
3	Steel	Shock mat at bottom of ballast	0.129
4	Steel	Shock mat at top and bottom of ballast	0.091
5	Sand	Without shock mat	0.080
6	Sand	Shock mat at top of ballast	0.055
7	Sand	Shock mat at bottom of ballast	0.056
8	Sand	Shock mat at top and bottom of ballast	0.028

mats would provide the optimum effect for subgrade of high impedance, i.e. rigid foundations, where the track is laid on a bridge deck with a reduced ballast thickness or a track supported on a rock foundation.

After each test, the ballast sample was sieved and the change in gradation was obtained. The breakage is quantified using the parameter, Ballast Breakage Index (BBI), proposed earlier by Indraratna et al., [23].

$$BBI = A/(A + B) \tag{5.4}$$

where, A is the shift in the PSD curve after the test, and B is the potential breakage or the area between the arbitrary boundary of maximum breakage and the final PSD.

The BBI values obtained from all the tests are presented in Table 5.4. The larger breakage of ballast particles can be attributed to the considerable non-uniform stress concentrations occurring at the corners of sharp angular fresh ballast particles under high impact induced contact stresses.

The ballast breakage is more pronounced for steel base than that for the relatively softer sand base. A lesser breakage is observed when shock mat is placed at bottom for steel base and at top for sand base. Placement of shock mats at the top and bottom of the ballast is the best combination that provides the minimum breakage.

However, use of concrete base is preferred to simulate the concrete bridge deck condition. Also use of capping and natural subgrade layer is encouraged to simulate real track substructure. As discussed in previous sections, prismoidal test chamber is best suited where independent major and minor principal stresses can be applied in mutually orthogonal directions permitting development of lateral strains in a direction parallel to the sleeper during loading. Therefore a large prismoidal test rig which can accommodate specimens 700 mm long, 600 mm wide, and 700 mm high can be used for more accurate analysis of composite track structures subjected to high impact loads (Fig. 5.37).

Pressure cells placed at top and bottom of ballast can record transient stresses transferred through the ballast during the impact event. The springs connected to lateral movable walls can provide the information of transient stresses developed during the loading. Settlement pegs and electronic potentiometers can monitor vertical and lateral movement of walls as discussed previously in section 4.3.1. A comprehensive research on this topic is currently in progress at the University of Wollongong.

Figure 5.37 (a) Top of Prismoidal test chamber showing rail-sleeper assembly instrumented with potentiometers, and (b) springs and potentiometers attached to a vertical wall of the chamber.

REFERENCES

1. Indraratna, B. and Salim, W.: Shear strength and degradation characteristics of railway ballast. *Proc. 14th Southeast Asian Geotechnical Conference*, Hong Kong, Vol. 1, 2001, pp. 521–526.
2. Salim, M.W.: Deformation and degradation aspects of ballast and constitutive modelling under cyclic loading. PhD Thesis, University of Wollongong, Australia, 2004.
3. Indraratna, B., Ionescu, D. and Christie, H.D.: Shear behaviour of railway ballast based on large-scale triaxial tests. *J. of Geotechnical and Geoenvironmental Engineering*, ASCE, Vol. 124, No. 5, 1998, pp. 439–449.
4. Indraratna, B., Salim, W., Ionescu, D. and Christie, D.: Stress-strain and degradation behaviour of railway ballast under static and dynamic loading, based on large-scale triaxial testing. *Proc. 15th Int. Conf. on Soil Mech. and Geotech. Engg*, Istanbul, Vol. 3, 2001, pp. 2093–2096.
5. Marsal, R.J.: Large scale testing of rockfill materials. *J. of the Soil Mech. and Found. Div.*, ASCE, Vol. 93, No. SM2, 1967, pp. 27–43.
6. Marachi, N.D., Chan, C.K. and Seed, H.B.: Evaluation of properties of rockfill materials. *J. of the Soil Mech. and Found. Div.*, ASCE, Vol. 98, No. SM1, 1972, pp. 95–114.
7. Charles, J.A. and Watts, K.S.: The influence of confining pressure on the shear strength of compacted rockfill. *Geotechnique*, Vol. 30, No. 4, 1980, pp. 353–367.
8. Indraratna, B., Wijewardena, L.S.S. and Balasubramaniam, A.S.: Large-scale triaxial testing of greywacke rockfill. *Geotechnique*, Vol. 43, No.1, 1993, pp. 37–51.
9. Raymond, G.P. and Davies, J.R.: Triaxial tests on dolomite railroad ballast. *J. of the Geotech. Engg. Div.*, ASCE, Vol. 104, No. GT6, 1978, pp. 737–751.
10. Marsal, R.J.: Mechanical properties of rockfill. In: *Embankment Dam Engineering*. Casagrande Volume, Wiley, New-York, 1973, pp. 109–200.
11. Indraratna, B. and Salim, W.: Modelling of particle breakage of coarse aggregates incorporating strength and dilatancy. *Geotechnical Engineering*, Proc. Institution of Civil Engineers, London, Vol. 155, Issue 4, 2002, pp. 243–252.
12. Bohac, J., Feda, J. and Kuthan, B.: Modelling of grain crushing and debonding. *Proc. 15th Int. Conf. on Soil Mech. and Geotech. Engg.*, Istanbul, Vol. 1, 2001, pp. 43–46.
13. McDowell, G.R. and Bolton, M.D.: On the micromechanics of crushable aggregates. *Geotechnique*, Vol. 48, No. 5, 1998, pp. 667–679.
14. Nakata, Y., Kato, Y. and Murata, H.: Properties of compression and single particle crushing of crushable soil. *Proc. 15th Int. Conf. on Soil Mech. and Geotech. Engg.*, Istanbul, Vol. 1, 2001, pp. 215–218.
15. Lade, P.V., Yamamuro, J.A. and Bopp, P.A.: Significance of particle crushing in granular materials. *J. of Geotech. Engg.*, ASCE, Vol. 122, No. 4, 1996, pp. 309–316.
16. Indraratna, B. and Salim, W.: Deformation and degradation mechanics of recycled ballast stabilised with geosynthetics. *Soils and Foundations*, Vol. 43, No. 4, 2003, pp. 35–46.
17. Festag, G. and Katzenbach, R.: Material behaviour of dry sand under cyclic loading. *Proc. 15th Int. Conf. on Soil Mech. and Geotech. Engg.*, Istanbul, Vol. 1, 2001, pp. 87–90.
18. Suiker, A.S.J.: *The mechanical behaviour of ballasted railway tracks*. PhD Thesis, Delft University of Technology, The Netherlands, 2002.
19. Indraratna, B., Khabbaz, H., Salim, W., Lackenby, J. and Christie, D.: Ballast characteristics and the effects of geosynthetics on rail track deformation. *Int. Conference on Geosynthetics and Geoenvironmental Engineering*, Mumbai, India, December 2004, (in press).
20. Lackenby, J.: *Factors affecting ballast degradation under cyclic loads*. PhD Thesis, University of Wollongong, Australia, 2005 (in preparation).

21. Esveld, C.: The significance of track resilience. *European Rail Review* Digital News Issue 10, 2009, pp. 14–20.
22. Bachmann, H., Ammann, W. and Deischl, F.: *Vibration Problems in Structures: Practical Guidelines.* Springer Verlag, 1997, Berlin.
23. Indraratna, B., Lackenby, J. and Christie, D.: Effect of confining pressure on the degradation of ballast under cyclic loading. *Geotechnique*, Vol. 55, No. 4, 2005, 325–328.

Chapter 6

Existing Track Deformation Models

Until today, the vast majority of railway design engineers have regarded ballast as an elastic granular medium. Although the accumulation of plastic deformation under cyclic traffic loading is evident, most researchers are primarily interested in modelling the dynamic resilient modulus of ballast. Limited research has been conducted on the constitutive modelling of ballast under cyclic loading, while some researchers have attempted to simulate the plastic deformation empirically. Despite spending a consider-able annual sum for the construction and maintenance of railway tracks, the design is still predominantly empirical in nature (Suiker, [1]). A large number of researchers have modelled the elasto-plastic deformation of sand and other granular media under monotonic and cyclic loadings. As ballast is comprised of coarse aggregates, these elasto-plastic deformation models are useful and may serve as a framework for devel-oping a specific model to simulate ballast behaviour including plastic deformation and particle breakage under cyclic loading.

6.1 PLASTIC DEFORMATION OF BALLAST

Various researchers have empirically modelled the permanent deformation of ballast under cyclic loading. Shenton [2] represented the ballast strain at any number of load cycles with the strain at the first cycle of loading and the logarithm of the number of load cycles, as given below:

$$\varepsilon_N = \varepsilon_1(1 + 0.2 \log_{10} N) \tag{6.1}$$

where, ε_N = average vertical strain of ballast at load cycle N, ε_1 = average vertical strain at load cycle 1, and N = number of load cycles.

A similar logarithmic function of load cycles was presented by Indraratna et al. [3, 4] when modelling the plastic deformation of ballast with/without geosynthetic reinforcement, where the settlement is given by:

$$S = a + b \log N \tag{6.2}$$

where, S = ballast settlement, N = number of load cycles, and a and b are empirical constants.

Stewart [5] conducted a series of variable amplitude cyclic triaxial tests on ballast and concluded that the predicted strains based on the superposition of ballast strains for the various load magnitudes using an equation similar to Equation 6.1 agreed well with the experimental results.

Shenton [6] presented an empirical model for the ballast settlement based on extensive field data and is given by:

$$S = K_1 N^{0.2} + K_2 N \tag{6.3}$$

where, S is the ballast settlement; K_1, K_2 are empirical constants, and N = total number of axles (or cycles). Shenton considered that the settlement of ballast is composed of two parts: the first component $(K_1 N^{0.2})$ predominates up to 1 million load cycles, and the second part $(K_2 N)$ is only a small portion of the settlement and becomes relatively insignificant above 1 million load cycles.

Raymond and Bathurst [7] correlated the track settlement to the logarithm of total tonnage based on the available field data, as shown below:

$$S_e(t) = a_r + a_0' \log\left(\frac{t}{t_r}\right) \tag{6.4}$$

where, $S_e(t)$ = mean ballast settlement over unit length at tonnage t, a_r = settlement at the reference tonnage, a_0' = slope of the semi-logarithmic relation, t_r = reference tonnage taken as 2 million ton, and t = total tonnage.

Chrismer and Selig [8] modelled ballast strain as a power function of the number of load cycles:

$$\varepsilon_N = \varepsilon_1 N^b \tag{6.5}$$

where, ε_N is the permanent strain after N load cycles, ε_1 is the strain at the first load cycle, b is a constant, and N is the number of load cycles. They concluded that the power equation represents ballast strain better than the logarithmic models.

Similarly, Indraratna et al. [9] and Ionescu et al. [10] reported that a power function best represented their ballast experimental data, as given by:

$$S = S_1 N^b \tag{6.6}$$

where, S = ballast settlement after N number of load cycles, S_1 = settlement after the first load cycle, b = empirical constant, and N = number of load cycles.

Recently, Suiker [1] and Suiker and de Borst [11] developed a plastic deformation model for ballast, where both plastic 'frictional sliding' and 'volumetric compaction' mechanisms have been considered during cyclic loading. They called it the 'Cyclic Densification Model', where the plastic flow rule has been decomposed into a frictional contribution and a compaction component, as given by:

$$\frac{d\varepsilon_{ij}^p}{dN} = \frac{d\kappa^p}{dN} m_{ij}^f + \frac{d\varepsilon_{vol,c}^p}{dN} m_{ij}^c \tag{6.7}$$

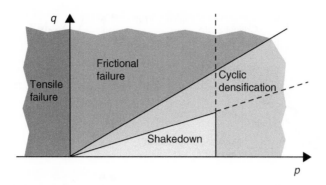

Figure 6.1 Four response regimes during cyclic loading (modified after Suiker, [1] and Suiker and de Borst, [11]).

where, $d\varepsilon_{ij}^{p}$ is the infinitesimal increment of plastic strain, $d\kappa^{p}$ is the increment of plastic distortional strain, $d\varepsilon_{vol,c}^{p}$ is the plastic volumetric strain increment due to cyclic compaction, m_{ij}^{f} and m_{ij}^{c} denote the flow directions for frictional sliding and volumetric compaction, respectively, and dN is the increment of load cycle.

Suiker and de Borst [11] divided the stress domain into four regimes:

- The shakedown regime where the cyclic response of ballast is fully elastic,
- The cyclic densification regime where progressive plastic deformation occurs under cyclic loading,
- The frictional failure regime where frictional collapse occurs due to cyclic stress level exceeding the static maximum strength, and
- Tensile failure regime where non-cohesive granular materials disintegrate due to induced tensile stresses.

These stress regimes are shown in Figure 6.1 in the *p-q* plane, where, *p* and *q* are the mean effective normal stress and deviator stress (invariants), respectively.

The cyclic densification model [1, 11] is an advanced step in modelling plastic deformation and plastic compaction of ballast under cyclic loading. However, particle breakage associated with cyclic loading, an important factor governing the plastic deformation and cyclic compaction of ballast, was not considered in their cyclic model. Therefore, a new constitutive model for ballast incorporating particle breakage has been developed by the authors and is presented in Chapter 7.

6.2 OTHER PLASTIC DEFORMATION MODELS

There are a number of other plasticity models available in the literature which were primarily developed to simulate the plastic deformation behaviour of clays, sands and gravels. However, being granular aggregates, sands and gravels deform in a similar way as of ballast (e.g. volumetric dilation under loading at low confining pressure).

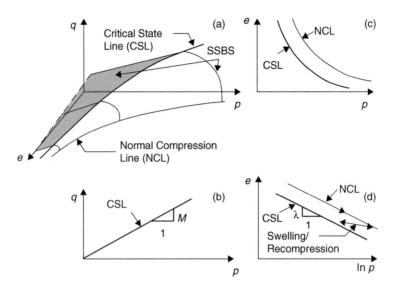

Figure 6.2 Critical state model, (a) state boundary surface, (b) projection of CSL in *q-p* plane, (c) projection of CSL and NCL on *e-p* plane, and (d) CSL and NCL plotted in *e-ln p* plane.

These plasticity models are expected to be useful in simulating the deformation and degradation of ballast under cyclic loading, and therefore, presented and discussed in the following Sections.

6.2.1 Critical state model

In the late 1950's and 1960's, Roscoe and his co-researchers developed a critical state model based on the theory of plasticity and soil behaviour at the critical states [12–15]. They were the first among others who successfully modelled the plastic deformation and the associated volume change behaviour of soils under shear stresses. Their mathematical model to simulate the plastic deformation of clay is known as 'Cam-clay' (Roscoe et al., [13]; Schofield and Wroth, [15]), which was subsequently modified by Roscoe and Burland [14] and is known as 'modified Cam-clay'.

The 'critical state' has been defined as the state at which soil continues to deform at constant stress and constant void ratio (Roscoe et al., [12]). The main features of the critical state model are:

- All possible states of a soil element form a stable state boundary surface (SSBS), as shown in Figure 6.2(a).
- Deformation of soil remains elastic until its stress state reaches the stable state boundary surface, i.e. yielding of soil initiates when a stress path meets the SSBS.
- At the critical state, the energy transmitted to a soil element across its boundary is dissipated within the soil element as frictional heat loss without changing the

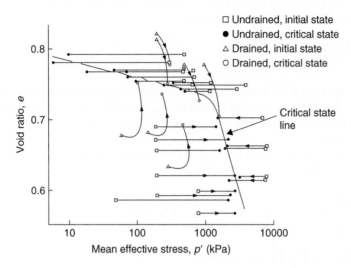

Figure 6.3 Bi-linear critical state line of sands (modified after Been et al., [16]).

stress or volume. Thus, at the critical state, $q = Mp$ (Fig. 6.2b), where, M is the coefficient of friction at the critical state.

- The projection of the critical state line (CSL) on e-p plane is parallel to the Normal Compression Line (NCL) obtained under isotropic compression (Fig. 6.2c). The NCL and the projection of CSL become parallel straight lines when plotted in a semi-logarithmic e-ln$p$ scale (Fig. 6.2d). The swelling and recompression lines are also assumed to be linear in e-ln$p$ plane.

Been et al. [16] studied the critical state/steady state of sands and concluded that the critical state line is approximately bilinear in the e-log p plane, as shown in Figure 6.3. They found an abrupt change in the slope of the critical state line for Leighton Buzzard sand and Erksak sand at about 1 MPa, and attributed this sudden change in the slope of the critical state line to the breakage of particles.

Although the original critical state model [12, 13] was based on extensive laboratory test results of remoulded clay, some researchers attempted to model the deformation behaviour of sands and gravels similar to the critical state (Cam-clay) model. In this respect, Schofield and Wroth [15] presented a critical state model for gravels (Granta-gravel) neglecting elastic component of the volumetric strain.

Jefferies [17] stated that the Cambridge-type models (e.g. Granta-gravel) could not reproduce softening and dilatancy of sands, which are on the dense (dry) side of the critical state line. It was pointed out that the inability of Cambridge-models to dilate is a large deficiency in modelling sand behaviour, as virtually all sands are practically denser than the critical and dilate during shearing. He proposed a critical state model for sand (Nor-sand) assuming associated flow (normality) and infinity of the isotropic normal compression line (NCL). The initial density of sand was incorporated through the state parameter ψ, as defined by Been and Jefferies [18]. Jefferies [17] employed the following dilatancy rule in his model:

$$D = \frac{M - \eta}{1 - N} \tag{6.8}$$

where, $D = \dot{\varepsilon}_p / \dot{\varepsilon}_q$ is a dilatancy function, ε_p and ε_q are strains corresponding to the stresses p and q, a dot superscript represents incremental change, M is the critical state friction coefficient, η is the shear stress ratio $(= q/p)$ and N is a density dependent material property.

Using Equation (6.8) and the normality condition, Jefferies [17] formulated the yield surface for Nor-sand, as given by:

$$\eta = \frac{M}{N}\left[1 + (N - 1)\left(\frac{p}{p_i}\right)^{N/(1-N)}\right] \quad \text{if } N \neq 0 \tag{6.9a}$$

$$\eta = M\left[1 + \ln\left(\frac{p_i}{p}\right)\right] \quad \text{if } N = 0 \tag{6.9b}$$

where, p_i is the mean stress at the image state defined by the condition $\dot{\varepsilon}_p = 0$. A simple hardening rule was used by Jefferies, as given below:

$$\frac{\dot{p}_i}{\dot{\varepsilon}_q} = h(p_{i,\max} - p_i) \tag{6.10}$$

where, h is a proportionality constant and $p_{i,\max}$ is the maximum value of p_i.

The Nor-sand [17] adequately modelled the deformation behaviour of sand including dilatancy, post-peak strain softening, the effects of confining pressure and initial density. However, researchers question the assumption of normality (associated flow) in sand, and therefore, most other researchers used non-associated flow in their formulations [19–23].

6.2.2 Elasto-plastic constitutive models

Lade [19] developed an elasto-plastic constitutive model for cohesionless soils based on the theory of plasticity, non-associated flow, an empirical work-hardening law and curved yield surfaces. He assumed that the total strain increments $d\varepsilon_{ij}$, are composed of three components, (a) elastic increments $d\varepsilon_{ij}^e$, (b) plastic collapse components $d\varepsilon_{ij}^{pc}$, and (c) plastic expansive increments $d\varepsilon_{ij}^{pe}$, such that:

$$d\varepsilon_{ij} = d\varepsilon_{ij}^e + d\varepsilon_{ij}^{pc} + d\varepsilon_{ij}^{pe} \tag{6.11}$$

The elastic strain increments were computed from pressure dependent unloading-reloading elastic modulus, as given by:

$$E_{ur} = K_{ur}p_a\left(\frac{\sigma_3}{p_a}\right)^n \tag{6.12}$$

where, $E_{ur} =$ unloading-reloading elastic modulus, $K_{ur} =$ dimensionless modulus number (constant), $p_a =$ atmospheric pressure, $\sigma_3 =$ confining pressure and n is an exponent.

Lade [19] expressed various yield surfaces and plastic potentials as functions of the stress invariants. Lade used identical formulation for the yield function and the plastic potential in modelling the plastic collapse component of strain, which is given by:

$$f_c = g_c = I_1^2 + 2I_2 \tag{6.13}$$

where, f_c is the yield surface, g_c is the plastic potential, the subscript c indicates plastic collapse, and I_1 and I_2 are the 1st and 2nd invariants of stresses, respectively. In modelling the plastic expansive strain component, Lade [19] employed two different functions for the yield surface and the plastic potential (i.e. non-associated flow), as given by:

$$f_p = (I_1^3/I_3 - 27)(I_1/p_a)^m \tag{6.14a}$$

$$g_p = I_1^3 - [27 + \eta_2(p_a/I_1)^m]I_3 \tag{6.14b}$$

where, I_3 is the third invariant of stresses, η_2 is a constant for the given values of f_p and σ_3, and m is an exponent.

Lade [19] also employed an isotropic work-hardening and softening law, as given by:

$$W_p = F_p(f_p) \tag{6.15}$$

where, $W_p =$ plastic work done and F_p is a monotonically increasing or decreasing positive function. The behaviour of cohesionless soils including dilatancy, strain-hardening and post-peak strain-softening was predicted very well by Lade's model. However, the capability of Lade's model to predict shear behaviour from an anisotropic initial stress state was neither verified nor discussed. This model was verified only for shearing from isotropic initial stress state. For employing a stress-strain constitutive model for the case of a complicated cyclic loading, where stresses are often changing from non-isotropic stress states, the model must be capable of predicting shear behaviour from both isotropic and anisotropic initial stress states.

Pender [20] successfully overcame the limitations of Lade's formulation and developed a constitutive model for the shear behaviour of overconsolidated soils based on the critical state framework, non-associated flow, and the theory of plasticity. He assumed constant stress ratio yield loci and parabolic undrained stress paths, as given by:

$$f = q - \eta_i p = 0 \tag{6.16}$$

$$\left(\frac{\eta - \eta_o}{AM - \eta_o}\right)^2 = \frac{p_{cs}}{p}\left[\frac{1 - \frac{p_o}{p}}{1 - \frac{p_o}{p_{cs}}}\right] \tag{6.17}$$

where, $f =$ yield function,
$\eta_i =$ a given stress ratio $(= q/p)$,

η_o is the initial stress ratio,

A is +1 for loading towards the critical state in compression, and −1 for extension,

p_{cs} is the value of p on the critical state line corresponding to the current void ratio,

p_o is the intercept of the undrained stress path with the initial stress ratio line, and

M is the slope of the critical state line in p-q plane.

Pender [20] assumed the ratio of plastic distortional strain increment ($d\varepsilon_s^p$) to plastic volumetric strain increment ($d\varepsilon_v^p$), as given by:

$$\frac{d\varepsilon_s^p}{d\varepsilon_v^p} = \frac{(AM - \eta_o)^2}{(AM)^2 \left(\frac{p_o}{p_{cs}} - 1\right) \left\{(AM - \eta_o) - (\eta - \eta_o)\frac{p}{p_{cs}}\right\}} \tag{6.18}$$

The general constitutive relationship for the incremental plastic strain is given by Hill [24] as:

$$d\varepsilon_{ij}^p = b\frac{\partial g}{\partial \sigma_{ij}}df \tag{6.19}$$

Combining Equations (6.16–6.19), Pender [20] formulated the following expression for the incremental plastic strains:

$$d\varepsilon_s^p = \frac{2\kappa\left(\frac{p}{p_{cs}}\right)(\eta - \eta_o)d\eta}{(AM)^2(1 + e)\left(\frac{2p_o}{p} - 1\right)\left[(AM - \eta_o) - (\eta - \eta_o)\frac{p}{p_{cs}}\right]} \tag{6.20}$$

$$d\varepsilon_v^p = \frac{2\kappa\left(\frac{p_o}{p_{cs}} - 1\right)\left(\frac{p}{p_{cs}}\right)(\eta - \eta_o)d\eta}{(AM - \eta_o)^2(1 + e)\left(\frac{2p_o}{p} - 1\right)} \tag{6.21}$$

where, κ is the slope of the swelling/recompression line in e-ln p plot.

Pender's model was able to predict non-linear stress-strain behaviour, dilatancy, strain-hardening and post-peak strain-softening aspects of overconsolidated soils during shearing. His model can also be applied to shearing from an initial stress of either isotropic or anisotropic state, which is an essential criterion for modelling the deformation behaviour under cyclic loading. The deformation of ballast under monotonic loading has been modelled by the authors following Pender's simulation technique along with a new plastic dilatancy rule incorporating particle breakage. The new constitutive model is presented and explained in detail in Chapter 7.

Pender [25] also introduced a cyclic hardening parameter to capture the cyclic stress-strain behaviour of soils and extended his previous formulation, as given by:

$$d\varepsilon_s^p = \frac{2\kappa\left(\frac{p}{p_{cs}}\right)(\eta - \eta_o)^{1+\xi}d\eta}{(AM)^2(1+e)\left(\frac{2p_o}{p} - 1\right)(AM - \eta_o)^\xi\left[(AM - \eta_o) - (\eta - \eta_o)\frac{p}{p_{cs}}\right]} \tag{6.22}$$

$$\xi = \left(\frac{|q_p|}{p_{cs}}\right)^{\hat{\alpha}}(H^{\hat{\beta}} - 1) \tag{6.23}$$

where, ξ is the cyclic hardening index, q_p is the change in q in the previous half cycle, H is the number of half cycles, and $\hat{\alpha}, \hat{\beta}$ are soil parameters for cyclic hardening.

Pender [25] considered that the value of cyclic hardening index (ξ) would increase with an increase in the number of half cycles, and therefore, formulated the hardening index (ξ) in an empirical way. He did not relate the cyclic hardening index with cyclic compaction (i.e. densification), which is often observed in cyclic tests of granular aggregates. Ballast usually hardens under cyclic loading due to plastic volumetric compaction (Suiker, [1]) and this aspect of volumetric behaviour is absent in Pender's [25] cyclic model.

Tatsuoka et al. [26] presented a cyclic stress-strain model for sand in plane strain loading. They expressed the relationship between stress and strain of sand under plane strain compression and plane strain extension in terms of an empirical hyperbolic equation, as given by:

$$y = \frac{x}{\frac{1}{C_1} + \frac{x}{C_2}} \tag{6.24}$$

where, $y = \tau/\tau_{max}$
$x = \varepsilon_y/\varepsilon_{yref}$
$\tau = \sigma_{vertical} - \sigma_{horizontal} = $ shear stress, $\tau_{max} = $ maximum shear stress,
$\varepsilon_y = \varepsilon_{vertical} - \varepsilon_{horizontal} = $ shear strain, $\varepsilon_{yref} = $ reference shear strain, and
C_1 and C_2 are the fitting parameters, which also depend on the strain level, x.

Tatsuoka et al. [26] described a set of rules (e.g. proportional rule, external and internal rules, drag rule etc.) to simulate the hysteretic stress-strain relationship under cyclic loading. They proposed a drag parameter, which is a function of plastic shear strain. The drag parameter was employed to simulate the evolution of stress-strain hysteretic loop as the number of load cycle increases. Tatsuoka et al. [26] used the following equations to model plastic dilatancy in plane strain cyclic loading:

$$d = \frac{s(1 + 1/K') + (1 - 1/K')}{s(1 - 1/K') + (1 + 1/K')} \quad \text{for loading} \tag{6.25a}$$

$$d = \frac{s(1 + 1/K') - (1 - 1/K')}{-s(1 - 1/K') + (1 + 1/K')} \quad \text{for unloading} \tag{6.25b}$$

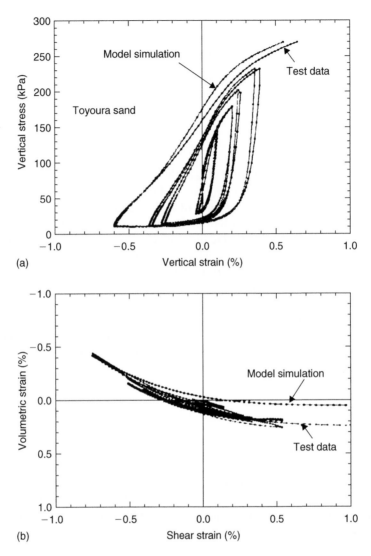

Figure 6.4 Model simulation of sand under plane strain cyclic loading, (a) stress-strain, and (b) volume change behaviour (modified after Tatsuoka et al., [26]).

where, $d = -d\varepsilon^p_{vol}/d\gamma^p$
$s = \sin\phi_{mob}$
$K' = $ model constant
$\phi_{mob} = $ mobilised friction angle.

Although the model was based on empirical formulations, Tatsuoka et al. [26] successfully simulated the stress-strain and volume change behaviour of sand under plane strain cyclic loading, as shown in Figure 6.4. One limitation of their model is that the hyperbolic stress-strain formulation (Equation 6.24) is independent of the plastic volumetric strain resulting from the dilatancy equations (Equations 6.25a and 6.25b), while many other researchers indicate that the volumetric strain significantly affects the stress-strain behaviour of soils including granular assembly [13, 15, 27].

6.2.3 Bounding surface plasticity models

To realistically model the stress-strain behaviour of soils under cyclic loading, some researchers introduced the concept of bounding surface plasticity in their formulations (Dafalias and Herrmann, [22, 23]; Mroz and Norris, [21]). The simple elasto-plastic or non-linear elastic models may be used to simulate the deformation behaviour of soils under monotonic loading with sufficient accuracy. However, for a complex loading system involving loading, unloading and repetitive actions of loads, more complex hardening rules should be examined to simulate cyclic deformation behaviour more realistically (Mroz and Norris, [21]).

The 'bounding surface' concept was originally introduced by Dafalias and Popov [28, 29], and simultaneously and independently by Krieg [30] in conjunction with an enclosed yield surface for metal plasticity. Both the name and concept were inspired from the observation of experimental results that the stress-strain curves converge to specific 'bounds' at a rate, which depends on the distance of the stress point from the bounds. Dafalias and Herrmann [22] presented two different direct bounding surface formulations within the framework of critical state soil plasticity for the quasi-elastic range in triaxial stress space. Dafalias and Herrmann [23] subsequently extended their previous formulations and presented a generalised bounding surface plasticity model in a three-dimensional stress space in terms of stress invariants. Figure 6.5 shows the schematic representation of the bounding surface.

Mroz and Norris [21] examined the qualitative response of a two surface plasticity model and a model with infinite number of loading surfaces under cyclic loading and then developed their formulations in triaxial stress space. The general expression of the plastic strain increment vector is given by [21]:

$$\dot{\varepsilon}^p = \frac{1}{K}\mathbf{n}_g(\mathbf{n}_f^T \cdot \dot{\sigma}) \tag{6.26}$$

where, $\dot{\varepsilon}^p$ is the plastic strain increment vector, $\dot{\sigma}$ is the stress increment vector, $\mathbf{n}_g$ and $\mathbf{n}_f$ are the unit vectors normal to the plastic potential and yield surface, respectively, and K is a scalar hardening modulus.

Mroz and Norris [21] considered that the hardening modulus K, (Equation 6.26) evolves from an initial value on the yield surface K_y, at point P (Fig. 6.6) to a bounding value K_R, at point R on the consolidation surface. The point R on the consolidation surface is a conjugate point of P such that the direction of the unit vector normal to the yield surface at point P is the same as the direction of unit vector normal to the consolidation surface at point R. The evolution of modulus K, depends on the distance between the current stress point P and its conjugate point R. The maximum distance between the yield and consolidation surfaces is given by:

$$K = K_R + (K_y - K_R)\left(\frac{\delta}{\delta_0}\right)^\gamma \tag{6.27}$$

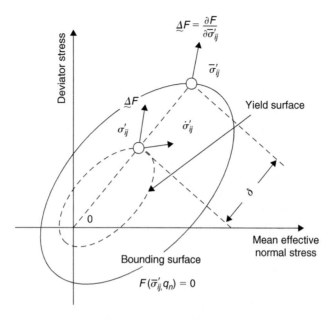

Figure 6.5 Schematic illustration of bounding surface (modified after Dafalias and Herrmann, [23]).

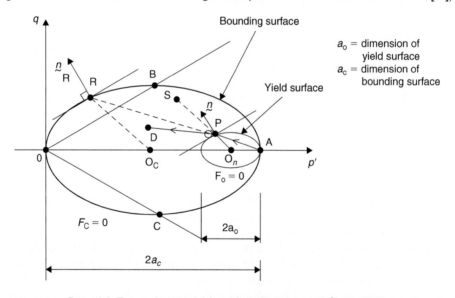

Figure 6.6 Two surface model (modified after Mroz and Norris, [21]).

$$\delta = f(\sigma'_R - \sigma'_P)^{1/2} \tag{6.28}$$
$$\delta_0 = 2(a_c - a_0) \tag{6.29}$$

where, a_c and a_0 are the semidiameters of the consolidation and yield surfaces, respectively (Fig. 6.6), and γ is a constant parameter. Mroz and Norris [21] indicated that the value of δ_0 changes only slightly due to change in density, while δ changes with

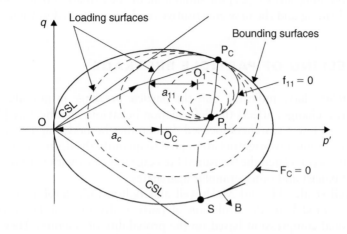

Figure 6.7 Model with infinite number of loading surfaces (modified after Mroz and Norris, [21]).

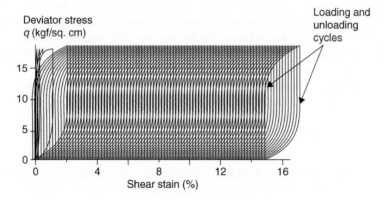

Figure 6.8 Model prediction for undrained cyclic triaxial loading by infinite loading surface hardening (modified after Mroz and Norris, [21]).

the change in stress and depends on the instantaneous positions of the yield and consolidation surfaces.

For the plastic model with infinite number of loading surfaces, Mroz and Norris [21] employed a plastic hardening modulus K, almost similar to Equation (6.27), as given by:

$$K = K_R + (K_y - K_R)(R_1)^\gamma \qquad (6.30)$$

$$R_1 = \frac{a_c - a_{l1}}{a_c} \qquad (6.31)$$

where, a_{l1} is the semidiameter of the first loading surface, $f_{l1} = 0$ (Fig. 6.7).

Although Mroz and Norris [21] had not quantitatively modelled any particular soil, the qualitative aspects of soil behaviour under cyclic loading were well predicted (Fig. 6.8). The authors have been inspired by the concept of varying hardening modulus within the bounding surface [21] for simulating the deformation behaviour of ballast under cyclic loading and the new constitutive model is presented in Chapter 7.

6.3 MODELLING OF PARTICLE BREAKAGE

Many researchers have indicated that the particle breakage in granular geomaterials due to stress changes affects the deformation behaviour significantly [27, 31–33]. However, only a few researchers focused their studies in modelling particle breakage under shearing. Some investigators attempted to quantify the degree of particle breakage, while others correlated the measured breakage indicator with various engineering properties of ballast and other granular aggregates.

McDowell et al. [34] and McDowell and Bolton [35] developed a conceptual and analytical model for the evolution of particle size in granular medium under one-dimensional compression based on the probability of fracture. They considered that the probability of grain fracture is a function of applied stress, particle size and co-ordination number (number of contacts with the neighbouring particles), and postulated that the plastic hardening is due to an increase in specific surface, which must accompany irrecoverable compression caused by particle breakage. McDowell and co-researchers indicated that when particles fracture, the smallest particles are geometrically self-similar in configurations under increasing stress (Fig. 6.9), and that a fractal geometry evolves with successive fracture of the smallest grains.

McDowell and co-researchers [34, 35] also added a fracture energy term to the well-known Cam-clay plastic work equation [13, 15] and is given by:

$$q\delta\varepsilon_q^p + p'\delta\varepsilon_v^p = Mp'\delta\varepsilon_q^p + \frac{\Gamma_s dS}{V_s(1+e)} \tag{6.32}$$

where, $\delta\varepsilon_q^p$ is the increment of plastic shear strain, $\delta\varepsilon_v^p$ is the increment of plastic volumetric strain, dS is the increase in surface area of volume V_s of solids distributed to a gross volume of $V_s(1+e)$, e is the void ratio and Γ_s is the 'surface free-energy'.

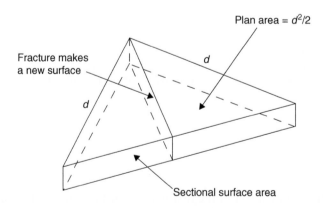

Figure 6.9 Crushing of a triangular particle into two geometrically similar particles (modified after McDowell et al., [34]).

Although McDowell and co-researchers added this surface energy term to the plastic work equation during shear deformation (Equation 6.32), they did not examine the applicability of their formulation nor verify the equation for shearing with available test data. They restricted their study to the volume change behaviour of aggregates caused by particle breakage in one-dimensional compression.

Ueng and Chen [36] particularly studied the effects of grain breakage on the shear behaviour of sands and formulated a useful relationship between the principal stress ratio, rate of dilation, angle of internal friction and the energy consumption due to particle breakage per unit volume during triaxial shearing. Their formulation is given by:

$$\frac{\sigma_1'}{\sigma_3'} = \left(1 + \frac{d\varepsilon_v}{d\varepsilon_1}\right) \tan^2\left(45° + \frac{\phi_f}{2}\right) + \frac{dE_B}{\sigma_3' d\varepsilon_1}(1 + \sin \phi_f) \tag{6.33}$$

where, σ_1' is the major principal stress, σ_3' is minor principal stress, $d\varepsilon_v$ is the volumetric strain increment, $d\varepsilon_1$ is the major principal strain increment, ϕ_f is the angle of internal friction and dE_B is the increment of energy consumption per unit volume caused by particle breakage during shearing.

Ueng and Chen [36] used the increase in specific surface area per unit volume (dS_v) as the indicator of particle breakage and correlated the rate of energy consumption due to particle breakage at failure $(dE_B/d\varepsilon_1)_f$, with the rate of increase in surface area at failure $(dS_v/d\varepsilon_1)_f$, as given by:

$$dE_B = kdS_v \tag{6.34}$$

where, k is a proportionality constant.

Ueng and Chen's [36] formulation is a significant development in modelling particle breakage under triaxial shearing. However, its application is limited to the strength of geomaterials in terms of principal stress ratio during triaxial loading. It cannot be used directly to predict the plastic deformation of ballast under monotonic and cyclic loadings and the associated particle breakage. In the new constitutive model presented in Chapter 7, the authors have employed part of Ueng and Chen's [36] techniques to incorporate particle breakage.

REFERENCES

1. Suiker, A.S.J.: *The mechanical behaviour of ballasted railway tracks.* PhD Thesis, Delft University of Technology, The Netherlands, 2002.
2. Shenton, M.J.: Deformation of railway ballast under repeated loading conditions. In: Kerr (ed.): *Railroad Track Mechanics and Technology.* Proc. of a symposium held at Princeton Univ., 1975, pp. 387–404.
3. Indraratna, B., Salim, W. and Christie, D.: Improvement of recycled ballast using geosynthetics. *Proc. 7th International Conference on Geosynthetics,* Nice, France, 2002, pp. 1177–1182.
4. Indraratna, B., Salim, W. and Christie, D.: Performance of recycled ballast stabilised with geosynthetics. *Conference on Railway Engineering (CORE 2002),* Wollongong, Australia, 2002, pp. 113–120.

5. Stewart, H.E.: Permanent strains from cyclic variable amplitude loadings. *J. of Geotechnical Engineering, ASCE*, Vol. 112, No. 6, 1986, pp. 646–660.
6. Shenton, M.J.: Ballast deformation and track deterioration. In: *Track Technology*. Proc. of a Conf. organized by the Inst. of Civil Engineers and held at the Univ. of Nottingham, 1984, pp. 242–252.
7. Raymond, G.P. and Bathurst, R.J.: Repeated-load response of aggregates in relation to track quality index. *Canadian Geotech. Journal*, Vol. 31, 1994, pp. 547–554.
8. Chrismer, S. and Selig, E.T.: Computer model for ballast maintenance planning. *Proc. 5th International Heavy Haul Railway Conference*, Beijing, 1993, pp. 223–227.
9. Indraratna, B., Salim, W., Ionescu, D. and Christie, D.: Stress-strain and degradation behaviour of railway ballast under static and dynamic loading, based on large-scale triaxial testing. *Proc. 15th Int. Conf. on Soil Mech. and Geotech. Engg*, Istanbul, Vol. 3, 2001, pp. 2093–2096.
10. Ionescu, D., Indraratna, B. and Christie, H.D.: Behaviour of railway ballast under dynamic loads. *Proc. 13th Southeast Asian Geotechnical Conference*, Taipei, 1998, pp. 69–74.
11. Suiker, A.S.J. and de Borst, R.: A numerical model for the cyclic deterioration of railway tracks. *Int. Journal for Numerical Methods in Engineering*. Vol. 57, 2003, pp. 441–470.
12. Roscoe, K.H., Schofield, A.N. and Wroth, C.P.: On yielding of soils. *Geotechnique*, Vol. 8, No. 1, 1958, pp. 22–53.
13. Roscoe, K.H., Schofield, A.N. and Thurairajah, A.: Yielding of clays in states wetter than critical. *Geotechnique*, Vol. 13, No. 3, 1963, pp. 211–240.
14. Roscoe, K.H. and Burland, J.B.: On the generalized stress-strain behaviour of wet clay. In: *Engineering Plasticity*, 1968, pp. 535–609.
15. Schofield, A.N. and Wroth, C.P.: *Critical State Soil Mechanics*. McGraw Hill, 1968.
16. Been, K., Jefferies, M.G. and Hachey, J.: The critical state of sands. *Geotechnique*, Vol. 41, No. 3, 1991, pp. 365–381.
17. Jefferies, M.G.: Nor-sand: a simple critical state model for sand. *Geotechnique*, Vol. 43, No. 1, 1993, pp. 91–103.
18. Been, K. and Jefferies, M.G.: A state parameter for sands. *Geotechnique*, Vol. 35, No. 2, 1985, pp. 99–112.
19. Lade, P.V.: Elasto-plastic stress-strain theory for cohesionless soil with curved yield surfaces. *International Journal of Solids and Structures*, Vol. 13, 1977, pp. 1019–1035.
20. Pender, M.J.: A model for the behaviour of overconsolidated soil. *Geotechnique*, Vol. 28, No. 1, 1978, pp. 1–25.
21. Mroz, Z. and Norris, V.A.: Elastoplastic and viscoplastic constitutive models for soils with application to cyclic loading. In: *Soil Mechanics-Transient and Cyclic Loads* (edited by Pande and Zienkiewicz), John Wiley & Sons, 1982, pp. 173–217.
22. Dafalias, Y.F. and Herrmann, L.R.: A bounding surface soil plasticity model. *Proc. Int. Symp. on Soils Under Cyclic and Transient Loading*, Swansea, U.K., 1980, pp. 335–345.
23. Dafalias, Y.F. and Herrmann, L.R.: Bounding surface formulation of soil plasticity. In: *Soil Mechanics – Transient and Cyclic Loads* (edited by Pande and Zienkiewicz), John Wiley & Sons, 1982, pp. 253–282.
24. Hill, R.: *The Mathematical Theory of Plasticity*. Oxford University Press, 1950.
25. Pender, M.J.: A model for the cyclic loading of overconsolidated soil. In: *Soil Mechanics – Transient and Cyclic Loads* (edited by Pande and Zienkiewicz), John Wiley & Sons, 1982, pp. 283–311.
26. Tasuoka, F., Masuda, T., Siddiquee, M.S.A. and Koseki, J.: Modeling the stress-strain relations of sand in cyclic plane strain loading. *J. of Geotechnical and Geoenvironmental Engineering, ASCE*, Vol. 129. No. 6, 2003, pp. 450–467.

27. Indraratna, B., Ionescu, D. and Christie, H.D.: Shear behaviour of railway ballast based on large-scale triaxial tests. *J. of Geotechnical and Geoenvironmental Engineering, ASCE,* Vol. 124. No. 5, 1998, pp. 439–449.

28. Dafalias, Y.F. and Popov, E.P.: A model of nonlinearly hardening materials for complex loadings. *Acta Mech.* Vol. 21, 1975, pp. 173–192.

29. Dafalias, Y.F. and Popov, E.P.: Plastic internal variables formalism of cyclic plasticity. *J. of Applied Mechanics, ASME,* Vol. 98, No. 4, 1976, pp. 645–650.

30. Krieg, R.D.: A practical two surface plasticity theory. *J. of Applied Mechanics, ASME,* Vol. 42, 1975, pp. 641–646.

31. Marsal, R.J.: Large scale testing of rockfill materials. *J. of the Soil Mech. and Found. Div., ASCE,* Vol. 93, No. SM2, 1967, pp. 27–43.

32. Hardin, B.O.: Crushing of soil particles. *J. of Geotechnical Engineering, ASCE,* Vol. 111, No. 10, 1985, pp. 1177–1192.

33. Lade, P.V., Yamamuro, J.A. and Bopp, P.A.: Significance of particle crushing in granular materials. *J. of Geotech. Engg., ASCE,* Vol. 122, No. 4, 1996, pp. 309–316.

34. McDowell, G.R., Bolton, M.D. and Robertson, D.: The fractal crushing of granular materials. *J. Mech. Phys. Solids,* Vol. 44, No. 12, 1996, pp. 2079–2102.

35. McDowell, G.R. and Bolton, M.D.: On the micromechanics of crushable aggregates. *Geotechnique,* Vol. 48, No. 5, 1998, pp. 667–679.

36. Ueng, T.S. and Chen, T.J.: Energy aspects of particle breakage in drained shear of sands. *Geotechnique,* Vol. 50, No. 1, 2000, pp. 65–72.

Chapter 7

A Constitutive Model for Ballast

Researchers and practicing engineers have long recognised that the ballast bed accumulates plastic deformation under cyclic loading. Despite this, little or no effort has been made to develop realistic constitutive stress-strain relationships, particularly modelling plastic deformation and particle degradation of ballast under cyclic loading. Several researchers attempted to model the constitutive behaviour of soils and granular aggregates under monotonic loading (e.g. Roscoe et al., [1]; Schofield and Wroth, [2]; Lade, [3]; Pender, [4]), and various approaches were made to simulate the cyclic response of granular media. Some are quite innovative and successful to a limited extent. Nevertheless, constitutive modelling of geomaterials under cyclic loading still remains a challenging task.

In the case of railway ballast, the progressive change in particle geometry due to internal attrition, grinding, splitting and crushing (i.e. degradation) under cyclic traffic loads further complicates the stress-strain relationships. There is a lack of realistic constitutive modelling, which includes the effect of particle breakage during shearing. In this respect, the authors have developed a new stress-strain and particle breakage model first for monotonic loading[1], and then extended for the more complex cyclic loading. In the following Sections, modelling of particle breakage and the formulations of new stress-strain relationships for monotonic and cyclic loadings are described in detail.

7.1 MODELLING OF PARTICLE BREAKAGE

Since triaxial testing is considered to be one of the most versatile and useful laboratory methods for evaluating the fundamental strength and deformation properties of geomaterials, a triaxial specimen has been considered as the basis for formulating the relationship between stress, strain and particle breakage (Indraratna and Salim, [5]). The axisymmetric triaxial specimen has one advantage that two of its principal stresses (and also strains) are equal, which reduces the number of independent stress-strain parameters governing the shear behaviour. Figure 7.1(a) shows an axisymmetric ballast specimen subjected to drained triaxial compression loading, while Figure 7.1(b) shows

---

[1] It is acknowledged that this Chapter also includes the essense of technical papers written by the authors, [5, 6].

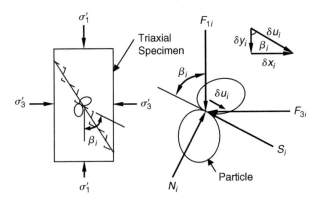

Figure 7.1 Triaxial compression of ballast, (a) specimen under stresses and saw-tooth deformation model, (b) details of contact forces and deformations of two particles at contact (modified after Indraratna and Salim, [5]; Salim and Indraratna, [6]).

the details of contact forces and the relative deformation between two typical particles in an enlarged scale.

The vertical force F_{1i}, and the horizontal force F_{3i}, are acting at contact i between the two particles, which are sliding relative to each other under the applied stresses (major effective principal stress σ_1', and minor effective principal stress σ_3'). It is assumed that the sliding plane makes an angle of β_i with the major principal stress, σ_1' (Fig. 7.1a). If N_i and S_i are the normal force and shear resistance, respectively, then by resolving the forces F_{1i} and F_{3i}, it can be shown that:

$$N_i = F_{1i} \sin \beta_i + F_{3i} \cos \beta_i \tag{7.1}$$

$$S_i = F_{1i} \cos \beta_i - F_{3i} \sin \beta_i \tag{7.2}$$

Assuming no cohesion (i.e. $c = 0$) between the ballast particles, the shear resistance S_i, can be expressed by the Mohr-Coulomb theory, as given by:

$$S_i = N_i \tan \phi_\mu \tag{7.3}$$

where, ϕ_μ is the friction angle between the two particles. Assuming δu_i is the incremental displacement at contact i in the direction of sliding, the horizontal and vertical displacement components δx_i and δy_i, can be expressed as:

$$\delta x_i = \delta u_i \sin \beta_i \tag{7.4}$$

$$\delta y_i = \delta u_i \cos \beta_i \tag{7.5}$$

$$\delta x_i = \delta y_i \tan \beta_i \tag{7.6}$$

If any particle breakage is accompanied by sliding during shear deformation, it is reasonable to assume that the total work done by the applied forces F_{1i} and F_{3i} at contact i, is spent on overcoming frictional resistance and particle breakage, hence:

$$F_{1i}\delta y_i - F_{3i}\delta x_i = N_i \tan \phi_\mu \delta u_i + \delta E_{bi} \tag{7.7}$$

where, δE_{bi} is the incremental energy spent on particle breakage at contact i due to the deformation δu_i. The energy term $(F_{3i}\delta x_i)$ on the left hand side of Equation 7.7 is shown to be negative due to the fact that the direction of the displacement component δx_i is opposite to the direction of applied force F_{3i}.

Substituting Equations 7.1, 7.5 and 7.6 into Equation 7.7 gives:

$$F_{1i}\delta y_i - F_{3i}\delta y_i \tan \beta_i = F_{1i}\delta y_i \tan \beta_i \tan \phi_\mu + F_{3i}\delta y_i \tan \phi_\mu + \delta E_{bi} \tag{7.8}$$

If the average number of contacts per unit length in the directions of three principal stresses σ'_1, σ'_2 and σ'_3 are denoted by n_1, n_2 and n_3, respectively, then the average contact forces and the vertical displacement component can be expressed as:

$$F_{1i} = \frac{\sigma'_1}{n_2 n_3} \tag{7.9}$$

$$F_{3i} = \frac{\sigma'_3}{n_1 n_2} \tag{7.10}$$

$$\delta y_i = \frac{\delta \varepsilon_1}{n_1} \tag{7.11}$$

where, $\delta \varepsilon_1$ is the major principal strain increment.

Replacing Equations 7.9–7.11 into Equation 7.8 gives:

$$\left(\frac{\sigma'_1}{n_2 n_3}\right)\left(\frac{\delta \varepsilon_1}{n_1}\right) - \left(\frac{\sigma'_3}{n_1 n_2}\right)\left(\frac{\delta \varepsilon_1}{n_1}\right) \tan \beta_i$$

$$= \left(\frac{\sigma'_1}{n_2 n_3}\right)\left(\frac{\delta \varepsilon_1}{n_1}\right) \tan \beta_i \tan \phi_\mu + \left(\frac{\sigma'_3}{n_1 n_2}\right)\left(\frac{\delta \varepsilon_1}{n_1}\right) \tan \phi_\mu + \delta E_{bi} \tag{7.12}$$

Multiplying both sides by $n_1 n_2 n_3$ gives:

$$\sigma'_1 \delta \varepsilon_1 - \sigma'_3 \delta \varepsilon_1 \left(\frac{n_3}{n_1}\right) \tan \beta_i = \sigma'_1 \delta \varepsilon_1 \tan \beta_i \tan \phi_\mu + \sigma'_3 \delta \varepsilon_1 \left(\frac{n_3}{n_1}\right) \tan \phi_\mu$$

$$+ \delta E_{bi} (n_1 n_2 n_3) \tag{7.13}$$

where, the product $n_1 n_2 n_3$ represents the total number of contacts in a unit volume of ballast.

Let $\delta E_B = \delta E_{bi}(n_1 n_2 n_3)$ represent the incremental energy spent on particle breakage per unit volume of ballast during the strain increment $\delta\varepsilon_1$, and $r_n = (n_3/n_1)$. Then, Equation 7.13 can be re-written as:

$$\sigma_1' \delta\varepsilon_1 - \sigma_3' \delta\varepsilon_1 r_n \tan\beta_i = \sigma_1' \delta\varepsilon_1 \tan\beta_i \tan\phi_\mu + \sigma_3' \delta\varepsilon_1 r_n \tan\phi_\mu + \delta E_B \tag{7.14}$$

The conventional triaxial stress invariants, p' (mean effective normal stress) and q (deviator stress), are:

$$p' = \frac{(\sigma_1' + 2\sigma_3')}{3}, \quad \text{and} \tag{7.15}$$

$$q = q' = \sigma_1' - \sigma_3' \tag{7.16}$$

Solving Equations 7.15 and 7.16, the stresses σ_1' and σ_3' can be written as:

$$\sigma_1' = p' + \frac{2q}{3} \tag{7.17}$$

$$\sigma_3' = p' - \frac{q}{3} \tag{7.18}$$

Substituting Equations 7.17 and 7.18 into Equation 7.14 gives:

$$\left(p' + \frac{2q}{3}\right)\delta\varepsilon_1 - \left(p' - \frac{q}{3}\right)\delta\varepsilon_1 r_n \tan\beta_i$$

$$= \left(p' + \frac{2q}{3}\right)\delta\varepsilon_1 \tan\beta_i \tan\phi_\mu + \left(p' - \frac{q}{3}\right)\delta\varepsilon_1 r_n \tan\phi_\mu + \delta E_B \tag{7.19}$$

Re-arranging Equation 7.19, the deviator stress ratio becomes:

$$\frac{q}{p'} = \frac{r_n \tan(\beta_i + \phi_\mu) - 1}{\left[\frac{2}{3} + \frac{1}{3} r_n \tan(\beta_i + \phi_\mu)\right]} + \frac{\delta E_B}{p' \delta\varepsilon_1 \left[\frac{2}{3} + \frac{1}{3} r_n \tan(\beta_i + \phi_\mu)\right][1 - \tan\beta_i \tan\phi_\mu]} \tag{7.20}$$

In case of infinitesimal increments (e.g. $\delta\varepsilon_1 \to 0$), the major principal strain increment $\delta\varepsilon_1$, can be replaced by the differential increment $d\varepsilon_1$. Similarly, the other finite increments δE_B, δy_i and δx_i can be substituted by the corresponding differentials dE_B, dy_i and dx_i, respectively. Thus, for the limiting case ($\delta\varepsilon_1 \to 0$), the term ($\delta E_B/\delta\varepsilon_1$) on the right hand side of Equation 7.20 becomes the derivative $dE_B/d\varepsilon_1$, which represents the rate of energy consumption due to particle breakage during shear deformation.

Rowe [7] studied the effect of dilatancy on the friction angle of granular aggregates and concluded that the interparticle friction angle ϕ_μ, should be replaced by ϕ_f, which is the friction angle of aggregates after correction for dilatancy. The friction angle ϕ_f, varies from ϕ_μ at very dense state to ϕ_{cv} at very loose condition, where deformation takes place at a constant volume. The energy spent on the rearrangement of particles during shearing has been attributed to the difference between ϕ_f and ϕ_μ. Rowe [7] also concluded that the dense assemblies of cohesionless particles deform in such a way that the minimum rate of internal energy (work) is absorbed in frictional heat. According to this principle, shear deformation occurs in ballast when at each contact i, the energy ratio (ER_i) of the work done by F_{1i} to that by F_{3i} (i.e. $ER_i = F_{1i}\delta y_i/F_{3i}\delta x_i$) is the minimum. By expanding the expression of ER_i and letting the derivative $d(ER_i)/d\beta_i = 0$, one can determine the critical direction of sliding at contact i (i.e. $\beta_i = \beta_c$) for the minimum energy ratio condition. In other words, $\beta_i = \beta_c$, when $ER_i = ER_{min}$ (minimum energy ratio).

Using the minimum energy ratio principle, Ueng and Chen [8] showed the following two expressions for the ratio r_n ($= n_3/n_1$) and the critical sliding angle β_c:

$$r_n = \frac{1 - \frac{d\varepsilon_v}{d\varepsilon_1}}{\tan \beta_c} \tag{7.21}$$

$$\beta_c = 45° - \frac{\phi_f}{2} \tag{7.22}$$

where, $d\varepsilon_v$ is the volumetric strain increment (compression is taken as positive) for the triaxial specimen corresponding to $d\varepsilon_1$.

Substituting Equations 7.21–7.22, ϕ_μ by ϕ_f and $\beta_i = \beta_c$ into Equation 7.20, and using the differential increment terms, the deviator stress ratio becomes (Indraratna and Salim, [5]):

$$\frac{q}{p'} = \frac{\left(1 - \frac{d\varepsilon_v}{d\varepsilon_1}\right) \tan^2\left(45° + \frac{\phi_f}{2}\right) - 1}{\left[\frac{2}{3} + \frac{1}{3}\left(1 - \frac{d\varepsilon_v}{d\varepsilon_1}\right) \tan^2\left(45° + \frac{\phi_f}{2}\right)\right]} + \frac{dE_B(1 + \sin \phi_f)}{p'd\varepsilon_1\left[\frac{2}{3} + \frac{1}{3}\left(1 - \frac{d\varepsilon_v}{d\varepsilon_1}\right) \tan^2\left(45° + \frac{\phi_f}{2}\right)\right]} \tag{7.23}$$

In the above model, ϕ_f is considered as the basic friction angle, which excludes the effects of both dilatancy and particle breakage.

It is interesting to note that Equation 7.23 simplifies to the well-known critical state equation when particle breakage is ignored. In critical state soil mechanics (Schofield and Wroth, [2]), particle breakage during shearing was not taken into account. At the critical state, soil mass deforms continuously at constant stress and constant volume. If the breakage of particles is ignored (i.e. $dE_B = 0$) at the critical state (i.e. $dp' = dq = d\varepsilon_v = 0$ and $\phi_f = \phi_{cs}$), then Equation 7.23 is reduced to the following critical state relationship:

$$\left(\frac{q}{p'}\right)_{cs} = \frac{\tan^2\left(45° + \frac{\phi_{cs}}{2}\right) - 1}{\frac{2}{3} + \frac{1}{3}\tan^2\left(45° + \frac{\phi_{cs}}{2}\right)} = \frac{6 \sin \phi_{cs}}{3 - \sin \phi_{cs}} = M \tag{7.24}$$

7.1.1 Evaluation of ϕ_f for ballast

In order to evaluate the basic friction angle (ϕ_f) for the ballast used by the authors, the last term of Equation 7.23 containing the energy consumption due to particle breakage is set to zero. The resulting apparent (equivalent) friction angle is denoted by ϕ_{fb}, which naturally includes the contribution of particle breakage but excludes the effect of dilation. Thus, Equation 7.23 is simplified to [5]:

$$\frac{q}{p'} = \frac{\left(1 - \frac{d\varepsilon_v}{d\varepsilon_1}\right)\tan^2\left(45° + \frac{\phi_{fb}}{2}\right) - 1}{\frac{2}{3} + \frac{1}{3}\left(1 - \frac{d\varepsilon_v}{d\varepsilon_1}\right)\tan^2\left(45° + \frac{\phi_{fb}}{2}\right)} \qquad (7.25)$$

Using the laboratory experimental results of deviator stress ratio at failure $(q/p')_f$, and the corresponding value of $(1 - d\varepsilon_v/d\varepsilon_1)_f$ into Equation 7.25, the value of ϕ_{fb} can be easily computed. The calculated values of ϕ_{fb} are plotted against the effective confining pressure (Fig. 7.2), and also against the rate of particle breakage at failure $(dB_g/d\varepsilon_1)_f$, as shown in Figure 7.3. The values of $(dB_g/d\varepsilon_1)_f$ for the fresh ballast used by the authors were obtained from the laboratory experimental results (Fig. 5.19, Chapter 5).

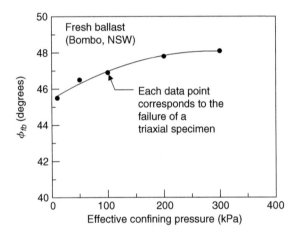

Figure 7.2 Effect of confining pressure on ϕ_{fb} (modified after Indraratna and Salim, [5]).

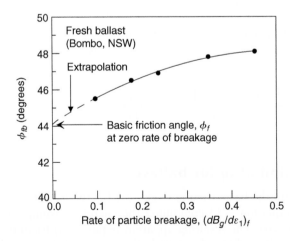

Figure 7.3 Estimation of ϕ_f from laboratory test data (modified after Indraratna and Salim, [5]).

It is evident that the angle ϕ_{fb}, increases at a decreasing rate with the increasing confining pressure (Fig. 7.2). At an elevated confining pressure, the degree of particle breakage is higher (see Fig. 5.19), which means increased energy consumption for higher particle breakage, which is clearly reflected in the increased values of ϕ_{fb}. Figure 7.3 reveals that ϕ_{fb} also increases non-linearly with the rate of particle breakage at failure $(dB_g/d\varepsilon_1)_f$. By extrapolating this relationship back to zero rate of particle breakage [i.e. $(dB_g/d\varepsilon_1)_f = 0$], the basic friction angle ϕ_f, excluding the effect of particle breakage, can be estimated. Based on the current test results, the value of ϕ_f for the fresh ballast (latite basalt) used by authors is found to be approximately 44° (Fig. 7.3).

7.1.2 Contribution of particle breakage to friction angle

The peak friction angle (ϕ_p) of ballast and other granular aggregates is conveniently calculated from the triaxial test results of peak principal stress ratio by re-arranging the Mohr-Coulomb failure criterion, as given in the following relationship:

$$\left(\frac{\sigma'_1}{\sigma'_3}\right)_p = \frac{1 + \sin\phi_p}{1 - \sin\phi_p} \tag{7.26}$$

Equation 7.26 relates the peak friction angle (ϕ_p) with the peak value of principal stress ratio $(\sigma'_1/\sigma'_3)_p$, hence provides an obvious upper bound for the internal friction angle of aggregates. In contrast, the basic friction angle (ϕ_f) evaluated at zero dilatancy and at zero particle breakage, provides a lower bound (Fig. 7.4), and is considered to be independent of the confining pressure (Indraratna and Salim, [5]). Therefore, the basic friction angle (ϕ_f) may be considered to be the same as the angle of repose of the material. As explained earlier, the apparent friction angle ϕ_{fb}, includes the effect of particle breakage, but excludes dilatancy.

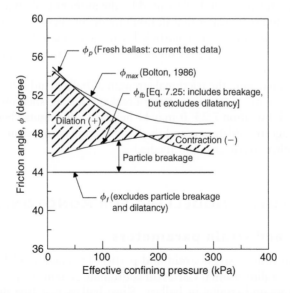

Figure 7.4 Effect of particle breakage and dilatancy on friction angle (modified after Indraratna and Salim, [5]).

Figure 7.4 illustrates the various angles of friction (ϕ_f, ϕ_{fb} and ϕ_p) computed for the fresh ballast and these friction angles were plotted against increasing confining pressure. This figure shows that the difference between ϕ_p (Equation 7.26) and ϕ_{fb} (Equation 7.25) at low confinement is very high because of higher dilatancy. At low stresses, the degree of particle breakage is also low, and therefore, the difference between ϕ_{fb} and ϕ_f is also small. As confining pressure increases, the difference between ϕ_{fb} and ϕ_f increases, which is attributed to the higher rate of particle degradation (i.e. increased energy consumption for higher particle breakage). At increased confining pressure, a higher rate of particle breakage contributes to an increase in friction angle; however, dilatancy is suppressed, and volumetric contraction adversely affects the friction angle. The peak friction angle (ϕ_p) computed from the laboratory triaxial test results can be viewed as the summation of basic friction angle ϕ_f, and the effects of dilatancy and particle breakage, as illustrated in Figure 7.4. It is noted that the peak friction angle decreases with increasing confining pressure, an observation consistent with the previous studies (Marsal, [9]; Charles and Watts, [10]; Indraratna et al., [11]).

Bolton [12] studied the strength and dilatancy of sand, and modelled the dilatancy-related component of friction angle ($\phi_{max} - \phi_{crit}$) as a function of relative dilatancy index, which depends on the initial density and effective mean stress at failure. Bolton used the notation ϕ_{crit} to indicate the friction angle at the critical state (i.e. at zero dilation). If the value of ϕ_f estimated in Figure 7.3 is considered as the value of ϕ_{crit} for the fresh ballast, then Bolton's model can be used to predict its maximum friction angle (ϕ_{max}). The predicted ϕ_{max} can be obtained by adding the dilatancy component to ϕ_{crit}. It should be mentioned here that Bolton's model does not incorporate particle breakage. While this is acceptable for fine granular media such as sand, where particle breakage may be insignificant, Bolton's dilatancy model is not appropriate for coarser, angular aggregates like ballast, where particle degradation can be significant. Nevertheless, the predicted ϕ_{max} for the fresh ballast used by the authors is shown in Figure 7.4 for comparison. This figure indicates that Bolton's model predicts ϕ_{max}, which agrees with ϕ_p at low confining pressure where particle breakage is small. However, it seems that Bolton's model overpredicts ϕ_{max} (or dilatancy-related friction component) for ballast at higher confining pressures.

The mechanism behind the frictional strength of ballast and other granular aggregates, particularly with regard to particle breakage during shearing is explained in Figure 7.4 through Equation 7.23. It may be helpful to distinguish between the effects of particle breakage and dilatancy, and the basic friction component of shear strength for ballast and other coarse granular media.

7.2 CONSTITUTIVE MODELLING FOR MONOTONIC LOADING

7.2.1 Stress and strain parameters

To develop a constitutive stress-strain and particle breakage model in a generalised stress space, a three-dimensional Cartesian coordinate system (x_j, $j = 1, 2, 3$) was used to define the stress and strains in ballast. Since ballast is a free draining granular medium, all the stresses used in the current model are considered to be effective.

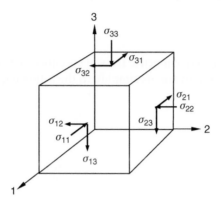

Figure 7.5 Three-dimensional stresses and index notations.

For a three-dimensional ballast element under stresses (Fig. 7.5), the following stress and strain invariants were used to formulate a relationship between the stress, strain, and particle breakage:

$$q = \sqrt{\frac{3}{2}s_{ij}s_{ij}} = \sqrt{\frac{1}{2}[(\sigma_{11} - \sigma_{22})^2 + (\sigma_{22} - \sigma_{33})^2 + (\sigma_{33} - \sigma_{11})^2] + 3(\sigma_{12}^2 + \sigma_{23}^2 + \sigma_{31}^2)}$$

(7.27)

$$p = \frac{1}{3}\sigma_{kk} = \frac{1}{3}(\sigma_{11} + \sigma_{22} + \sigma_{33})$$

(7.28)

where, q is the distortional stress (invariant), p is the mean effective normal stress (invariant), σ_{ij} is the stress tensor ($i = 1, 2, 3$, and $j = 1, 2, 3$) and s_{ij} is the stress deviator tensor, as defined below:

$$s_{ij} = \sigma_{ij} - \frac{1}{3}\sigma_{kk}\delta_{ij}$$

(7.29)

In the above, δ_{ij} is the Kronecker delta (i.e. $\delta_{ij} = 1$ if $i = j$, and $\delta_{ij} = 0$ if $i \neq j$). The usual summation convention over the repeated indices is adopted in these notations.

The complementary strain invariants are the distortional strain ε_s, and volumetric strain ε_v, respectively, as defined below:

$$\varepsilon_s = \sqrt{\frac{2}{3}e_{ij}e_{ij}}$$

$$= \sqrt{\frac{2}{9}[(\varepsilon_{11} - \varepsilon_{22})^2 + (\varepsilon_{22} - \varepsilon_{33})^2 + (\varepsilon_{33} - \varepsilon_{11})^2] + \frac{4}{3}(\varepsilon_{12}^2 + \varepsilon_{23}^2 + \varepsilon_{31}^2)}$$

(7.30)

$$\varepsilon_v = \varepsilon_{kk} = \varepsilon_{11} + \varepsilon_{22} + \varepsilon_{33}$$

(7.31)

where, ε_{ij} is the strain tensor, and e_{ij} is the strain deviator tensor, which is defined as:

$$e_{ij} = \varepsilon_{ij} - \frac{1}{3}\varepsilon_{kk}\delta_{ij} \tag{7.32}$$

For the special case of an axisymmetric triaxial specimen (where $\sigma_2 = \sigma_3$ and $\varepsilon_2 = \varepsilon_3$), the above stress and strain invariants simplify to the following well-known functions:

$$q = \sigma_1 - \sigma_3 \tag{7.33}$$

$$p = \frac{1}{3}(\sigma_1 + 2\sigma_3) \tag{7.34}$$

$$\varepsilon_s = \frac{2}{3}(\varepsilon_1 - \varepsilon_3) \tag{7.35}$$

$$\varepsilon_v = \varepsilon_1 + 2\varepsilon_3 \tag{7.36}$$

7.2.2 Incremental constitutive model

In classical soil plasticity, the total strains ε_{ij}, are usually decomposed into elastic (recoverable) and plastic (irrecoverable) components ε_{ij}^e and ε_{ij}^p, respectively:

$$\varepsilon_{ij} = \varepsilon_{ij}^e + \varepsilon_{ij}^p \tag{7.37}$$

where, the superscript e denotes the elastic component, and p represents the plastic component. Accordingly, the strain increments are also divided into elastic and plastic components:

$$d\varepsilon_{ij} = d\varepsilon_{ij}^e + d\varepsilon_{ij}^p \tag{7.38}$$

Similarly, the increments of strain invariants are also separated into elastic and plastic components, as given below:

$$d\varepsilon_s = d\varepsilon_s^e + d\varepsilon_s^p \tag{7.39}$$

$$d\varepsilon_v = d\varepsilon_v^e + d\varepsilon_v^p \tag{7.40}$$

The elastic components of a strain increment can be computed using the theory of elasticity, where the elastic distortional strain increment $(d\varepsilon_s^e)$ is given by:

$$d\varepsilon_s^e = \frac{dq}{2G} \tag{7.41}$$

where, G is the elastic shear modulus.

The elastic volumetric strain increment $d\varepsilon_v^e$, can be determined using the swelling/recompression constant κ, and is given by [1, 2]:

$$d\varepsilon_v^e = \frac{\kappa}{1+e_i}\left(\frac{dp}{p}\right) \tag{7.42}$$

where, e_i is the initial void ratio at the start of shearing.

In formulating Equation 7.23, the special case of axisymmetric triaxial shearing (i.e. $\sigma_2 = \sigma_3$ and $\varepsilon_2 = \varepsilon_3$) was considered (see Fig. 7.1) and only the plastic components of strain increment were taken into account. Thus, the strain increments $d\varepsilon_v$ and $d\varepsilon_1$ in Equation 7.23 refer to the plastic strain increments $d\varepsilon_v^p$ and $d\varepsilon_1^p$, respectively. Equation 7.23 can be extended to a generalised stress-strain formulation by replacing the principal strain increment with a combination of strain invariants. The principal strain increments of an axisymmetric specimen can easily be replaced with the incremental strain invariants using Equations 7.35–7.36; hence, it can be shown that:

$$1 - \frac{d\varepsilon_v^p}{d\varepsilon_1^p} = -2\frac{d\varepsilon_3^p}{d\varepsilon_1^p} = \frac{d\varepsilon_s^p - \frac{2}{3}d\varepsilon_v^p}{d\varepsilon_s^p + \frac{1}{3}d\varepsilon_v^p} \tag{7.43}$$

Substituting $d\varepsilon_1$ by $d\varepsilon_1^p$, $d\varepsilon_v$ by $d\varepsilon_v^p$ and Equation 7.43 into Equation 7.23 gives:

$$\frac{q}{p} = \frac{\left(\frac{d\varepsilon_s^p - \frac{2}{3}d\varepsilon_v^p}{d\varepsilon_s^p + \frac{1}{3}d\varepsilon_v^p}\right)\tan^2\left(45° + \frac{\phi_f}{2}\right) - 1}{\left[\frac{2}{3} + \frac{1}{3}\left(\frac{d\varepsilon_s^p - \frac{2}{3}d\varepsilon_v^p}{d\varepsilon_s^p + \frac{1}{3}d\varepsilon_v^p}\right)\tan^2\left(45° + \frac{\phi_f}{2}\right)\right]}$$
$$+ \frac{dE_B\left(1 + \sin\phi_f\right)}{p\left(d\varepsilon_s^p + \frac{1}{3}d\varepsilon_v^p\right)\left[\frac{2}{3} + \frac{1}{3}\left(\frac{d\varepsilon_s^p - \frac{2}{3}d\varepsilon_v^p}{d\varepsilon_s^p + \frac{1}{3}d\varepsilon_v^p}\right)\tan^2\left(45° + \frac{\phi_f}{2}\right)\right]} \tag{7.44}$$

Critical State Line (CSL) and the critical state parameters are often employed in modelling plastic deformations of soils. Critical state parameters are the fundamental properties of a soil including a granular assembly. In case of a granular medium where progressive particle breakage occurs under imparted loading, the critical state line of the aggregates also changes gradually. However, in the current formulation, it is assumed that the critical state line of ballast remains unchanged (i.e. fixed) in the p-q-e space to serve as a reference state. Considering the small change in particle size distribution after testing (see Figs. 5.16 and 5.28), it is expected that the change in the critical state line under working loads would be small and the corresponding errors in model computation resulting from this simplified assumption will be negligible.

Using the critical state friction ratio $M = 6\sin\phi_f/(3 - \sin\phi_f)$, it can be shown that,

$$\tan^2\left(45° + \frac{\phi_f}{2}\right) = \frac{1 + \sin\phi_f}{1 - \sin\phi_f} = \frac{3 + 2M}{3 - M} \tag{7.45}$$

$$1 + \sin\phi_f = \frac{6 + 4M}{6 + M} \tag{7.46}$$

Replacing Equations 7.45–7.46 and $q/p = \eta$ (stress ratio) into Equation 7.44 gives:

$$\eta \left[\frac{2}{3} + \frac{1}{3} \left(\frac{d\varepsilon_s^p - \frac{2}{3}d\varepsilon_v^p}{d\varepsilon_s^p + \frac{1}{3}d\varepsilon_v^p} \right) \left(\frac{3 + 2M}{3 - M} \right) \right] = \left(\frac{d\varepsilon_s^p - \frac{2}{3}d\varepsilon_v^p}{d\varepsilon_s^p + \frac{1}{3}d\varepsilon_v^p} \right) \left(\frac{3 + 2M}{3 - M} \right) - 1$$
$$+ \frac{dE_B}{p \left(d\varepsilon_s^p + \frac{1}{3}d\varepsilon_v^p \right)} \left(\frac{6 + 4M}{6 + M} \right) \quad (7.47)$$

Re-arrangement of Equation 7.47 gives:

$$\frac{\eta}{3}[9d\varepsilon_s^p - 2Md\varepsilon_v^p] = -3d\varepsilon_v^p + 3Md\varepsilon_s^p - Md\varepsilon_v^p + \frac{dE_B}{p}\left[\frac{(3 - M)(6 + 4M)}{6 + M} \right] \quad (7.48)$$

Equation 7.48 can be further re-arranged to give the ratio between the plastic volumetric and distortional strain increments, as given below:

$$\frac{d\varepsilon_v^p}{d\varepsilon_s^p} = \frac{9(M - \eta)}{9 + 3M - 2\eta M} + \frac{dE_B}{pd\varepsilon_s^p}\left(\frac{9 - 3M}{9 + 3M - 2\eta M} \right)\left(\frac{6 + 4M}{6 + M} \right) \quad (7.49)$$

Equation 7.49 captures: (a) plastic volumetric strain increment associated with plastic distortional strain increment, and (b) corresponding energy consumption for particle breakage during shear deformation. It is relevant to note here that Equation 7.49 becomes undefined when $d\varepsilon_s^p$ becomes zero under isotropic stress condition; hence, it is only valid for shearing where stresses are anisotropic [6]. In Equation 7.49, the rate of energy consumption per unit volume of ballast $(dE_B/d\varepsilon_s^p)$ must be determined first. The incremental energy consumption due to particle breakage per unit volume dE_B (Equation 7.49), can be related to the increment of breakage index dB_g, where the breakage index can be measured in the laboratory, as explained earlier.

The experimental values of (q/p'), $(1 - d\varepsilon_v/d\varepsilon_1)$, and the basic friction angle of fresh ballast (ϕ_f) were substituted into Equation 7.23, and the values of $dE_B/d\varepsilon_1$ were then back calculated. From the experimental results (Fig. 5.19, Chapter 5), the rates of particle breakage $dB_g/d\varepsilon_1$, at various axial strains and confining pressures, were determined. The computed $dE_B/d\varepsilon_1$ values are plotted against these experimental $dB_g/d\varepsilon_1$ values, as shown in Figure 7.6. This figure indicates that $dE_B/d\varepsilon_1$ and $dB_g/d\varepsilon_1$ are linearly related to each other. Therefore, it can be assumed that the incremental energy consumption due to particle breakage per unit volume is proportional to the corresponding increment of breakage index (i.e., $dE_B = \beta dB_g$, where β is a constant of proportionality). Therefore, Equation 7.49 becomes:

$$\frac{d\varepsilon_v^p}{d\varepsilon_s^p} = \frac{9(M - \eta)}{9 + 3M - 2\eta M} + \frac{\beta dB_g}{pd\varepsilon_s^p}\left(\frac{9 - 3M}{9 + 3M - 2\eta M} \right)\left(\frac{6 + 4M}{6 + M} \right) \quad (7.50)$$

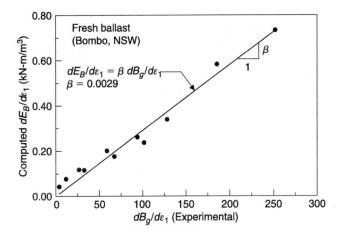

Figure 7.6 Relationship between the rate of energy consumption and rate of particle breakage (after Salim and Indraratna, [6]).

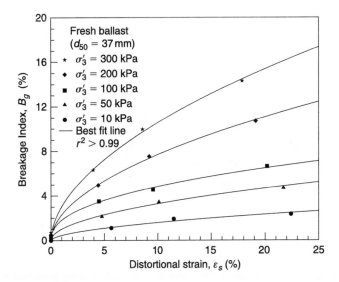

Figure 7.7 Variation of particle breakage of fresh ballast with distortional strain and confining pressure (re-plotted from Fig. 5.19).

The experimental data of Figure 5.19 were re-plotted as B_g versus distortional strain ε_s, as shown in Figure 7.7. These breakage data are re-plotted in a modified scale as $\ln\{p_{cs(i)}/p_{(i)}\}B_g$ versus ε_s, as shown in Figure 7.8, where, p_{cs} is the value of p on the critical state line at the current void ratio and the subscript (i) indicates the initial value at the start of shearing. The definition of p_{cs} is illustrated in Figure 7.9 for clarity. Figure 7.8 shows that the wide variations of B_g values (Fig. 7.7) due to varying confining pressures are practically eliminated in this technique and that all breakage

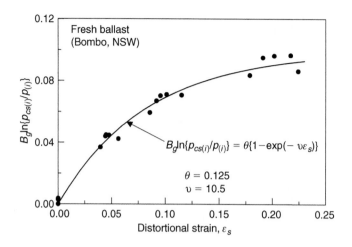

Figure 7.8 Modelling of ballast breakage during triaxial shearing (modified after Salim and Indraratna, [6]).

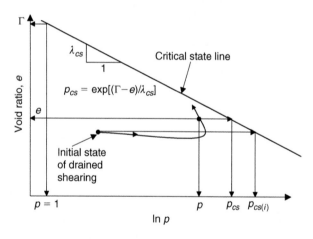

Figure 7.9 Definition of p_{cs} and typical e-ln p plot in a drained shearing (modified after Salim and Indraratna, [6]).

data fall close to a single line (non-linear). Thus, the breakage of particles under triaxial shearing may be represented by a single non-linear function, as given by:

$$B_g = \frac{\theta\{1 - \exp(-\upsilon\varepsilon_s)\}}{\ln\left\{\frac{p_{cs(i)}}{p_{(i)}}\right\}} \tag{7.51}$$

where, θ and υ are two material constants relating to the breakage of ballast.

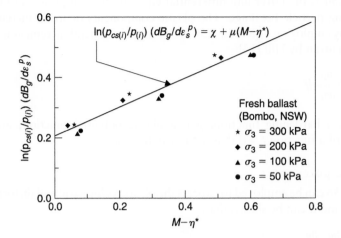

Figure 7.10 Modelling of rate of ballast breakage (modified after Salim and Indraratna, [6]).

The values of $dB_g/d\varepsilon_s^p$ at various distortional strains and confining pressures can be obtained readily from Figure 7.7. These breakage rates are then plotted as $\ln\{p_{cs(i)}/p_{(i)}\}dB_g/d\varepsilon_s^p$ versus $(M - \eta^*)$, as shown in Figure 7.10, where $\eta^* = \eta(p/p_{cs})$.

Figure 7.10 indicates that the values of $\ln\{p_{cs(i)}/p_{(i)}\}dB_g/d\varepsilon_s^p$ are related to $(M - \eta^*)$ linearly, irrespective of the confining pressures. Thus, a linear relationship between the rate of particle breakage $(dB_g/d\varepsilon_s^p)$ and $(M - \eta^*)$ is proposed [6], as follows:

$$\frac{dB_g}{d\varepsilon_s^p} = \frac{\chi + \mu(M - \eta^*)}{\ln\left(\frac{p_{cs(i)}}{p_{(i)}}\right)} \tag{7.52}$$

where, χ and μ are two material constants relating to the rate of ballast breakage (Fig. 7.10). Substituting Equation 7.52 into Equation 7.50 gives:

$$\frac{d\varepsilon_v^p}{d\varepsilon_s^p} = \frac{9(M - \eta)}{9 + 3M - 2\eta M} + \left(\frac{\beta}{p}\right)\left[\frac{\chi + \mu(M - \eta^*)}{\ln\left(\frac{p_{cs(i)}}{p_{(i)}}\right)}\right]\left(\frac{9 - 3M}{9 + 3M - 2\eta M}\right)\left(\frac{6 + 4M}{6 + M}\right) \tag{7.53}$$

Equation 7.53 can be re-written as:

$$\frac{d\varepsilon_v^p}{d\varepsilon_s^p} = \frac{9(M - \eta)}{9 + 3M - 2\eta M} + \left(\frac{B}{p}\right)\left[\frac{\chi + \mu(M - \eta^*)}{9 + 3M - 2\eta M}\right] \tag{7.54}$$

where,

$$B = \frac{\beta}{\ln\left(\frac{p_{cs(i)}}{p_{(i)}}\right)}\left[\frac{(9 - 3M)(6 + 4M)}{6 + M}\right] = \text{constant} \tag{7.55}$$

Equation 7.54 is the governing differential equation for the plastic strain increment incorporating particle breakage. The plastic components of strain increment can be computed by employing Equation 7.54 along with the general incremental constitutive relationship given by Hill [13]:

$$d\varepsilon_{ij}^p = h\frac{\partial g}{\partial \sigma_{ij}}df \tag{7.56}$$

where, h is a hardening function, g is a plastic potential function, and df is the differential of a function $f = 0$ that defines yield locus.

The plastic potential, g

Equation 7.56 can be employed to express the plastic volumetric and distortional strain increments and it can be shown that:

$$\frac{d\varepsilon_v^p}{d\varepsilon_s^p} = \frac{\partial g}{\partial p}\Big/\frac{\partial g}{\partial q} \tag{7.57}$$

By definition, the plastic strain increment vector is normal to the plastic potential surface. Thus, at any point (p, q) on the plastic potential $g = g(p, q)$,

$$\frac{d\varepsilon_v^p}{d\varepsilon_s^p} = -\frac{dq}{dp} \tag{7.58}$$

Substituting Equation 7.58 into Equation 7.54 gives:

$$-\frac{dq}{dp} = \frac{9(M - \eta)}{9 + 3M - 2\eta M} + \left(\frac{B}{p}\right)\left[\frac{\chi + \mu(M - \eta^*)}{9 + 3M - 2\eta M}\right] \tag{7.59}$$

Equation 7.59 can be re-written in the following form:

$$\frac{dq}{dp} + \frac{9(Mp - q) + B\{\chi + \mu(M - q/p_{cs})\}}{(9 + 3M)p - 2qM} = 0 \tag{7.60}$$

This is a first-order linear differential equation of q. The solution of Equation 7.60 gives the plastic potential function $g(p, q)$. It is pertinent to mention here that it only requires the partial derivatives of g with respect to p and q, rather than the explicit function of g, to derive expressions for the plastic strain increments. Since Equation 7.60 is linear in q,

$$\frac{\partial g}{\partial q} = 1 \tag{7.61}$$

$$\frac{\partial g}{\partial p} = \frac{9(Mp - q) + B\{\chi + \mu(M - q/p_{cs})\}}{(9 + 3M)p - 2qM} = \frac{9(M - \eta) + (B/p)\{\chi + \mu(M - \eta^*)\}}{9 + 3M - 2\eta M} \tag{7.62}$$

The derivation technique of Equations 7.61 and 7.62 from the differential equation (Equation 7.60) is shown in Appendix A, based on a simple example. It is relevant to note that substitution of Equations 7.61 and 7.62 into Equation 7.57 satisfies the governing differential equation (Equation 7.54).

To formulate the yield and hardening functions, the following assumption and postulates are made with regard to railway ballast:

Assumption: As shear deformation increases, the material (ballast) moves towards the critical state.

The critical state has been defined earlier in Chapter 6. In critical state soil mechanics, it is commonly assumed that the projections of the critical state line on e-ln p and p-q planes are straight lines and this is also implied in the current formulation. Indraratna and Salim [14] presented experimental evidence that ballast, like other soils, moves towards a common (critical) state as the shear deformation increases, irrespective of the initial states and confining pressures (see Figs. 5.20–5.21).

Postulate A: The material (ballast) deforms plastically, if and only when there is a change in the stress ratio, $q/p(=\eta)$.

A hypothesis similar to the above postulate was made by Pender [4] for overconsolidated soils. The implication of this postulate is that it specifies the yield function f, for ballast. Within the common range of stresses (<1 MPa) encountered in railway tracks, Postulate A is only valid for time-independent situations (i.e. no creep effects).

Plastic deformation occurs in ballast resulting from grain slippage, particle rolling, grain attrition, fracture and crushing, and the resulting rearrangement of particles. Under isotropic stress (i.e. $q=0$, $\eta=0$), it is believed that the above mentioned mechanisms of grain rearrangement are insignificant in coarse aggregates like ballast, hence no apparent plastic deformation (Salim and Indraratna, [6]). However, a small increase in stress ratio (and corresponding distortional stress, q) brings the ballast specimen closer to its critical state, activates the grain rearrangement mechanisms and leads to incremental shear distortion (irrecoverable) and associated plastic volumetric change. It is believed that under stress levels approaching the crushing strength of aggregates, time-dependent (creep) effects will also lead to additional particle breakage and associated plastic deformation. At very high values of p where the grains may crush and even pulverise, Postulate A needs to be modified to incorporate a capped-type yield surface, which is more appropriate for clays and sands. However, within the scope of this study, creep has not been incorporated, rather the behaviour is focussed on ballast deformation and particle breakage alone under imparted loading.

In the current model (non-capped), the yield loci are represented by constant stress ratio ($\eta=$ constant) lines in the p-q plane (Fig. 7.11). The yield locus moves kinematically along with its current stress ratio as the stress changes. Mathematically, the yield function f, specifying the yield locus for the current stress ratio η_j, was expressed by Pender [4] as:

$$f = q - \eta_i p = 0 \tag{7.63}$$

Figure 7.12 shows the direction of plastic strain vectors (Equation 7.49) for different yield loci. Each plastic strain increment vector can be separated into a volumetric component and a distortional component, as mentioned earlier. It is usually assumed that the plastic distortional strain increment ($d\varepsilon_s^p$) is positive when $d\eta$ is positive. If the effect of particle breakage on the direction of plastic strain increment is small and $d\varepsilon_s^p$ is positive, then according to Equation 7.49, the plastic volumetric strain increment

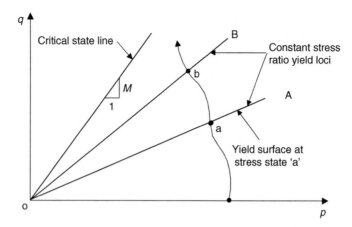

Figure 7.11 Yield loci represented by constant stress ratio lines in p, q plane (inspired by Pender, [4] and modified after Salim and Indraratna, [6]).

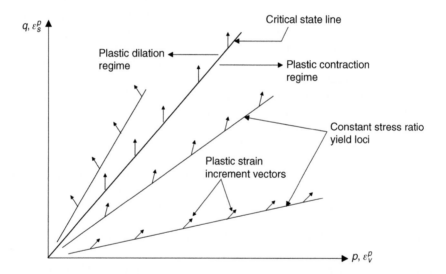

Figure 7.12 Plastic strain increment vectors for different yield loci (modified after Salim and Indraratna, [6]).

will be either positive, zero, or negative, depending primarily on the sign of the term $(M - \eta)$, i.e. on the position of current yield locus relative to the critical state line (CSL) in the p-q plane.

If $\eta < M$ (i.e. the current yield locus is below the CSL), the direction of the plastic strain increment will be such that its volumetric component becomes positive (i.e. contraction, see Fig. 7.12). In contrast, if $\eta > M$ (i.e. the current yield locus is above the CSL), the increment of plastic volumetric strain will be negative (i.e. dilation). Thus, the p-q plane is considered to be divided into two distinct regimes by the CSL.

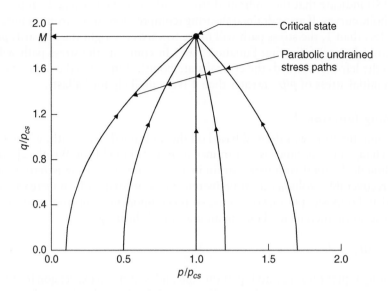

Figure 7.13 Parabolic undrained stress paths (inspired by Pender, [4] and modified after Salim and Indraratna, [6]).

The area above the CSL is the plastic dilation regime, and the area below the CSL is the plastic contraction regime (Fig. 7.12).

Differentiating Equation 7.63, and substituting $dq = \eta_i dp + p d\eta$,

$$df = dq - \eta_i dp = p d\eta \qquad (7.64)$$

The hardening function (h) is formulated based on an undrained stress path where the volumetric strain is constrained to zero. The second postulate is made regarding the shape of the undrained stress path, as described below.

Postulate B: The undrained stress paths are parabolic in the $p - q$ plane and are expressed by the following relationship (Pender, [4]):

$$\left(\frac{\eta}{M}\right)^2 = \frac{p_{cs}}{p}\left[\frac{1 - \frac{p_o}{p}}{1 - \frac{p_o}{p_{cs}}}\right] \qquad (7.65)$$

where, p_{cs} is the value of p on the critical state line corresponding to the current void ratio, as illustrated in Figure 7.9. Thus, $p_{cs} = \exp\{(\Gamma - e)/\lambda_{cs}\}$, $\Gamma =$ void ratio on the CSL at $p = 1$, and λ_{cs} is the slope of the projection of CSL on the e-$\ln p$ plane, and p_o is the value of p at the intersection of the undrained stress path with the initial stress ratio line.

Figure 7.13 shows the parabolic undrained stress paths (Equation 7.65) in q/p_{cs} and p/p_{cs} plane. In this figure, p and q are normalised by p_{cs}. No undrained test on ballast was carried out by the authors to verify Postulate B. However, previous experimental results reported by other researchers (e.g. Roscoe et al., [1]; Ishihara et al., [15]) indicate that the undrained stress paths may be reasonably approximated by parabolic curves. If an undrained shearing (compression) starts from an initial stress of p/p_{cs} less than 1, the stress path will move towards the right (i.e. towards $p/p_{cs} = 1$ at the critical state) following Equation 7.65. In contrast, the stress path will move towards the left (i.e. towards the critical state point), if undrained compression starts from an initial stress of p/p_{cs} greater than 1 (very unlikely for ballast).

Hardening function, h

A hardening function was derived based on the undrained stress path, where the total volume change of a specimen is constrained to zero. Schofield and Wroth [2] explained that although the total volumetric strain in an undrained shearing is zero, there is an elastic (recoverable) volumetric strain increment associated with an increase in p, and an equal and opposite plastic volumetric strain component compensates for the elastic volumetric strain increment. Thus, in an undrained shearing,

$$d\varepsilon_v^p + d\varepsilon_v^e = d\varepsilon_v = 0 \qquad (7.66)$$

Substituting Equation 7.42 into Equation 7.66 and writing an expression for the plastic volumetric strain increment following Equation 7.56, it can be shown that:

$$h \frac{\partial g}{\partial p} p d\eta + \frac{\kappa dp}{p(1 + e_i)} = 0 \qquad (7.67)$$

Differentiating Equation 7.65 and simplifying, an alternative differential form of the undrained stress path is obtained:

$$\frac{2p}{M^2} \left(\frac{p_o}{p_{cs}} - 1 \right) \eta d\eta + \left(\frac{2p_o}{p} - 1 \right) p_{cs} \left(\frac{dp}{p} \right) = 0 \qquad (7.68)$$

Substituting Equations 7.62 and 7.68 into Equation 7.67 and re-arranging, the hardening function becomes:

$$h = \frac{2\kappa \left(\frac{p_o}{p_{cs}} - 1 \right) (9 + 3M - 2\eta M)\eta}{M^2(1 + e_i) \left(\frac{2p_o}{p} - 1 \right) p_{cs} \left[9(M - \eta) + \frac{B}{p} \{ \chi + \mu(M - \eta^*) \} \right]} \qquad (7.69)$$

It should be mentioned here that the above hardening function (Equation 7.69) clearly depends on p, p_o and p_{cs}, besides other parameters. The parameter p represents the current mean stress, while p_{cs} is the image of current void ratio in terms of stress on the critical state line (see Fig. 7.9). Thus, the hardening function (Equation 7.69) correctly incorporates the effect of current void ratio (or density) relative to the critical state void ratio. The above expression of hardening function h, gives a positive value if a ballast specimen is in a state looser than the critical (i.e. $p_o > p_{cs}$). In the normal range of stresses, ballast and other coarse aggregates remain in states denser than the critical (i.e. $p_o < p_{cs}$), and therefore, the sign of the hardening function (Equation 7.69) should be reversed (Salim and Indraratna, [6]). Substituting Equation 7.69 with a negative

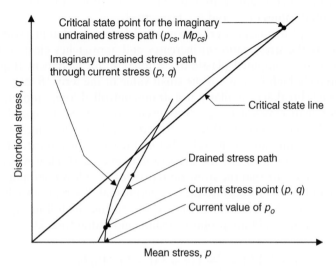

Figure 7.14 Definition of p_o in a drained shearing (inspired by Pender, [4] and modified after Salim and Indraratna, [6]).

sign and also Equations 7.61 and 7.64 into Equation 7.56, the plastic distortional strain increment becomes:

$$d\varepsilon_s^p = \frac{2\kappa\left(\frac{p}{p_{cs}}\right)\left(1 - \frac{p_o}{p_{cs}}\right)(9 + 3M - 2\eta M)\eta d\eta}{M^2(1 + e_i)\left(\frac{2p_o}{p} - 1\right)\left[9(M - \eta) + \frac{B}{p}\{\chi + \mu(M - \eta^*)\}\right]} \quad (7.70)$$

Equation 7.70 is based on the strain hardening function derived from an undrained stress path where both p_o and p_{cs} remain constant throughout. Therefore, the factor $(1 - p_o/p_{cs})$ in the numerator remains a constant during an undrained test and may be considered as a function of the initial state of ballast at the start of shearing. In drained shearing, the value of p_{cs} varies as the void ratio (e) changes. The parameter p_o is re-defined for a drained test as the value of p at the intersection of the initial stress ratio line with an imaginary undrained stress path, which passes through the current stress (p, q) point and current (p_{cs}, Mp_{cs}) point corresponding to the current void ratio (Pender, [4]). This definition of p_o in a drained test is graphically illustrated in Figure 7.14.

Since the void ratio (e) varies during drained shearing, the corresponding p_{cs} (see Fig. 7.9) changes, as mentioned earlier. Therefore, the imaginary undrained stress path (Equation 7.65), which is a function of p_{cs}, also varies during a drained test, resulting in a variable p_o value (Fig. 7.14). For drained shearing, the plastic distortional strain increment may be expressed by modifying Equation 7.70, as given below:

$$d\varepsilon_s^p = \frac{2\alpha\kappa\left(\frac{p}{p_{cs}}\right)\left(1 - \frac{p_{o(i)}}{p_{cs(i)}}\right)(9 + 3M - 2\eta M)\eta d\eta}{M^2(1 + e_i)\left(\frac{2p_o}{p} - 1\right)\left[9(M - \eta) + \frac{B}{p}\{\zeta + \mu(M - \eta^*)\}\right]} \quad (7.71)$$

where, α is a model constant relating to the initial stiffness of ballast, and $p_{o(i)}$ and $p_{cs(i)}$ are the initial values of p_o and p_{cs}, respectively.

Numerical implementation of the above model indicates that in a stress-controlled computation, as the stress ratio (η) increases and approaches close to the value of M (i.e. $\eta \approx M$), the computed plastic distortional strain increment (Equation 7.71) becomes extremely high because of the small value of the term $(B/p\{\chi + \mu(M - \eta^*)\})$ related to particle breakage. Similarly, in a strain-controlled computation, as the plastic distortional strain increases at a stress ratio (η) close to M, the corresponding increment in stress ratio becomes very small (close to zero), and the resulting total stress ratio practically remains the same as its value before the strain increment. Thus, it is clear that Equation 7.71 doesn't allow the stress ratio to exceed M. However, experimental results of ballast indicate that the stress ratio exceeds M at low confining pressure (see Figs. 5.5, 5.6 and 5.20). To capture these experimental observations where the stress ratio (η) may exceed the value of M at low confinement, the following modifications to Equations 7.71 and 7.54 are proposed (Salim and Indraratna, 2004):

$$d\varepsilon_s^p = \frac{2\alpha\kappa \left(\frac{p}{p_{cs}}\right)\left(1 - \frac{p_{o(i)}}{p_{cs(i)}}\right)(9 + 3M - 2\eta^*M)\eta d\eta}{M^2(1 + e_i)\left(\frac{2p_o}{p} - 1\right)\left[9(M - \eta^*) + \frac{B}{p}\{\chi + \mu(M - \eta^*)\}\right]} \tag{7.72}$$

$$\frac{d\varepsilon_v^p}{d\varepsilon_s^p} = \frac{9(M - \eta)}{9 + 3M - 2\eta^*M} + \left(\frac{B}{p}\right)\left[\frac{\chi + \mu(M - \eta^*)}{9 + 3M - 2\eta^*M}\right] \tag{7.73}$$

The term $(M - \eta^*)$ in the denominator of Equation 7.72 will now vary from a positive value to zero as the distortional strain increases. The stress ratio (η) may increase to a value equal to or higher than M (at small strain), but the value of $(M - \eta^*)$ remains substantially greater than zero, providing an acceptable value of $d\varepsilon_s^p$.

It is often necessary to conduct a strain-controlled computation to predict the post-peak behaviour of ballast. For the strain-controlled prediction, Equation 7.72 can be re-written in the following form:

$$d\eta = \frac{M^2(1 + e_i)\left(\frac{2p_o}{p} - 1\right)\left[9(M - \eta^*) + \frac{B}{p}\{\chi + \mu(M - \eta^*)\}\right]d\varepsilon_s^p}{2\alpha\kappa\left(\frac{p}{p_{cs}}\right)\left(1 - \frac{p_{o(i)}}{p_{cs(i)}}\right)(9 + 3M - 2\eta^*M)\eta} \tag{7.74}$$

7.3 CONSTITUTIVE MODELLING FOR CYCLIC LOADING

In Section 7.2, a new constitutive model for ballast incorporating particle breakage has been presented for monotonic loading, where the shear stress is increased from an isotropic initial stress state (i.e. initial stress ratio is zero). In the case of cyclic loading, stress can increase or decrease from any state, isotropic or even anisotropic. Therefore, in order to formulate a constitutive model for cyclic loading, a stress-strain and particle breakage model must be developed first for shearing from an anisotropic initial stress state, where shearing may commence from an initial stress ratio, η_i. The model should cover shearing from both isotropic ($\eta_i = 0$) and anisotropic ($\eta_i \neq 0$) initial stress states.

7.3.1 Shearing from an anisotropic initial stress state

To extend the above constitutive model (described in Section 7.2) for shearing from an anisotropic initial stress state where the initial stress ratio is represented by η_i, Postulate B needs to be amended as follows:

Postulate B1:   The generalised undrained stress path from an initial stress ratio of η_i, is assumed to be parabolic, and is given by:

$$\left(\frac{\eta - \eta_i}{M - \eta_i}\right)^2 = \frac{p_{cs}}{p}\left[\frac{1 - \frac{p_o}{p}}{1 - \frac{p_o}{p_{cs}}}\right] \tag{7.75}$$

where, p_o and p_{cs} are the same as defined earlier.

   Postulate B1 is a modified form of a hypothesis proposed by Pender [4]. Differentiating Equation 7.75 with respect to p and re-arranging gives:

$$\frac{2(\eta - \eta_i)\left(1 - \frac{p_o}{p_{cs}}\right)\left(\frac{p}{p_{cs}}\right)d\eta}{(M - \eta_i)^2\left(\frac{2p_o}{p} - 1\right)} = \frac{dp}{p} \tag{7.76}$$

The plastic potential function (g) used for shearing from an isotropic initial stress state is also used for shearing from an anisotropic initial stress state. Substituting Equations 7.62 and 7.76 into Equation 7.67 and solving for the hardening function, it can be shown that:

$$b = \frac{2\kappa\left(\frac{1}{p_{cs}}\right)\left(\frac{p_o}{p_{cs}} - 1\right)(9 + 3M - 2\eta M)(\eta - \eta_i)}{(M - \eta_i)^2(1 + e_i)\left(\frac{2p_o}{p} - 1\right)[9(M - \eta) + \frac{B}{p}\{\chi + \mu(M - \eta^*)\}]} \tag{7.77}$$

Substituting Equations 7.61, 7.64 and 7.77 into Equation 7.56, the plastic distortional strain increment can now be written as:

$$d\varepsilon_s^p = \frac{2\kappa\left(\frac{p}{p_{cs}}\right)\left(1 - \frac{p_o}{p_{cs}}\right)(9 + 3M - 2\eta M)(\eta - \eta_i)d\eta}{(M - \eta_i)^2(1 + e_i)\left(\frac{2p_o}{p} - 1\right)[9(M - \eta) + \frac{B}{p}\{\chi + \mu(M - \eta^*)\}]} \tag{7.78}$$

In Section 7.2.2, it was pointed out that the theoretical formulation of plastic distortional strain increment (Equation 7.71, which is similar to Equation 7.78) could not predict the stress-strain behaviour of ballast well, especially at low confining pressures, where the theoretical model (Equation 7.71) underpredicted shear stress and the shear strength. To capture the experimental observations that the stress ratio η can exceed the critical state value (M) at low confining pressures, Equation 7.78 has also been amended with a modified stress ratio η^* (similar to Equation 7.72), and the following modified form of the plastic distortional strain increment is proposed:

$$de_s^p = \frac{2\alpha\kappa\left(\frac{p}{p_{cs}}\right)\left(1 - \frac{p_{o(i)}}{p_{cs(i)}}\right)(9 + 3M - 2\eta^*M)(\eta - \eta_i)d\eta}{(M - \eta_i)^2(1 + e_i)\left(\frac{2p_o}{p} - 1\right)[9(M - \eta^*) + \frac{B}{p}\{\chi + \mu(M - \eta^*)\}]} \tag{7.79}$$

where, $\eta^* = \eta(p/p_{cs})$, as shown earlier.

The relationship between the plastic volumetric strain increment and plastic distortional strain increment remains the same, as given by Equation 7.73, and the particle breakage is also simulated as before (Equation 7.51). The modified plastic hardening function corresponding to Equation 7.79 is given by:

$$h = \frac{2\alpha\kappa\left(\frac{1}{p_{cs}}\right)\left(1 - \frac{p_{o(i)}}{p_{cs(i)}}\right)(9 + 3M - 2\eta^*M)(\eta - \eta_i)}{(M - \eta_i)^2(1 + e_i)\left(\frac{2p_o}{p} - 1\right)[9(M - \eta^*) + \frac{B}{p}\{\chi + \mu(M - \eta^*)\}]} \tag{7.80}$$

7.3.2 Cyclic loading model

A common shortcoming of many stress-strain constitutive models for geomaterials is that these were developed for specific requirements and applicable only to specific loading conditions. This limitation in constitutive modelling becomes pronounced when an artificial distinction is made between monotonic and cyclic loadings for practical purposes (Dafalias and Herrmann, [16]). In reality, cyclic loading is a sequence of several monotonic ones, a combination of loading, unloading, and reloading. Therefore, the realistic constitutive laws should be based on a more fundamental framework so that they are applicable to all types of loading, whether monotonic, cyclic or any other combination.

The classical theory of plasticity provides such a framework and significant advances have been made in the past 4 decades, especially after the development of the critical state theory by Roscoe and co-researchers [1, 2]. These theories can adequately and accurately simulate the deformation response of geomaterials under monotonic loading. However, some important aspects of deformation behaviour, particularly in cyclic loading, cannot be adequately modelled with these theories. One of the main reasons is that in the classical concept of yield surface, there is little flexibility in varying the plastic modulus when the loading directions are changed. This implies a purely elastic stress domain, which is contrary to reality for many geomaterials (Dafalias and Herrmann, [16]). Therefore, the classical theory of plasticity is unable to simulate, even qualitatively, the accumulation of plastic strains with increasing load cycles.

To overcome these limitations, a new concept of plasticity called 'bounding surface plasticity' was introduced by Dafalias and Popov [17, 18] and Krieg [19], as mentioned earlier in Chapter 6. The salient features of the bounding surface plasticity theory are: (a) plastic deformation may occur for stress changes within the bounding surface, and (b) the possibility of having a very flexible plastic modulus. These are the clear advantages over the classical yield surface plasticity theory.

The most difficult part of constitutive modelling, especially in cyclic loading, is the mathematical description of the appropriate evolution of hardening modulus. The memory of particular loading events and progressive cyclic hardening or softening phenomena should be included in the model (Mroz and Norris, [20]). One possibility is to consider a smaller yield surface within a larger bounding surface and vary the plastic (or hardening) modulus depending on the distance of the current stress point relative to its conjugate point on the bounding surface, as was examined by Mroz and Norris [20] and Dafalias and Herrmann [16], among others. Their novel approach was considerably successful, at least qualitatively, for predicting different aspects of soil behaviour under loading, both monotonic and cyclic.

To simulate the response of ballast under cyclic loading, the concept of bounding surface plasticity along with varying hardening function was adopted by the authors, as described in the following Sections.

7.3.2.1 Conceptual model

In the current formulation, it is assumed that under cyclic reloading, ballast deforms plastically but at a smaller scale. These small plastic deformations are also governed by linear kinematic yield surfaces (same as Equation 7.63) within a larger bounding yield surface. During a virgin loading where the stress state remains on the bounding surface, plastic deformations are the same as in case of monotonic shearing (Section 7.2). The plastic deformations under cyclic loading are generally computed by formulating an appropriate plastic hardening function that varies with the state of geomaterials (i.e. p, q and e) and the previous stress history.

Before formulating an appropriate varying hardening function for the generalised cyclic loading, the evolution of plastic hardening function during a simple loading-unloading and reloading path 'a-b-c-d' (Fig. 7.15) is considered first. The constant

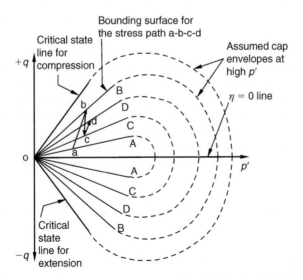

Figure 7.15 Bounding surface for a simplified stress path 'a-b-c-d' under cyclic loading.

stress ratio lines (OA, OB, OC, OD etc.) shown in Figure 7.15 represent yield loci, as mentioned earlier in Postulate A. The dotted curves shown in Figure 7.15 represent possible caps of the yield loci at very high stress levels. However, it is anticipated that at very high stress levels, ballast will yield in both isotropic compression (i.e. $\eta = 0$) and shearing (i.e. $d|\eta| > 0$), and that the degree of particle breakage will be very high at those stresses.

According to Postulate A, the line OA (Fig. 7.15) connecting the initial stress point 'a' and the origin of stresses 'O', represents the initial yield locus. The line OA' represents a similar yield locus for the negative q (i.e. in extension). In triaxial extension, q is often considered to be negative. If the stress ratio at point 'a' represents the maximum past stress ratio of ballast, then the line OA forms its current bounding surface. It is assumed that if the stress state is on the current bounding surface and the change of stress is directed towards the exterior of the bounding surface (i.e. away from the $\eta = 0$ line, or $d|\eta| > 0$), it represents 'loading', which causes plastic deformation, in addition to elastic strain. The plastic deformation associated with this 'loading' will be governed by the bounding hardening function h_{bound}, which is the same as given by Equation 7.80.

In contrast, if any change of stress is directed towards the interior of bounding surface (i.e. towards the $\eta = 0$ line, or $d|\eta| < 0$), it represents 'unloading' and causes only elastic recovery of strains. There is no plastic deformation associated with 'unloading'. If the change of stress commences from a point interior to the bounding surface and is directed towards the bounding surface ($d|\eta| > 0$), it represents 'reloading' and also causes plastic deformation, but at a considerably smaller scale. The plastic deformation associated with this 'reloading' will be governed by a new hardening function, h_{int}. The mathematical formulation of h_{int}, is given later in Section 7.3.2.2.

In stress path 'a-b' (Fig. 7.15), since the stress point 'a' is on the current bounding surface (OA) and the direction of stress change is towards the exterior of the current bounding surface, the plastic hardening function for this 'loading' is given by Equation 7.80. At the end of stress path 'a-b', a new bounding surface is formed by the line OB, (connecting the stress point 'b' and the origin 'O'). During the stress path 'b-c', since the stress change is directed towards the interior of current bounding surface (i.e. towards $\eta = 0$ line), the deformation corresponding to this 'unloading' is purely elastic.

During the stress path 'c-d', since the stress change starts from a point ('c'), which is inside the current bounding surface (OB), and the stress change is directed towards the bounding surface, the plastic hardening function for this 'reloading' will be h_{int}. It is also assumed that the hardening function h_{int}, starts with an initial value at the beginning of reloading (e.g. point 'c' in Fig. 7.15), and gradually evolves to the bounding value as the stress path meets the current bounding surface.

The essential features of the current cyclic constitutive model are summarised below:

- Plastic deformations are associated with all 'loading' and 'reloading', in addition to elastic strains
- 'Unloading' causes only elastic recovery of strain
- 'Loading' is defined by: $\eta = \eta_{bound}$ and $d|\eta| > 0$
- 'Unloading' is defined by: $d|\eta| < 0$
- 'Reloading' is defined by: $|\eta| < |\eta_{bound}|$ and $d|\eta| > 0$

- If $\eta = \eta_{bound}$, $h = h_{bound}$
- If $|\eta| < |\eta_{bound}|$, $h = h_{int}$

where, $h_{bound} =$ hardening function at the bounding surface given by Equation 7.80 and $\eta_{bound} =$ stress ratio at the bounding surface.

7.3.2.2 Mathematical model

The mathematical expressions of the initial hardening function $h_{int(i)}$ and the evolution of plastic hardening function h_{int}, within the bounding surface are given by:

$$h_{int(i)} = h_i e^{-\xi_1 \varepsilon_v^p} \tag{7.81}$$

$$h_{int} = h_{int(i)} + (h_{bound} - h_{int(i)}) R^\gamma e^{-\xi_2 \varepsilon_v^p} \tag{7.82}$$

$$R = \frac{\eta - \eta_i}{\eta_{bound} - \eta_i} \tag{7.83}$$

where, $h_i =$ initial hardening function at the start of cyclic loading (e.g. $h_i = h$ at point 'a' in Figure 7.15), $h_{int} =$ hardening function at the interior of bounding surface (for 'reloading'), $h_{int(i)} =$ initial value of h_{int} for 'reloading', ξ_1, ξ_2 and γ are dimensionless parameters and the first two are related to cyclic hardening.

The function h_{int} for the first 'reloading' is modelled by Equation 7.82. For the second and subsequent 'reloadings', h_{int} is given by:

$$h_{int} = h_{int(i)} + (h_{bound} - h_{int(i)}) R^\gamma e^{-\xi_3 \varepsilon_{v1}^p} \tag{7.84}$$

where, ξ_3 is another dimensionless parameter related to cyclic hardening and ε_{v1}^p is the accumulated plastic volumetric strain since the end of the first load cycle.

The plastic distortional strain increment corresponding to any 'loading' is given by Equation 7.79 and for a 'reloading', $d\varepsilon_s^p$ is given by:

$$d\varepsilon_s^p = h_{int} p \, d\eta \tag{7.85}$$

Equation 7.73 gives the plastic volumetric strain increment, as in monotonic shearing, and Equation 7.51 gives the particle breakage. Although actual breakage process depends on the cyclic loading and the fatigue failure of ballast grains, the particle breakage has been modelled in the current formulation as a function of distortional strain ε_s, initial mean stress $p_{(i)}$ and the initial void ratio represented by the parameter $p_{cs(i)}$, based on the experimental findings (see Fig. 7.7). Each load increment during loading and reloading causes an increase in stress ratio $d\eta$, resulting in an increase in plastic distortional and volumetric strains (Equations 7.85 and 7.73, respectively). These strains are accumulated with increasing load cycles, although there is no net change in q for a system of cyclic loading with a constant load amplitude. The increase in distortional strains and the induced internal stresses cause attrition, grinding, breakage of sharp corners and asperities, and even splitting and crushing of weaker grains. All these degradation aspects are included together in the breakage index (B_g), as modelled by Equation 7.51. Thus, the effect of cyclic loading on the particle breakage process has been adequately simulated in the authors' model.

The implementation of the above constitutive model has been carried out numerically and the verification of the model is discussed in the following Section.

7.4 MODEL VERIFICATION AND DISCUSSION

The new stress-strain and particle breakage constitutive model has been examined and verified by comparing the model response with the laboratory experimental data for both monotonic and cyclic loadings. The model parameters were evaluated using the triaxial test results. Additionally, ballast specimens under triaxial stresses were analysed by finite element method (FEM) employing a computer code ABAQUS[2], and the numerical predictions were also compared with the analytical model predictions. This Section describes the numerical techniques adopted to implement the authors' constitutive model, the evaluation of model parameters, and the comparison of analytical and numerical predictions with the test data. The analytical predictions using the monotonic loading model (Section 7.2) were compared with the triaxial test results of fresh ballast, while the predictions using the cyclic loading model (Section 7.3) were verified against the prismoidal triaxial test results of fresh ballast.

7.4.1 Numerical method

To implement the current constitutive model, a simple numerical procedure was adopted to solve the differential Equations 7.41–7.42, 7.73, 7.79 and 7.85, which could not be integrated directly. For monotonic model predictions, a strain-controlled computation was conducted adopting the following equation:

$$(\eta)_{n+1} = (\eta)_n + \left(\frac{d\eta}{d\varepsilon_s^p}\right)_n \delta\varepsilon_s^p \tag{7.86}$$

where, the subscript 'n' represents a current value and the subscript 'n + 1' indicates a value after the increment.

For cyclic model predictions, a stress-controlled computation was carried out following the equation:

$$(\varepsilon_s^p)_{n+1} = (\varepsilon_s^p)_n + \left(\frac{d\varepsilon_s^p}{d\eta}\right)_n \delta\eta \tag{7.87}$$

For both monotonic and cyclic model predictions, the numerical values of ε_v^p, ε_s^e, and ε_v^e were computed by:

$$(\varepsilon_v^p)_{n+1} = (\varepsilon_v^p)_n + \left(\frac{d\varepsilon_v^p}{d\varepsilon_s^p}\right)_n \delta\varepsilon_s^p \tag{7.88}$$

---

[2] ABAQUS software is commercialized by Hibbit, Karlsson & Sorensen, Inc., 1080 Main Street, Pawtucket, RI 02860-4847, USA.

$$\left(\varepsilon_s^e\right)_{n+1} = \left(\varepsilon_s^e\right)_n + \left(\frac{d\varepsilon_s^e}{dq}\right)_n \delta q \tag{7.89}$$

$$\left(\varepsilon_v^e\right)_{n+1} = \left(\varepsilon_v^e\right)_n + \left(\frac{d\varepsilon_v^e}{dp}\right)_n \delta p \tag{7.90}$$

Equation 7.79 was used for the derivatives $d\eta/d\varepsilon_s^p$ and $d\varepsilon_s^p/d\eta$ of Equations 7.86 and 7.87, respectively. Equations 7.73, 7.41, and 7.42 were used for the derivatives $d\varepsilon_v^p/d\varepsilon_s^p$, $d\varepsilon_s^e/dq$, and $d\varepsilon_v^e/dp$ of Equations 7.88, 7.89, and 7.90, respectively, for both monotonic and cyclic model predictions.

7.4.2 Evaluation of model parameters

The monotonic shearing model (Section 7.2) contains 11 parameters, which can be evaluated using conventional drained triaxial test results together with the measurements of particle breakage, as explained below. The critical state parameters (M, λ_{cs}, Γ and κ) can be determined from a series of drained triaxial compression tests conducted at various effective confining pressures. The slope of the line connecting the critical state points in the p-q plane gives the value of M, and that in the e-$\ln p$ plane gives λ_{cs}. The void ratio (e) of the critical state line at $p = 1$ kPa is the value of Γ. The parameter κ can be determined from an isotropic (hydrostatic) loading-unloading test with the measurements of volume change. The slope of the unloading part of isotropic test data plotted in the e-$\ln p$ plane gives the value of κ. The elastic shear modulus G, can be evaluated from the unloading part of stress-strain (q-ε_s) plot in triaxial shearing.

The model parameter β (Equation 7.55) can be evaluated by measuring the particle breakage (B_g) at various strain levels, as explained earlier in Section 7.2.2 (Fig. 7.6). The parameters θ and υ can be determined by replotting the breakage data as $\ln\{p_{cs(i)}/p_{(i)}\}B_g$ versus ε_s^p (Fig. 7.8), and finding the coefficients of the non-linear function (Equation 7.51) that best represent the test data. The parameters χ and μ can be evaluated by plotting the rate of particle breakage data in terms of $\ln\{p_{cs(i)}/p_{(i)}\}dB_g/d\varepsilon_s^p$ versus $(M - \eta^*)$ (Fig. 7.10) and determining the values of the intercept and slope of the best-fit line. The parameter α is used in the current model to match the initial stiffness of the analytical predictions with the experimental results and can be evaluated by a regression analysis or a trial and error process comparing the model predictions with a set of experimental data.

The cyclic loading model (Section 7.3) has 4 parameters in addition to the above. These 4 parameters can be evaluated from the stress-strain measurements for a number of load cycles during a cyclic test. The parameter ξ_1 can be determined from the initial re-loading data, while the parameters ξ_2 and ξ_3 can be evaluated from the remaining parts of the first re-loading and the following re-loading data, respectively. The model parameter γ can also be evaluated from any re-loading stress-strain data. The determination of the above model parameters (both for monotonic and cyclic models) from laboratory experimental test results are explained further in Appendix B.

7.4.3 Model predictions for monotonic loading

The deformation response of ballast under monotonic loading was predicted using the new constitutive model (Section 7.2), and then compared with the experimental

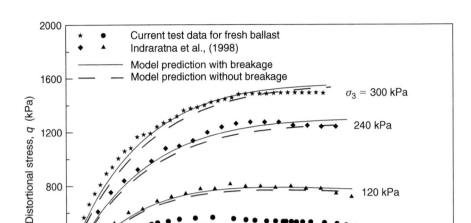

Figure 7.16 Analytical prediction of stress-strain of ballast with and without particle breakage compared to test data (modified after Salim and Indraratna, [6] and Salim, [21]).

results. In predicting ballast behaviour using the authors' model, the following model parameters were used: $M = 1.9$, $\lambda_{cs} = 0.188$, $\Gamma = 1.83$, $\kappa = 0.007$, $G = 80$ MPa, $\alpha = 28$, $\beta = 0.0029$ kN-m/m^3, $\chi = 0.21$, $\mu = 0.50$, $\theta = 0.125$, and $\nu = 10.5$. Ten of the above 11 parameters were evaluated from drained triaxial compression test results, as explained earlier in Section 7.4.2. The value $\alpha = 28$ was determined by initial stiffness matching of the analytical predictions with several test results of ballast (Salim and Indraratna, 2004).

The analytical predictions were made following a strain-controlled computation. For a given initial state of ballast (p, q and e), a small plastic distortional strain increment was assumed and the corresponding new stress ratio was computed as per the numerical procedure shown earlier (Section 7.4.1). The corresponding plastic and elastic volumetric strains were computed using Equations 7.88 and 7.90, while the elastic distortional strain increment was obtained using Equation 7.89. The breakage index (B_g) at the end of strain increment was computed by Equation 7.51.

Figure 7.16 shows the stress-strain predictions for ballast, while Figure 7.17 illustrates the volume change predictions compared to the authors' experimental data and the previous ballast test data, as reported by Indraratna et al. [11]. The analytical predictions without any particle breakage (i.e. using $\beta = 0$ in Equation 7.55) are also shown in these figures for comparison. Excellent agreement is found between the current model predictions and the experimental data, especially with particle breakage. Since the confining pressures used in the laboratory experiments were small (300 kPa maximum) compared to the compressive strength of the parent rock of about 130 MPa

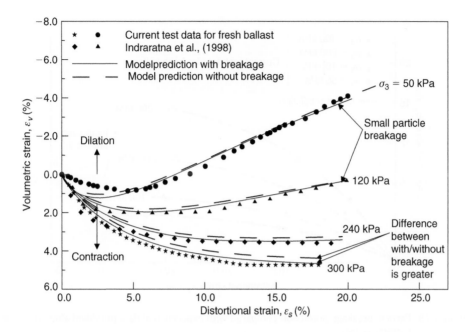

Figure 7.17 Volume change predictions with and without particle breakage compared to test data (modified after Salim and Indraratna, [6] and Salim, [21]).

(Indraratna et al., [11]), only a small fraction of the imparted energy was consumed in particle breakage. Therefore, the difference between the model predictions with and without particle breakage is small (Figs. 7.16–7.17). As seen in Figure 7.17, the gap between the predicted curves with and without breakage increases as the confining pressure increases (e.g. $\sigma_3 = 300$ kPa), where particle breakage becomes increasingly more significant. It is anticipated that at very high confining pressures (>1 MPa), particle breakage will be high and particle crushing will dominate the deformation behaviour of ballast, especially the volumetric changes.

Figure 7.18 shows the model prediction of particle breakage (B_g) compared to the experimental data. It shows that the predicted breakage values are close to the measured data. Figure 7.18 verifies that the authors' analytical model predicts the breakage of ballast to an acceptable accuracy.

As mentioned earlier, the postulates made in the current model are comparable to the hypotheses made by Pender [4] for overconsolidated soils. Despite these similarities, there are some significant differences between these two approaches. Pender [4] assumed that all soils, which are denser than the critical (i.e. $p_o < p_{cs}$), would exhibit plastic dilation during shear deformation. He adopted a function for the ratio between plastic strain increments, $d\varepsilon_v^p / d\varepsilon_s^p$, which makes the plastic volumetric strain increment negative (i.e. dilation) for all soils denser than the critical. However, Indraratna and Salim [14] reported that at a relatively high confinement (>200 kPa), plastic volumetric contraction occurs during shearing of ballast, which is still on the denser side of the

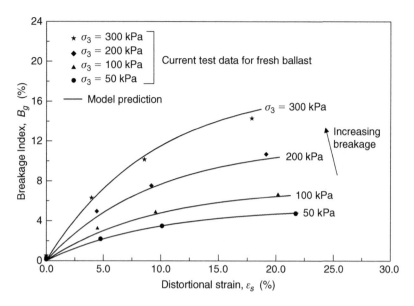

Figure 7.18 Particle breakage prediction compared with experimental data (modified after Salim and Indraratna, [6]).

critical state line (CSL). This aspect of ballast behaviour is well captured in the current model. Equation 7.73 provides positive plastic volumetric strain (i.e. contraction) for ballast, which is denser than the critical, as long as the stress ratio (η) does not exceed M.

In contrast, Pender's [4] hypothesis always provides plastic dilation (negative $d\varepsilon_v^p$) for all stress ratios if the soil is on the denser side of the CSL (i.e. $p_o < p_{cs}$). Other major difference between the two models is the incorporation of particle breakage, which is absent in Pender's [4] model. Any particle breakage will consume part of the imparted energy, and therefore, a reduced amount of energy will be spent on frictional deformation and the resulting plastic distortional strain increment will be smaller. This is clearly reflected in the denominators of Equations 7.72 and 7.79, which include the breakage term. Moreover, particle breakage will contribute to an increase in plastic volumetric strain (contraction), an aspect that is correctly represented in the current model (Equation 7.73).

An interesting point to note is that Equation 7.73 of the current model always governs the plastic volumetric strain (positive or negative) towards the critical state. At the initial stage of shearing ($\eta < M$), Equation 7.73 provides plastic volumetric contraction ($d\varepsilon_v^p$ positive) so that ballast hardens, and as a result, it can sustain additional shear stress (i.e. η increases towards M). If the stress ratio η exceeds M (under low confinement), Equation 7.73 provides negative $d\varepsilon_v^p$ (or dilation) when the value of the breakage related term is small, and therefore, the material softens, and the stress ratio gradually decreases towards the critical state value M.

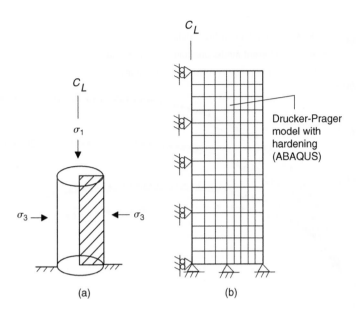

Figure 7.19 (a) Ballast specimen, (b) discretisation and mesh used in finite element modelling of the ballast specimen.

7.4.4 Analytical model compared to FEM predictions

The analytical model predictions are also compared with the results of finite element analysis employing ABAQUS. The finite element code ABAQUS is a powerful tool and commercially available for analysing a wide range of engineering problems including geomechanics. In this Section, the analytical model predictions and the ABAQUS finite element predictions are compared with the experimental data.

Finite element analyses were carried out for a cylindrical ballast specimen (Fig. 7.19a) using axisymmetric elements. As $\sigma_2 = \sigma_3$ and $\varepsilon_2 = \varepsilon_3$ in triaxial shearing (i.e. axisymmetric), the shaded area of the specimen (Fig. 7.19a) was discretised, as illustrated in Figure 7.19(b). The left boundary of Figure 7.19(b) represents the central specimen axis, which does not move laterally under triaxial loading, hence the roller supports to restrain lateral movement (i.e. vertical degree of freedom only).

In ABAQUS, the extended Drucker-Prager model with hardening was used to simulate inelastic deformation of granular materials (Hibbit, Karlsson and Sorensen, Inc., [22]). Figures 7.20(a) and (b) show the FEM stress-strain and volume change predictions compared to the analytical predictions. The experimental results are also plotted in these figures for convenience and comparison.

Figure 7.20(a) indicates that both the analytical and FEM models predict the stress-strain response of ballast fairly well, but the authors' constitutive model is slightly better. In contrast, Figure 7.20(b) clearly shows that the FEM model (ABAQUS) could not simulate the volumetric response of ballast well, especially at high confining pressures (e.g. 200 and 300 kPa). In particular, the finite element simulation

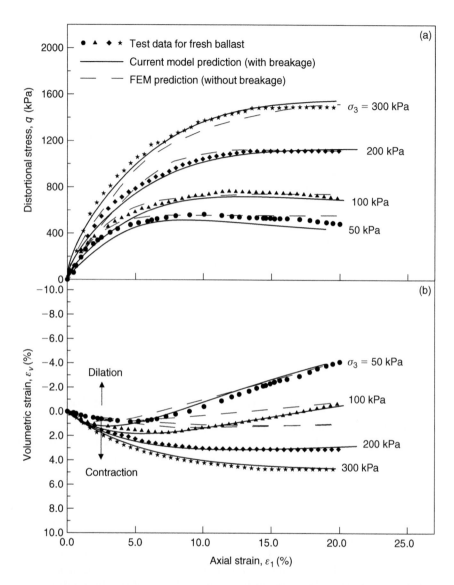

Figure 7.20 Analytical model predictions of ballast compared with FEM analysis results and experimental data, (a) stress-strain, and (b) volume change behaviour.

could not predict the specimen contraction at high stresses. Apart from restrained lateral displacements at high confining pressures, particle breakage is also increasingly more significant, as discussed earlier, hence, the subsequent overall contraction of the specimen is inevitable.

Particle breakage was not taken into account in the constitutive model of ABAQUS. Moreover, the plastic volumetric deformation of geomaterials is simulated in ABAQUS by a single value of dilation angle, which restricts the volumetric contraction in the

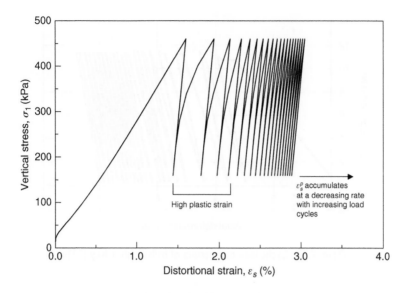

Figure 7.21 Qualitative model prediction of cyclic stress-strain of ballast.

finite element simulation. Therefore, it is not surprising that acceptable volumetric matching could not be achieved in ABAQUS simulation. As the authors' constitutive model incorporates the effect of particle breakage on both volumetric and distortional strains and also appropriately simulates the plastic volumetric response associated with shearing (Equation 7.73), better predictions of volumetric behaviour using the current model were achieved (Fig. 7.20a).

7.4.5 Model predictions for cyclic loading

The qualitative prediction of cyclic stress-strain using the authors' constitutive model is shown in Figure 7.21. In addition to 11 model parameters used in monotonic model, the following values of 4 additional cyclic model parameters were used: $\xi_1 = 1400$, $\xi_2 = 25$, $\xi_3 = 3400$, and $\gamma = 2$. Figure 7.22 shows the cyclic load-deformation test results of ballast as reported by Key [23]. Comparing Figures 7.21 and 7.22, it may be concluded that the qualitative stress-strain model prediction is comparable to the experimental data. The qualitative model prediction (Fig. 7.21) also shows that as the load cycle increases, the plastic strain accumulates at a decreasing rate, which is a key feature of cyclic deformation behaviour of many geomaterials. It also depicts that the plastic strain is high in the first cycle of loading, then gradually decreases with increasing load cycles, a typical behaviour of ballast under cyclic loading (Key, [23]).

The model predictions of distortional strain (ε_s) and volumetric strain (ε_v) of fresh ballast (wet) under a system of cyclic vertical stress and lateral confinement similar to that applied in the prismoidal triaxial tests are compared with the experimental data, as shown in Figures 7.23–7.24. Additionally, 2 other cyclic stress-strain models (Tatsuoka

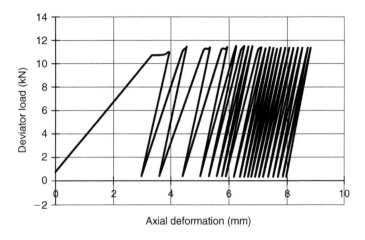

Figure 7.22 Cyclic load test results of ballast (after Key, [23]).

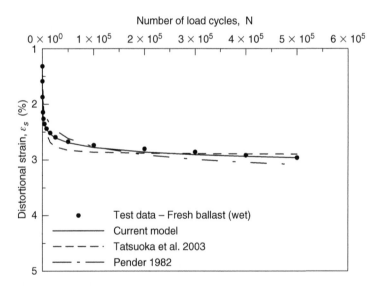

Figure 7.23 Model prediction of ballast distortional strain compared with experimental data.

et al., [24] and Pender, [25]) were also employed to predict the cyclic response of ballast and those predictions are also compared with the current model. Since the model parameters were evaluated from the triaxial test results of fresh ballast, which was saturated prior to drained shearing, cyclic model predictions using those parameters were compared with the results of fresh ballast tested in a wet state.

Tatsuoka et al. [24] simulated the stress-strain hysteretic loop in a plane strain cyclic loading based on an empirical hyperbolic relationship (Equation 6.24). The evolution of the stress-strain with increasing load cycles was governed by a set of rules in

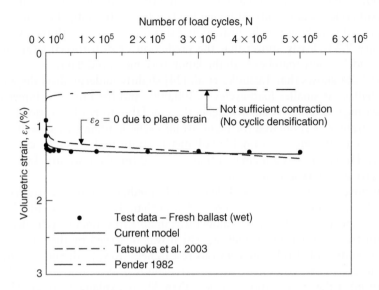

Figure 7.24 Model prediction of volumetric strain of ballast compared with test data.

their technique, as mentioned earlier in Chapter 6. In contrast, Pender's [25] model was formulated based on the critical state framework and the classical theory of plasticity. Since there is little flexibility in the classical plasticity theory in varying the plastic modulus when loading direction is reversed, as mentioned earlier in Section 7.3, Pender [25] adopted a cyclic hardening index ξ (Equation 6.23) in his model to overcome this limitation. On the other hand, the current model was developed based on the critical state framework and the bounding surface plasticity concept, rather than the classical plasticity theory. The current model also incorporates particle breakage under loading.

The following parameters were used for analysing ballast behaviour using Tatsuoka et al. [24]: $\gamma_{ref} = 1.61\%$, $\beta_{max} = 0.024$, $F = 0.14$, $M_o = 2000$, $K' = 0.45$ for loading in the first cycle, $K' = 0.24$ for reloading and $K' = 0.24106$ for unloading. The parameter γ_{ref} was evaluated from the monotonic shearing results of ballast (q_{max}/G). The parameter β_{max} represents the maximum drag in Tatsuoka et al. [24] and is related to the plastic shear strain in cyclic loading. The parameter M_o was evaluated from the initial stiffness of $\sin\phi_{mob}-\gamma$ relationship. As Tatsuoka et al. [24] did not indicate the evaluation technique for the model parameters F and K', the above values of these parameters were used in this study to give the best possible predictions.

The following parameters were used for the prediction of ballast behaviour using Pender's [25] model: $M = 1.90$, $\lambda = 0.188$, $\kappa = 0.007$, $G = 80\,\text{MPa}$, $\hat{\alpha} = 0.05$ and $\hat{\beta} = 0.10$. The first 4 parameters of Pender's [25] model (i.e. M, λ, κ and G) are the same as in the authors' model. Pender [25] did not show the evaluation technique for the model parameters $\hat{\alpha}$ and $\hat{\beta}$. The above values of $\hat{\alpha}$ and $\hat{\beta}$ were used by the authors to give the best possible predictions using Pender's [25] model.

Figure 7.23 shows that Pender's [25] model slightly underpredicts the distortional strain at smaller load cycles ($<100,000$) but overpredicts slightly at higher load cycles

(>200,000). In contrast, Tatsuoka et al. [24] slightly overpredicts distortional strain at smaller load cycles (<200,000). At higher load cycles (>200,000), Tatsuoka et al. [24] gives improved matching for the distortional strain prediction with the experimental data. Figure 7.23 clearly shows that the prediction of distortional strain using the authors' model closely matches with the laboratory measured data.

Figure 7.24 shows that Tatsuoka et al. [24] slightly underpredicts the volumetric strain of ballast at smaller load cycles (<300,000) and the rate of volumetric strain with increasing load cycles is slightly higher than the laboratory observations. Although the stress-strain was simulated for plane strain cyclic loading (i.e. $\varepsilon_2 = 0$), Tatsuoka et al. [24] generally gives reasonable volumetric strain under triaxial cyclic loading (Fig. 7.24). However, it is anticipated that as $\varepsilon_2 = 0$ (in plane strain), Tatsuoka et al. [24] will give excessive lateral strains (ε_3).

In contrast, Pender's [25] model clearly underpredicts volumetric strain of ballast. Since Pender [25] considered that all soils denser than the critical would dilate plastically during shear deformation, his model was unable to simulate cyclic densification (i.e. volumetric contraction) of ballast, which was observed by the authors in their laboratory study and also by the previous researchers (Key, [23]; Suiker, [26]). In the authors' model, plastic volumetric strain increment is positive (i.e. contraction, rather than dilation) if the stress ratio (η) is less than M, as explained earlier. This plastic volumetric contraction is accumulated with increasing load cycles, causing cyclic densification in ballast. Thus, the current model correctly simulates the volumetric response of ballast under cyclic loading, as revealed in Figure 7.24.

Figure 7.25 shows the predicted particle breakage (B_g) of ballast using the current authors' model compared to the experimental data. Since Tatsuoka et al. [24] and Pender [25] did not consider any breakage of particles under cyclic loading, these models were unable to simulate ballast breakage under cyclic loading, and therefore,

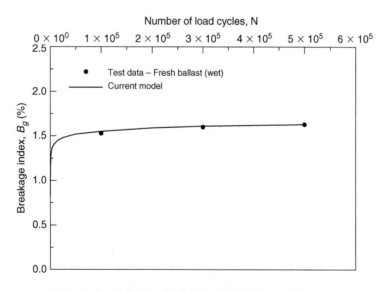

Figure 7.25 Prediction of ballast breakage under cyclic loading.

not shown in this figure. Tatsuoka et al. [24] and Pender [25] developed their models primarily for sands and overconsolidated fine-grained soils, where particle breakage is insignificant.

In rail tracks, particle breakage is the main source of ballast fouling, as mentioned earlier in Chapter 3, and also affects the strength and deformation behaviour of ballast. In the authors model, the particle breakage has been incorporated in the incremental stress-strain formulations appropriately. Figure 7.25 shows that the predicted breakage of ballast increases rapidly up to about 50,000 load cycles, beyond which the increase in breakage becomes marginal. The close agreement between the model predictions and the experimental data (Fig. 7.25) verifies that the authors' model can predict ballast breakage under cyclic loading to an acceptable accuracy.

REFERENCES

1. Roscoe, K.H., Schofield, A.N. and Thurairajah, A.: Yielding of clays in states wetter than critical. *Geotechnique*, Vol. 13, No. 3, 1963, pp. 211–240.
2. Schofield, A.N. and Wroth, C.P.: *Critical State Soil Mechanics*. McGraw Hill, 1968.
3. Lade, P.V.: Elasto-plastic stress-strain theory for cohesionless soil with curved yield surfaces. *International Journal of Solids and Structures*, Vol. 13, 1977, pp. 1019–1035.
4. Pender, M.J.: A model for the behaviour of overconsolidated soil. *Geotechnique*, Vol. 28, No. 1, 1978, pp. 1–25.
5. Indraratna, B. and Salim, W.: Modelling of particle breakage of coarse aggregates incorporating strength and dilatancy. *Geotechnical Engineering*, Proc. Institution of Civil Engineers, London, Vol. 155, Issue 4, 2002, pp. 243–252.
6. Salim, W. and Indraratna, B.: A new elasto-plastic constitutive model for coarse granular aggregates incorporating particle breakage. *Canadian Geotechnical Journal*, Vol. 41, No. 4, 2004, pp. 657–671.
7. Rowe P.W.: The stress-dilatancy relation for the static equilibrium of an assembly of particles in contact. *Proceedings Royal Society*, Vol. A269, 1962, pp. 500–527.
8. Ueng, T.S. and Chen, T.J.: Energy aspects of particle breakage in drained shear of sands. *Geotechnique*, Vol. 50, No. 1, 2000, pp. 65–72.
9. Marsal, R.J.: Large scale testing of rockfill materials. *J. of the Soil Mech. and Found. Div.*, ASCE, Vol. 93, No. SM2, 1967, pp. 27–43.
10. Charles, J.A. and Watts, K.S.: The influence of confining pressure on the shear strength of compacted rockfill. *Geotechnique*, Vol. 30, No. 4, 1980, pp. 353–367.
11. Indraratna, B., Ionescu, D. and Christie, H.D.: Shear behaviour of railway ballast based on large-scale triaxial tests. *J. of Geotechnical and Geoenvironmental Engineering*, ASCE, Vol. 124. No. 5, 1998, pp. 439–449.
12. Bolton M.D.: The strength and dilatancy of sands. *Geotechnique*, Vol. 36, No. 1, 1986, pp. 65–78.
13. Hill, R.: *The Mathematical Theory of Plasticity*. Oxford University Press, 1950.
14. Indraratna, B. and Salim, W.: Shear strength and degradation characteristics of railway ballast. *Proc. 14th Southeast Asian Geotechnical Conference*, Hong Kong, Vol. 1, 2001, pp. 521–526.
15. Ishihara, K., Tatsuoka, F. and Yasuda, S.: Undrained deformation and liquefaction of sand under cyclic stresses. *Soils and Foundations*, Vol. 15, No. 1, 1975, pp. 29–44.
16. Dafalias, Y.F. and Herrmann, L.R.: Bounding surface formulation of soil plasticity. In: *Soil Mechanics – Transient and Cyclic Loads* (edited by Pande and Zienkiewicz), 1982, pp. 253–282.

17. Dafalias, Y.F. and Popov, E.P.: A model of nonlinearly hardening materials for complex loadings. *Acta Mech*. Vol. 21, 1975, pp. 173–192.
18. Dafalias, Y.F. and Popov, E.P.: Plastic internal variables formalism of cyclic plasticity. *J. of Applied Mechanics, ASME*, Vol. 98, No. 4, 1976, pp. 645–650.
19. Krieg, R.D.: A practical two surface plasticity theory. *J. of Applied Mechanics, ASME*, Vol. 42, 1975, pp. 641–646.
20. Mroz, Z. and Norris, V.A.: Elastoplastic and viscoplastic constitutive models for soils with application to cyclic loading. In: *Soil Mechanics-Transient and Cyclic Loads* (edited by Pande and Zienkiewicz), 1982, pp. 173–217.
21. Salim, M.W.: Deformation and degradation aspects of ballast and constitutive modelling under cyclic loading. PhD Thesis, University of Wollongong, Australia, 2004.
22. Hibbit, Karlsson and Sorensen, Inc.: *ABAQUS/Standard User's Manual*. Version 6.3, Vol. 2, 2000.
23. Key, A.J.: Behaviour of Two Layer Railway Track Ballast under Cyclic and Monotonic Loading. PhD Thesis, University of Shefield, UK, 1998.
24. Tasuoka, F., Masuda, T., Siddiquee, M.S.A. and Koseki, J.: Modeling the stress-strain relations of sand in cyclic plane strain loading. *J. of Geotechnical and Geoenvironmental Engineering, ASCE*, Vol. 129. No. 6, 2003, pp. 450–467.
25. Pender, M.J.: A model for the cyclic loading of overconsolidated soil. In: *Soil Mechanics – Transient and Cyclic Loads* (edited by Pande and Zienkiewicz), 1982, pp. 283–311.
26. Suiker, A.S.J.: *The mechanical behaviour of ballasted railway tracks*. PhD Thesis, Delft University of Technology, The Netherlands, 2002.

Chapter 8

Track Drainage and Use of Geotextiles

Ballast layer is designed to be free draining but when the ballast voids are wholly or partially occupied due to the intrusion of fine particles, the ballast can be considered to be "fouled". During operation, ballast deteriorates due to the breakage of angular corners and sharp edges, infiltration of fines from the surface, and mud pumping from the subgrade under train loading. As a result of these actions ballast becomes fouled, less angular, and its shear strength is reduced. Fouling materials have traditionally been considered as unfavourable to track structure. According to Selig and Waters [1], ballast breakdown, on average, accounts for up to 76% of fouling, followed by 13% of infiltration from subballast, 7% infiltration from surface ballast, 3% from subgrade intrusion, and 1% from sleeper wear. However, Feldman and Nissen [2] reported that for tracks in Australia used predominantly for coal transport, coal dust accounts for 70%–95% of contaminants and ballast breakdown contributes from 5%–30%. To ensure acceptable track performance, it is necessary to maintain a good drainage condition within the ballast layer.

Drainage plays a significant role in the stability and safety of a track substructure. Fouling causes a reduction in the drainage capacity of ballast. In saturated tracks, poor drainage can lead to the build up of excess pore water pressure under train loading. If the permeability of substructure elements, especially the subballast layer becomes excessively low, the excess pore water pressure developed under an axle loading often cannot dissipate completely before the next load is imposed. Thus, the residual pore pressures accumulate with increasing load cycles. After a few load cycles, the total excess pore water pressure becomes very high and often causes 'clay pumping', as described earlier.

Geosynthetics are now being used in track successfully by various railway organisations worldwide to significantly improve substructure drainage characteristics. In Chapters 4 and 5, the stabilisation aspects of recycled ballast using various geosynthetics were discussed. In this Chapter, several key issues of track drainage are highlighted and the effectiveness of various commercially available geosynthetics for enhancing track drainage is discussed.

8.1 DRAINAGE

The primary purpose of track drainage is to remove water from the substructure as fast as possible and maintain the load bearing stratum relatively dry. To fulfil this objective, the load bearing layer (ballast) is usually composed of coarse and uniformly-graded aggregates with large-size voids ensuring sufficiently high permeability. Since the ballast is laid on fine-grained subgrade, a filtering layer (subballast) is usually placed in between these two media to prevent inter-penetration.

Water can enter into the load-bearing stratum from 4 different sources:

- Precipitation (rain and snow)
- Surface flow from adjacent hill slopes
- Upward seepage from subgrade, and
- High groundwater table in low-lying coastal regions.

Track substructure should be adequately designed and constructed in such a way that the infiltrated water is quickly and completely drained out from the load bearing layers to the nearby drainage ditches or pipes. Internal drainage is usually ensured by placing a subballast layer having appropriate gradation. In the case of inadequate drainage, the following problems may occur in track:

- Decrease in ballast shear strength, stiffness and load bearing capacity
- Increased track settlement
- Softening of subgrade
- Formation of slurry and clay pumping under cyclic loading
- Ballast attrition by jetting action and freezing of water
- Sleeper degradation by water jetting.

All these problems will degrade the track performance and demand additional maintenance. To prevent or minimise these problems, adequate drainage is imperative in ballasted tracks.

8.1.1 Subballast permeability

Flow through porous media such as ballast and subballast is usually determined using Darcy's law:

$$v = ki \tag{8.1}$$

where, v = average velocity of fluid, i = hydraulic gradient, and k is the coefficient of permeability (hydraulic conductivity). The best way to evaluate the value of k for a particular porous medium is by conducting field or laboratory experiments such as constant head or falling head permeability test. However, many researchers have presented several empirical formulae based on the characteristic grain size of the medium to model permeability.

Indraratna et al. [3] stated that the following Hazen's formula can be used to estimate the permeability of granular aggregates:

$$k = C(D_{10})^2 \tag{8.2}$$

Table 8.1 Permeability values for different fouled ballast (modified from Selig and Waters, [1]).

Fouling Category	Fouling Index	Representative k values (mm/sec)
Clean	<1	25–50
Moderately Clean	1–9	2.5–25
Moderately Fouled	10–19	1.5–2.5
Fouled	20–39	0.005–1.5
Highly Fouled	>39	<0.005

where, k is the coefficient of permeability and C is an empirical constant which varies in the range of 40–150. Indraratna and Vafai [4] suggested that the permeability of granular materials can be better represented by the following empirical formula:

$$k = \alpha(D_5 \times D_{10})^\beta \qquad\qquad (8.3)$$

where, α and β are two empirical constants.

Auvinet and Bouvard [5] explained that it is extremely difficult to measure the actual pore dimensions. They indicated that the 'pore size' is ambiguous because the voids of a granular assembly are irregular and continuous. Pores are a complicated network of voids interconnected by narrow passages. Craig [6] indicated that the presence of a small percentage of fines in coarse granular aggregates greatly influences the permeability. Marsal [7] reported that some of the smaller detached grains resulting from the degradation process became idle. These particles fill the voids between larger grains but do not constitute part of the ballast matrix. Instead they decrease the void ratio and permeability of the granular assembly.

Selig and Waters [1] presented the typical values of k for the ballast having different degrees of fouling, as given in Table 8.1. The definition of 'Fouling Index' for the classification of ballast fouling is given in the following Section.

8.1.2 Drainage requirements

To design an adequate and satisfactory drainage system, it is imperative to first examine the subsurface conditions, ground water and climatic conditions. Subsurface explorations must be carried out to characterise subgrade soils including type, layering and drainage properties. The proposed drainage system should have sufficient capacity to drain the highest expected inflow rate of water during the design life of the system.

The first requirement in achieving a satisfactory track drainage is to maintain the ballast clean enough to ensure sufficiently high permeability for quick drainage (Salig and Walter, [1]). Secondly, the surface of the subballast and subgrade should be sloped towards the sides. The third requirement in track drainage is to provide suitable means (channel or conduit) to carry away the water which emanates from the substructure, as shown in Figure 8.1.

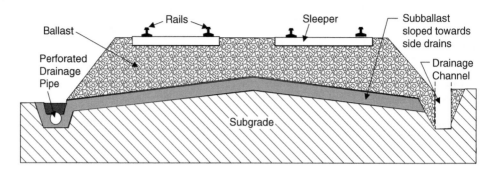

Figure 8.1 Schematic illustration of track drainage system.

8.2 FOULING INDICES

8.2.1 Fouling index and percentage of fouling

Selig and Waters [1] introduced the fouling index (*FI*) to describe ballast fouling based on gradations obtained for representative samples of ballast in North America (Fig. 8.2) as:

$$FI = P_4 + P_{200} \tag{8.4}$$

where, P_4 and P_{200} are percentages of ballast particles passing the No. 4 sieve (4.75 mm) and No. 200 sieve (0.075 mm) respectively. The categories of fouling based on *FI* are given in Table 8.1. The particles passing through the 0.075 mm sieve are included twice to emphasize the adverse influence of fine particles.

A related index to *FI* is the percentage of fouling (% fouling) which is the ratio of the dry weight of material passing 9.5 mm sieve to the dry weight of total sample. However, it should be noted that the above relationship is not applicable for all types of fouling due to the limited types of fouling materials used in this empirical development. Care should be taken when evaluating fouled ballast with a larger percent of particles finer than 0.075 mm.

8.2.2 Percentage void contamination

Feldman and Nissen [14] presented a parameter named Percentage Void Contamination (*PVC*) to capture the effect of void decrease in ballast as:

$$PVC = \frac{V_2}{V_1} \times 100\% \tag{8.5}$$

where, V_1 is the void volume between re-compacted ballast particles and V_2 is the total volume of re-compacted fouling material (particles passing 9.5 mm sieve), respectively. The samples for *PVC* tests are taken from the total depth of the ballast. Therefore V_1 represents the void volume of the entire ballast layer.

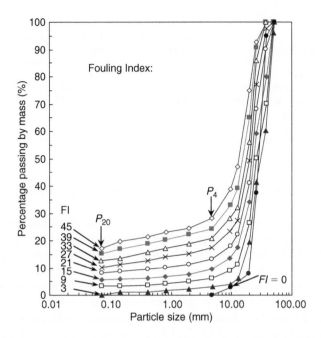

Figure 8.2 Gradations representing ballast conditions from clean to highly fouled (Modified from Selig and Waters, [1]).

Although the *PVC* method is a direct measure of percentage of voids occupied by fouling particles, the measurement of volume is time consuming. Furthermore, as the total volume of fouling particles is used, the gradation of fouling particles cannot be taken into account. For example, if the contaminates are all composed of coarse particles (4.75 mm to 9.5 mm), there should still be sufficient voids between the fouling particles, hence, the ballast drainage capacity would not be significantly reduced. In this regard, *PVC* may overestimate the extent of fouling. The authors suggest using the solid volume of fouling particles rather than the total volume in calculating the *PVC*. By using the solid volume, a smaller value of *PVC* will be obtained if there is insufficient quantity of fine particles within the contaminates, and vice versa.

8.2.3 Relative ballast fouling ratio

By comparing the % fouling and PVC, Indraratna et al. [15] proposed a new parameter, i.e. the Relative Ballast Fouling Ratio ($R_{b\text{-}f}$). It is a weighted ratio of the dry weight of fouling particles (passing 9.5 mm sieve) to the dry weight of ballast (particles retaining on 9.5 mm sieve).

The relative ballast fouling ratio can be defined as:

$$R_{b\text{-}f} = \frac{M_f \times \frac{G_{s\text{-}b}}{G_{s\text{-}f}}}{M_b} \times 100\% \qquad (8.6)$$

Table 8.2 Categories of fouling based on the R_{b-f}.

Category	Percentage of fouling (%)	Relative ballast fouling ratio (%)
Clean	<2	<2
Moderately clean	2 to <9.5	2 to <10
Moderately fouled	9.5 to <17.5	10 to <20
Fouled	17.5 to <34	20 to <50
Highly fouled	≥34	≥50

where, M_f and M_b, and G_{s-f} and G_{s-b} are mass and specific gravities of fouling materials and ballast, respectively.

In Equation (8.6), only the mass of ballast, and the mass and specific gravity of fouling material need to be measured. This will greatly speed up the measurements compared to the *PVC* method, and will be more attractive to the practicing track engineer. In comparison with *FI*, the magnitude of R_{b-f} can better represent the degree of fouling by various materials of different specific gravities. According to the relationship between *FI* and % fouling, categories of fouling based on the % fouling and R_{b-f} can be calculated from those based on *FI*. The calculated results are listed in Table 8.2.

A rate of contamination and a ballast life can be predicted for a track section given the value of R_{b-f}, a limit of allowable extent of fouling, a time period since undercutting and any changes in traffic volume. An average R_{b-f} can be calculated by performing tests every two kilometers along a track section. The rate of fouling (*FR*) can then be calculated by dividing the average R_{b-f} value (R_{b-f}^{Ave}) by the actual ballast life (BL_{ACT}) since last undercutting of the track section as follows:

$$FR = R_{b-f}^{Ave}/BL_{ACT} \tag{8.7}$$

With the above calculated *FR* and a prescribed allowable R_{b-f} limit (R_{b-f}^{All}), the allowable ballast life (BL_{ALL}) can be determined as:

$$BL_{ALL}(\text{years}) = R_{b-f}^{All}/FR \tag{8.8}$$

The value of BL_{ALL} can now be incorporated in track maintenance schedules as a quantitative index, in addition to standard track inspection routines and qualitative guidelines. Similarly, a relative fouling index can be defined using relative mass of fouling particles considering the specific gravities of ballast and fouling material.

8.3 GEOSYNTHETICS IN RAIL TRACK

Geosynthetics is the collective term applied to thin, flexible sheets manufactured from synthetic materials (e.g. polyethylene, polypropylene, polyester etc.), which are used in conjunction with soils and aggregates to enhance the soil properties (e.g. shear strength, hydraulic conductivity, filtration, separation etc.). Over the past few decades, various

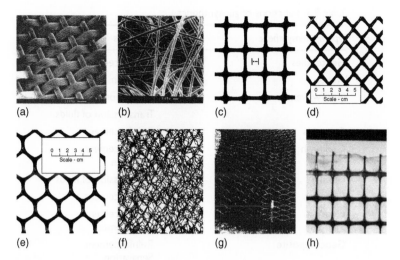

Figure 8.3 Types of geosynthetics, (a) woven-geotextile, (b) non-woven geotextile, (c) geogrid, (d) geonet, (e) geomesh, (f) geomat, (g) geocell, (after Ingold, [7]) and (f) geocomposite used by the authors.

types of geosynthetics have been tried out in track to minimise settlement and enhance drainage, but mainly as trial and error exercises. In the following Sections, different types of geosynthetics available for geotechnical applications and their effectiveness in harsh railway environment are discussed.

8.3.1 Types and functions of geosynthetics

Geosynthetics may be classified into two major groups: (a) geotextiles, and (b) geomembranes (Ingold, [7]). Geotextiles are basically textile fabrics, which are permeable to fluids (water and gas). There are some other synthetic products closely allied to geotextiles such as geogrids, geomeshes, geonets and geomats, which have all been used in geotechnical practice. All geotextiles and related products are permeable to fluids, whereas geomembranes are substantially impermeable to fluids and hence, primarily used for retention purposes. Figure 8.3 shows the common types of geosynthetics used in geotechnical engineering. The functions of these geosynthetics are summarised in Table 8.3.

Taking the functions into account, various types of geosynthetics have been used in different tracks depending on the specific requirements, cost and the engineering properties of the substructure materials. Geosynthetics generally minimise the track settlement by restricting lateral movement (through transferring lateral loads from ballast to geosynthetics by shear). Geotextiles dissipate excess pore pressure which is often developed in saturated subgrade under rapid cyclic loading. They also keep ballast relatively clean through the separation and filtering functions.

Geotextiles have been frequently used in track substructure, especially in localised mud problem areas, such as (a) wet cuts, (b) soft subgrade, (c) road grade crossings, (d) railroad track crossings, and (e) turnouts (Selig and Waters, [1]). Figure 8.4 shows

Table 8.3 Functions of geosynthetics.

Type of Geosynthetic	Functions
Geotextiles	Reinforcement
Woven	Filtration
Non-woven	Separation
	Transmission of fluids
Geogrids	Reinforcement
Geomesh	Reinforcement
	Filtration
Geonets	Transmission of fluids
Geomats	Reinforcement
Geocells	Reinforcement
	Confinement
Geocomposite	Reinforcement
	Separation
	Filtration
	Transmission of fluids
Geomembranes	Isolation
	Separation
	Reinforcement

Figure 8.4 Installation of geotextile and geogrid under the ballast layer (NSW, Australia).

a typical example of geosynthetics usage in track structure where a geotextile and geogrid was laid under the ballast layer.

Amsler [8] reported a case study in Geneva regarding the track performance with and without geosynthetics (Figs. 8.5a–b). In 1982, the left track (Fig. 8.4a) was completely reconstructed using a traditional design cross-section (without any geosynthetics). In 1983, the right track (Fig. 8.4b) was redone following a new design cross-section incorporating non-woven geotextiles at the subbase/subgrade interface.

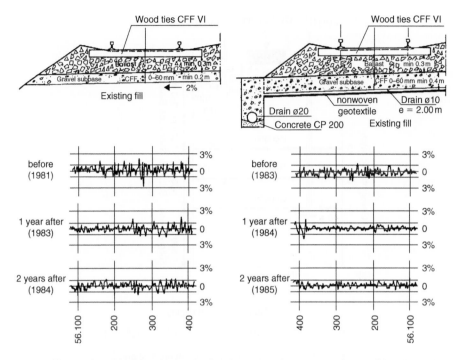

Figure 8.5 Effects of geosynthetics in track, (a) left track without geosynthetics, (b) right track with geosynthetics (after Amsler, [8]).

Both tracks were monitored by a track-quality measuring wagon before and after rehabilitation. The cross slope difference per millimetre between two rails of a track (warp) as a function of traveled distance was used as an indicator of stability and riding comfort. The pre- and post-renewal monitored data (warp) of both the tracks are presented in Figure 8.4 immediately below the design cross-sections. The smaller values of the measured data after installation of geotextiles on the right track (Fig. 8.4b) clearly show the benefits of using geosynthetics in rail track.

Amsler [8] concluded that the use of geotextiles significantly improved the track quality and that the improvement was maintained for a relatively long period. Other researchers reported similar improvement in track performance with the use of geosynthetics (Ashpiz et al., [9]; Selig and Waters, [1]). Track rejuvenation without geosynthetics, however, improved the performance for a shorter period of time and deteriorated almost to the pre-renewal level within about 1–2 years (Amsler, [8]).

Atalar et al. [10] studied the effects of geogrids on the settlement behaviour of track foundation in a large-scale model apparatus. Their test equipment and the settlement results are shown in Figure 8.6. They reported that the subbase settlement decreased significantly when only one layer of geogrid and geotextile combination was included at the subbase/subgrade interface. Settlement decreased further when additional layers of geogrid were placed inside and on top of the subbase (Fig. 8.6b). Bathurst and Raymond [11] also reported a significant decrease in permanent settlement when a geogrid was included at different elevations within the ballast layer.

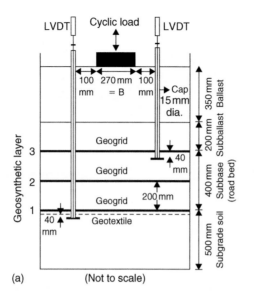

(a)   (Not to scale)

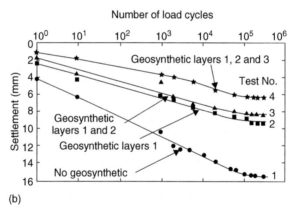

(b)

Figure 8.6 Use of geogrids in ballast bed, (a) test set-up, (b) settlement results (after Atalar et al., [10]).

Railway engineers often express concerns about the durability of geosynthetics in the harsh track environment due to the close contact with sharp angular ballast and heavy cyclic traffic loading. In this respect, several investigators studied the durability of geosynthetics in ballast bed. Most of them reported favourably. Selig and Waters [1] found that even after 3 years of service in a British Rail site, the extracted geogrid and geotextiles were in reasonably good condition.

Ashpiz et al. [9] investigated the durability of spunbonded geotextiles used in St. Petersburg–Moscow line. They reported only 0.2% and 0.3% surface damage after 1 year and 5 years of service, respectively. The retained strength was found to be about 74% and 72% after 1 year and 5 year periods, respectively. They reported some contamination of geotextiles when extracted from a track after 5 years of service.

A visual inspection revealed that the contamination was mainly due to fines generated by abrasion and breakage of upper ballast. Based on the laboratory testing of uncleaned geotextiles within the ballast layer, they concluded that the contamination of geotextiles had little influence on the drainage capacity of the ballast-geotextile system.

Nancey et al. [12] reported similar findings regarding the durability of a thick geotextile tested at 50 Hz frequency eccentric wheel loads in the 'vibrogir' model device. After 200 hours of cyclic loading (equivalent to 730 MGT loading), they found that the flow capacity, permeability, and puncture resistance of the thick geotextile were almost unaffected by the simulated traffic. Raymond and Bathurst [13] however, reported evidence of minor particle penetration holes in used geotextiles extracted from 175 mm depth of a rail track.

8.4 USE OF GEOSYNTHETIC VERTICAL DRAINS AS A SUBSURFACE DRAINAGE

Low-lying areas having large volumes of plastic clays can sustain high excess pore water pressures during both static and cyclic (repeated) loading. In poorly drained situations, the increase in excess pore pressures will decrease the effective load bearing capacity of the formation soil. Under certain circumstances, slurrying of clay beneath rail tracks may initiate pumping of the soil upwards, thereby clogging the ballast bed and promoting undrained shear failure (Chang, [16]; Indraratna et al., [17]). Geosynthetic prefabricated vertical drains (PVDs) can be installed to dissipate excess pore pressures by radial consolidation before they can build up to critical levels. These PVDs continue to dissipate excess pore water pressures even after the cyclic load stops (Indraratna et al., [18]).

8.4.1 Apparatus and test procedure

The tests were carried out using the large-scale cylindrical triaxial equipment. The equipment was further modified to measure the excess pore water pressure at different locations inside the specimen. Miniature 3 pore pressure transducers were fitted through the base of the triaxial rig and then inserted through the specimen pedestal and positioned at the locations shown in Figure 8.7.

The clay used in this study was kaolinite of specific gravity G_s 2.7. The liquid limit w_L was 55% and the plastic limit w_P was 27%. The compression index c_c was 0.42 and the swelling index c_s was 0.06. The soil specimens (with and without PVD) were subjected to anisotropic consolidation under an effective vertical stress of 40 kPa ($K_o = 0.60$ representing the in-situ stress), where K_o is the ratio of the effective horizontal to the effective vertical stress. Double drainage via the top and bottom of the specimen (in addition to radial drainage for tests with PVD) was permitted during the anisotropic consolidation to attain a degree of consolidation of 95%, (approximately 5 weeks with PVD and 9 weeks without PVD).

Three separate series of tests were conducted: (a) cyclic partially drained with PVD, (b) cyclic consolidated undrained (cyclic CK_oU) without PVD and (c) cyclic unconsolidated undrained (cyclic UU) without PVD. A series of conventional monotonic triaxial tests was conducted first according to ASTM (2002) to obtain the maximum deviator

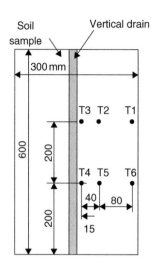

Figure 8.7 Locations of the pore pressure transducers at different positions from the PVD inside the soil sample (Indraratna et al., [18]).

stress at failure (q_f) during static loading. Then a cyclic stress ratio (CSR) of 0.65 was selected, where CSR is defined as the ratio of the cyclic deviator stress q_{cyc} to the static deviator stress at failure q_f (Brown et al., 1975; Zhou and Gong, 2001). Larew and Leonards (1962) defined the term critical cyclic stress ratio as the level of cyclic deviator stress above which a sample would experience failure after a certain number of loading cycles. Failure denotes a condition of rapidly accumulating non-recoverable (permanent) deformations with increasing number of cycles, and this can be represented in a semi-logarithmic plot at the point where the deformation curve starts to concave downwards indicating rapid increase in displacement. It is reported by various researchers that this critical cyclic stress ratio is between 0.6 and 0.7 (Ansal and Erken, [19]; Miller et al., [20]; Zhou and Gong, [21]). A sinusoidal cyclic load was applied to the specimen under one-way stress-controlled conditions at a frequency of 5 Hz simulating a 100 km/hr train speed. The applied cyclic amplitude was 25 kPa. The cyclic load application with PVD was carried out under radial and top drainage with no bottom drainage, in order to simulate the field boundary condition. The tests without PVD were carried out under totally undrained conditions. Membrane corrections were applied according to ASTM (22). The axial and volumetric deformations were measured using linear variable differential transformers (LVDT). Also, the measurements of excess pore pressures for all test series were made at the locations shown in Figure 8.7.

8.4.2 Test results and analysis

Excess pore ware pressure ratio (u^*) is defined as the excess pore water pressure normalised to the initial effective pressure (Miller et al., [20]; Zhou and Gong, [21]). Figure 8.8 shows the excess pore pressures ratio (u^*) versus the number of loading

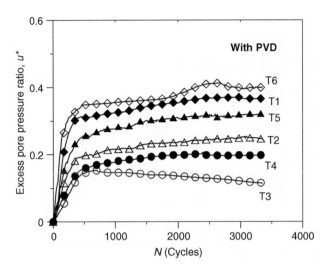

Figure 8.8 Excess pore pressures generated inside the soil sample at different locations from the PVD with the application of cyclic loads (Indraratna et al., [18]).

cycles (N) under the partially drained condition with PVD. The response of the six transducers shows the effect of the drainage path length on the development of the excess water pore pressures. Measured excess pore pressures and the corresponding excess pore water pressure ratio (u^*) versus the number of loading cycles (N) under the three separate series of tests are shown in 3(a). Without PVD, the excess pore pressure increased rapidly ($u^* \approx 0.9$), and undrained failure occurred very quickly. In contrast, with PVD, the excess pore pressure increased to a constant value ($u^* < 0.4$) after 500 loading cycles. As expected, the excess pore pressures measured at T3 and T4 were the lowest (i.e. shortest drainage path length), while the highest values were observed at T1 and T6 (Fig. 8.8). The data confirm the effectiveness of PVD in reducing the rapidly induced excess pore water pressures under cyclic loads, thereby mitigating potential undrained failure. Figure 8.9(b) presents the development of volumetric strains (compression) with the number of cycles associated with the dissipation of the excess pore pressures with PVD. As expected, the volumetric strains approach a constant level at 1.7% in compliance with the relatively constant u^*. For the tests without PVD, the measured volumetric strains were almost zero (Fig. 8.9b) as the cyclic load applications were carried out under totally undrained conditions.

During the application of cyclic loads, the PVD significantly decreases the build up of excess pore water pressure, and also accelerates its dissipation during any rest period. In reality, the dissipation of the pore water pressure during the rest period will make the track more stable for the next loading stage (i.e. subsequent passage of train). Soft formation beneath rail track stabilised by radial drainage (PVD) can be subjected to cyclic stress levels higher than the critical cyclic stress ratio without causing undrained failure.

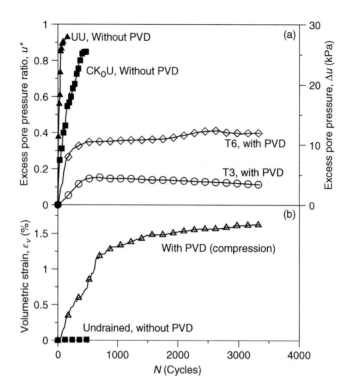

Figure 8.9 (a) Excess pore pressures generated with and without PVD under cyclic loads, (b) Volumetric compressive strains developed under cyclic loads with PVD (Indraratna et al., [18]).

REFERENCES

1. Selig, E.T. and Waters, J.M.: *Track Technology and Substructure Management*. Thomas Telford, London, 1994.
2. Feldman, F. and Nissen, D.: Alternative testing method for the measurement of ballast fouling: percentage void contamination. *Conference on Railway Engineering, Wollongong, Australia*, 2002, pp. 101–109.
3. Indraratna, B., Khabbaz, H., Lackenby, J. and Salim, W.: *Engineering behaviour of railway ballast – a critical review*, Technical Report 1, Rail-CRC Project No. 6, 2002, Cooperative Research Centre for Railway Engineering and Technologies, University of Wollongong, NSW, Australia.
4. Indraratna, B., and Vafai, F.: Analytical Model for Particle Migration Within Base Soil-Filter System. *Journal of Geotechnical and Geoenvironmental Engineering*, ASCE, Vol. 123, No. 3, 1997, pp. 100–109.
5. Auvinet, G. and Bouvard, D.: Pore Size Distribution of Granular Media. Powders and Grains, Biarez & Gourves (eds), Balkema, *Netherlands*, 1989, pp. 35–40.
6. Craig, R.F.: Soil Mechanics. E & FN Spon, London, 1997.
7. Marsal, R.J.: *Mechanical Properties of Rockfill. In: Embankment Dam Engineering*. Wiley, New York, 1973, pp. 109–200.

8. Ingold, T.S.: *Geotextiles and Geomembranes Manual*. Elsevier, UK, 1994.
9. Amsler, P.: Railway track maintenance using geotextile. *Procedings of 3rd International Conference on Geotextiles*, Vienna, 1986, pp. 1037–1041.
10. Ashpiz, E.S., Diederich, R. and Koslowski, C.: The use of spunbonded geotextile in railway track renewal St. Petersburg-Moscow. *Procedings of 7th Int. Conference on Geosynthetics*, Nice, France, 2002, pp. 1173–1176.
11. Atalar, C., Das, B.M., Shin, E.C. and Kim, D.H.: Settlement of geogrid-reinforced railroad bed due to cyclic load. *Procedings of 15th Int. Conf. on Soil Mech. Geotech. Engg.*, *Istanbul*, Vol. 3, 2001, pp. 2045–2048.
12. Bathurst, R.J. and Raymond, G.P.: Geogrid reinforcement of ballasted track. Transportation Research Record, TRB, Vol. 1153, 1987, pp. 8–14.
13. Nancey, A., Imbert, B. and Robinet, A.: Thick and abrasion resistant geotextile for use under the ballast in railways structure. *Procedings of 7th Int. Conf. on Geosynthetics*, *Nice, France*, 2002, pp. 1183–1189.
14. Raymond, G.P. and Bathurst, R.J.: Test results on exhumed railway track geotextiles. *Procedings of 4th Intl. Conference on Geotextiles*, Geomembranes and Related Products, 1990, pp. 197–202.
15. Indraratna, B., Su, L.J. and Rujikiatkamjorn C.: A new parameter for classification and evaluation of railway ballast fouling. *Canadian Geotechnical Journal*, 2010 (Accepted).
16. Chang, C.S.: Residual undrained deformation from cyclic loading. *Journal of Geotechnical Engineering Division*, ASCE, 108(GT4), 1982, pp. 637–646.
17. Indraratna, B., Balasubramaniam, A. and Balachandran, S.: Performance of test embankment constructed to failure on soft marine clay. *Journal of Geotechnical Engineering*, ASCE, 118(1), 1992, pp. 12–33.
18. Indraratna, B., Attya, A., and Rujikiatkamjorn, C.: Experimental Investigation on effectiveness of a vertical drain under cyclic loads. *Journal of Geotechnical and Geoenvironmental Engineering*, ASCE, 135(6), 2009, pp. 835–839.
19. Ansal, A.M. and Erken, A.: Undrained behaviour of clay under cyclic shear stresses. *Journal of Geotechnical Engineering*, ASCE, 115(7), 1989, pp. 968–983.
20. Miller, G.A., Teh, S.Y., Li, D. and Zaman, M.M.: Cyclic shear strength of soft railroad subgrade. *Journal of Geotechnical and Geoenvironmental Engineering*, ASCE, 126(2), 2000, 139–147.
21. Zhou, J. and Gong, X.: Strain degradation of saturated clay under cyclic loading. *Canadian Geotechnical Journal*, 38, 2001, pp. 208–212.
22. ASTM: *Standard test method for consolidated undrained triaxial compression test for cohesive soils*. ASTM D4767-02. 2002.

Chapter 9

Role of Subballast, its Drainage and Filtration Characteristics

The salient design feature of the subballast layer (sometimes called capping) is to protect the natural subgrade soil or embankment fill from excessive load that can lead to unacceptable settlement or bearing capacity failure under extreme conditions. The use of elastic theory (Chapter 2) to design the subballast layer (about 100 to 150 mm thick) as a relatively stiff medium of compacted broadly graded granular fills is common practice. However, the key role of subballast is to act as a drainage layer and as an effective filtration medium. Drainage plays a significant role in the stability and safety of a track substructure. Saturated tracks can lead to a build up of excess pore water pressure under train loading. If the hydraulic conductivity of the substructure elements becomes excessively low, especially the subballast layer, the excess pore water pressure developed under axle loading often cannot dissipate completely before the next load is imposed. Thus, the residual pore pressures accumulate with increasing load cycles. After a few load cycles, the total excess pore water pressure becomes very high and often causes clay pumping [1]. Thus the subballast layer plays two major roles in track substructure, (a) act as a permeable medium to transmit water laterally into the drainage channels, and (b) dissipate excess pore water pressure from saturated subgrades by allowing upward flow. The subballast, therefore, must have greater permeability than the subgrade soils.

This Chapter firstly explains the existing subballast selection criteria with reference to filtration and drainage. In an effort to enhance the selection criteria further, a critical review is given for the past empirical and mathematical investigations on filtration and the subsequent development of geometric-probabilistic methods[1]. Locke et al. [2] highlighted that the evaluation of filter effectiveness based on the constriction size distribution (CSD) can be more appropriate than the sole use of particle sizes. The development and effectiveness of constriction-based retention criteria, valid for both uniform and well-graded materials based on experimental evidence, are presented here. Furthermore, implications to current design guidelines are given and the need for further investigation is discussed. Lastly, experimental investigations undertaken at University of Wollongong into subballast filtration behaviour under cyclic conditions are detailed with the analysis of the results obtained.

---

[1] It is acknowledged that this Chapter also includes the essence of technical papers written by the authors [3, 4].

9.1 SUBBALLAST SELECTION CRITERIA

Research has been conducted to establish the grading requirements of granular filters for drains associated with seepage of water from soil under steady conditions. The mechanism of seepage associated with the combination of subgrade and ballast in rail track environment is governed mainly by the cyclic nature of the load produced by the passing traffic. However, little research has been done to establish any gradation criteria for repeated load situations. The selection criteria currently used in the industry are mainly based on filtration studies using static loading [5, 6]. These design criteria were developed based on steady seepage force rather than the usual cyclic conditions prevalent in rail tracks.

In rail track environments, the three sources of water entering the substructure are precipitation, surface flow, and subsurface seepage. Because ballast has an open surface, any precipitation falling onto the track penetrates the ballast rather than run off the surface. Water flowing down adjacent slopes also goes through the ballast and the underlying layers, unless diverted. Finally, in regions with a high groundwater table, water can seep upward from the subsurface and enter the substructure zone. Adequate drainage for these sources of water is of the utmost importance in order to prevent or minimise substructure problems related to excess water.

9.1.1 Filtration and drainage criteria

The subballast must prevent the intermixing of ballast and subgrade and the upward migration of subgrade particles into the ballast. Intermixing results from progressive penetration of the coarse ballast particles into the finer subgrade, accompanied by the upward displacement of the subgrade particles into the ballast voids. This process can occur when the subgrade is at any moisture condition. Upward migration of subgrade particles develops from at least three sources [7]:

(a)   subgrade seepage carrying soil particles;
(b)   hydraulic pumping of slurry from subgrade attrition; and
(c)   pumping of slurry through opening and closing of subgrade cracks and fissures.

Preventing intermixing and migration may be achieved by using a proper subballast gradation. This is known as a separation function. The filter criteria were first developed by Bertram in 1940 [6] with advice from Terzaghi and Casagrande. Subsequent studies were made by the U.S. Army Corps of Engineers and the U.S. Bureau of Reclamation. The two separation gradation criteria are:

$$D_{15} \leq 5 \cdot d_{85} \tag{9.1}$$

And

$$D_{50} \leq 25 \cdot d_{50} \tag{9.2}$$

where D_n is the filter grain size and d_n is the base particle size, which passes n percent by weight of the total filter and base, respectively.

The criterion in Equation (9.1) causes the particles at the coarsest end of the protected soil (d_{85}) to be blocked by the particles at the finest end of the filter (D_{15}). Assuming that no gaps exist in the grading of either the soil or the filter, the blocking action extends through the entire grading of both materials and a stable network of particles exists. The criterion in Equation (9.2) helps to avoid gap graded filters and create a filter gradation that is mostly parallel to the protected soil.

For medium to highly plastic clays without silt and sand, the criteria in Equations (9.1) and (9.2) are relaxed for seepage applications to permit easier filter selection. In these cases, the D_{15} size of the filters may be as large as 0.4 mm and Equation (9.2) may be ignored. To minimise the chance of filter particle segregation the coefficient of uniformity (Eq. (9.3)) must not exceed 20.

$$C_u = \left[\frac{D_{60}}{D_{10}} \leq 20 \right] \tag{9.3}$$

Deviation from the above recommendations may be desired in some cases because obtaining a suitable subballast gradation may prove difficult. In such cases laboratory tests can be conducted to test the filter capability of the subballast under repeated loading.

Not only must the subballast satisfy the criteria in Equations (9.1) and (9.2) in relation to the subgrade, but the criteria must also be satisfied in relation to the ballast. This condition simultaneously places an upper and lower limit on the acceptable subballast gradation. In case a single subballast material cannot be found to fit this range of sizes for a particular subgrade and ballast, then a two layer subballast may be used. The upper layer would be coarser to match with the ballast, while the lower layer would be finer to match with the subgrade. The relationship between these two layers of subballast must also satisfy Equations (9.1) and (9.2). A properly graded layer of sand and gravel subballast combined with adequate external drainage would prevent slurry forming by eliminating subgrade attrition [8]. One reason is that the high stresses at the ballast contact points on the subgrade surface are eliminated by the cushioning effect of the subballast.

As an intermediate layer, the drainage design of the subballast must consider both the underlying subgrade and the overlying ballast. The general guideline dictates that the subballast hydraulic conductivity should be at least an order of magnitude smaller than that the ballast; and have a surface sloped towards the outside of the track. In order to drain water seeping from the subgrade, including that produced by excess pore pressure generated from cyclic stresses, the subballast should also have a hydraulic conductivity greater than the subgrade. The exceptions are when the subgrade is relatively permeable, such as a layer of natural sand or sand-gravel, or when no upward seepage is expected, such as on an embankment.

Therefore, the subballast must generally have a hydraulic conductivity between that of the subgrade and that of the ballast. This requirement probably is achieved just by satisfying the separation criteria of Equations (9.1) and (9.2). However, an additional established criterion is used to ensure adequate hydraulic conductivity to drain an adjacent layer given as:

$$D_{15} > (4 \sim 5)d_{15} \tag{9.4}$$

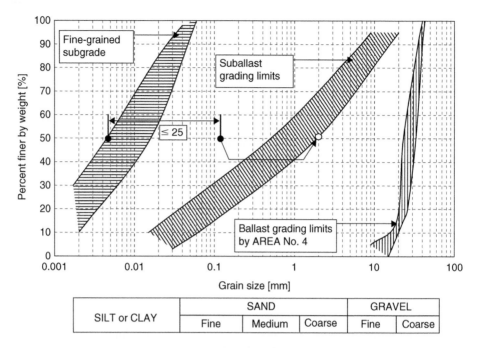

Figure 9.1 A 1-layer subballast system in relation to AREA No. 4 ballast grading and fine grained subgrade (modified after Selig and Waters, [7]).

Each subballast layer of a different material or gradation should have a nominal compacted thickness of at least 150 mm, to allow for construction variability and some subsequent compression under traffic. To serve as a structural material, the subballast must also be permeable enough to avoid a significant positive pore pressure build up under repeated load, must consist of durable particles, and must not be sensitive to changes in moisture content. Such a material is represented by mixtures of sand and gravel particles composed of crushing and abrasion resistant minerals. These materials may be available in natural deposits or may be produced by crushing rock or durable slags. Furthermore, soil susceptible to frost must be insulated by a sufficiently thick covering layer of non-frost susceptible subballast soil, which also limits freezing temperature. The combined thickness of the ballast and the subballast insulates the subgrade and good drainage helps limit the source of water that feeds the growth of ice lens.

9.1.2 Case studies of subballast selection

Figure 9.1 shows an example of a one layer subballast gradation relative to a typical ballast gradation specified by AREA No. 4 and a fine grained subgrade. A broad gradation ranging from fine gravel to silt size is required to satisfy Equation (9.2). However, the uniformity criterion specified in Equation (9.3) was not simultaneously met.

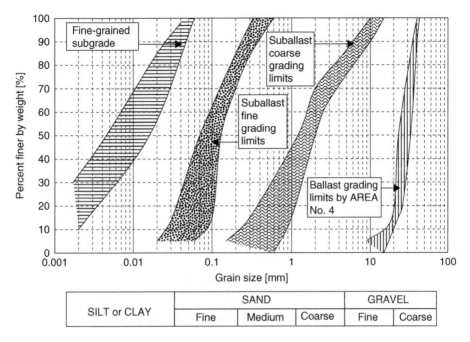

	SAND			GRAVEL	
SILT or CLAY	Fine	Medium	Coarse	Fine	Coarse

Figure 9.2 A 2-layer subballast system in relation to AREA No. 4 ballast grading according to and a fine grained subgrade (modified after Selig and Waters, [7]).

In case a single subballast material cannot be found to fit the desired range of sizes for a particular ballast and subgrade, then a two layer subballast may be used. In this type of subballast arrangement, the lower layer is the capping layer or blanket layer. An example of both a two layer subballast gradation in relation to a typical ballast gradation (AREA No. 4) and a fine grained subgrade is given in Figure 9.2. The blanketing sand layer recommended by British Rail and ASTM D1241 [9] is broadly graded, which is very close to the subballast layer shown in Figure 9.1.

The particle size distribution (PSD) of the capping material shown in Figure 9.3 closely resembles the subballast presently used in New South Wales, Australia. Also shown is the PSD of ballast used in NSW rail tracks according to the specification of TS 3402 [10].

A study conducted by Haque et al. [11] on the filtration behaviour of granular media under cyclic loading used two filter material gradations similar to the typical capping material gradation usually placed underneath the ballast layer in a railway track in the state of Victoria, Australia. The base soil used was locally available clayey silt with a typical gradation shown in Figure 9.4, together with the gradations of filter materials and ballast as prescribed by Australian Standard 2758.7 [12].

In Queensland, Australia, Queensland Rail (QR) occasionally uses a material described as MRD Type 2.4 Unbound soil as a capping layer in railway substructure. The Department of Main Roads uses this material as a base or subbase layer in road pavements. Figure 9.5 illustrates the PSD of this material used in the laboratory testing, showing a well graded (GW) soil.

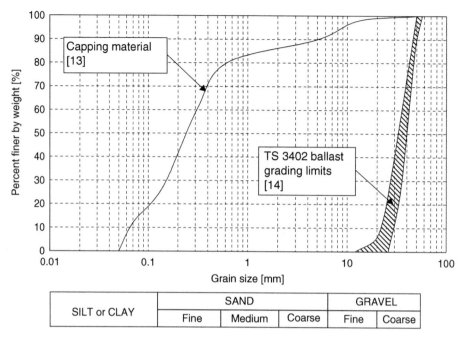

	SAND			GRAVEL	
SILT or CLAY	Fine	Medium	Coarse	Fine	Coarse

Figure 9.3 PSD of ballast and capping layer in NSW, Australia (after Trani, [15]).

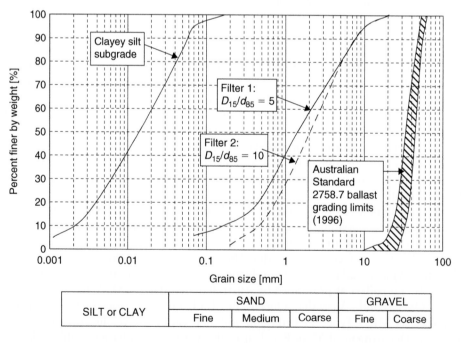

	SAND			GRAVEL	
SILT or CLAY	Fine	Medium	Coarse	Fine	Coarse

Figure 9.4 Gradation of filter material in relation to the PSD of clayey silt subgrade and AS 2758.7 ballast (after Trani, [15]).

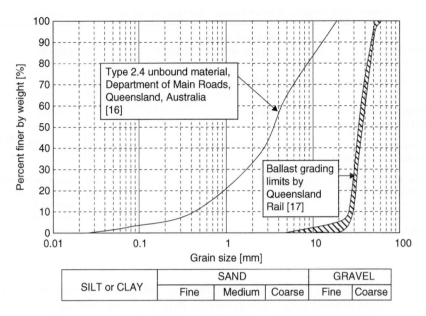

Figure 9.5 PSD of MRD Type 2.4 unbound material (after Trani, [15]).

9.2 EMPIRICAL STUDIES ON GRANULAR FILTRATION

While experimentally developed filter design criteria do not explain the fundamental mechanics of filtration and include over simplified assumptions and procedural bias, they are simple to use and have an implicit consideration for all the major factors affecting filtration (i.e., biological, chemical, geometric, hydraulic, and physical). Design criteria based on experimental studies are usually given in the form of one or more grain size ratios for the base and filter materials. Using metal sieves as filters, Vaughan and Soares [18] and Kwang [19] showed that the use of d_{85} to represent base soil stability would be acceptable. On the other hand, studies conducted by Kenney et al. [20] indicated that filter particles within the range of D_5 to D_{15} seem to govern the constriction size, which is largely independent of the shape of the filter PSD curve and layer thickness. If the subballast is able to retain the finer particles, Terzaghi's retention ratio D_{15}/d_{85} is a good representation of the stability of a subgrade – subballast combination. Honjo and Veneziano [21] validated this claim through a statistical analysis on extensive data on previous laboratory results and practical experience confirming that the grain size ratio of D_{15}/d_{85} is the most suitable parameter in designing filters for cohesionless base soils. Other grain size ratios ($D_{50}/d_{50}, D_{15}/d_{15}$), as proposed by some researchers, do not correlate well with filter performance [21, 22].

9.2.1 Natural resources conservation service (NRCS) method

This design procedure is mainly based on the results of laboratory tests carried out by Sherard and colleagues [22–25] through the Natural Resources Conservation Service

Table 9.1 Recommended empirical filter retention criteria (modified after Indraratna and Locke, [26]).

Base soil category	Base soil % passing 75 μm sieve ($<$4.75 mm)*	Filter criterion
I. Fine silt or clay	$>$85	$\dfrac{D_{15}}{d_{85R}} \leq 9$
II. Sandy silts/clays and silty/clayey sands	40–85	$D_{15} \leq 0.7$ mm
III. Sands, sandy gravels with few fines	$<$15 and (d_{95R}/d_{75R}) $\leq$ 7	$\dfrac{D_{15}}{d_{85R}} \leq \left[5 - 0.5\left(\dfrac{d_{95R}}{d_{75R}}\right)\right]$
IV. Soils intermediate between previous two categories	5–39	Extrapolate between the two previous values based on % passing 75 μm sieve

*of portion passing 4.75 mm sieve size.

(NRCS). The guidelines require classifying the base soils into four categories, depending on the fines content (i.e., fraction smaller than US #200 sieve size, 0.075 mm), determined after regrading the base soil PSD curves for the particle size larger than US #4 sieve size. Subsequently, the maximum D_{15} size of effective filters for each group is determined by the design criteria. With some modifications from the original NRCS tabulated guidelines, Indraratna and Locke [26] presented a retention criterion for the four categories of base soil, as shown in Table 9.1.

The NRCS guidelines also impose constraints on the maximum size of filter particles and the C_u of the filter bands in order to prevent segregation during installation and to avoid the selection of gap graded filters. Foster and Fell [27] suggested that the lower limit of fines content for Category II base soil should be changed from 40 to 35%, while the maximum D_{15} for dispersive soil in the same group should be lowered to 0.5 mm. Although studies such as Sherard and Dunnigan [22, 25] and Foster and Fell [27] found that tests on fine silts and clays failed with retention ratios from 6 to 14, they still recommended $D_{15}/d_{85R} \leq 9$ as the most appropriate filter criterion for the soils in the first group by considering the average value.

The finer particles of internally unstable, broadly graded soils (i.e., $C_u \geq 20$) can move into the voids between the coarser particles leading to erosion even when the coarser base particles are retained by a filter. In order to have successful filtration within this soil type, the process of self-filtration where a base soil – granular filter interface would prevent further erosion of base soil, is important. Lafleur et al. [28] indicated that the extent of mass loss is greater for broadly graded cohesionless base soils before self-filtration occurs. Locke and Indraratna [29] introduced the Reduced PSD method to determine the self-filtering stable fraction of a broadly graded base soil for both Categories I and II. The self-filtering fine fraction is determined by dividing the PSD at a point n (where n is the percentage passing diameter d_n) to define d_{15} of the coarse fraction and d_{85} of the fine fraction as given in Equations (9.5) and (9.6). These new design criteria often allow coarser filters for self-filtering base

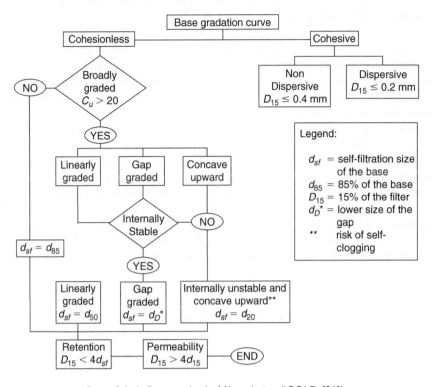

Figure 9.6 Lafleur method of filter design (ICOLD, [31]).

soil, while significantly finer filters may be necessary to protect some broadly graded materials.

$$d_{15coarse} = d_{n+0.15(100-n)} \qquad (9.5)$$

$$d_{85fine} = d_{0.85n} \qquad (9.6)$$

9.2.2 Self filtration method

The concept of self-filtration in relation to broadly graded cohesionless base soils was further studied by Lafleur [30] and Lafleur et al. [28]. It was reported that the classic Terzaghi's retention criterion leads to unsafe filter designs when applied to this soil group. Furthermore, Lafleur et al. [30] found out that the base particle size in the case of broadly graded and gap graded base soils is invariably smaller than d_{85} in comparison to the size of filter opening suggested by Kenney et al. [20]. Based on filtration test results on broadly graded cohesionless tills, Lafleur [30] suggested a design procedure involving the original Terzaghi's criterion where d_{85} is replaced by the appropriate indicative base particle size. This procedure, as depicted in Figure 9.6, separates the crack susceptible materials, i.e., cohesive, from cohesionless base soils. Considering the latter, the initial step is to determine if the soil is broadly graded. Soils with $C_u < 20$ should be considered as broadly graded if segregation occurs during placement but if the soil is not broadly graded, the self-filtration size, d_{sf}, is equal to d_{85}.

9.3 MATHEMATICAL FORMULATIONS IN DRAINAGE AND FILTRATION

Deficiencies in empirical investigations are addressed in rigorous analytical modelling and numerical simulations. Mathematical modelling of the filtration behaviour of base particles provides useful predictions on time dependent changes in filters as well as an indication of the required thickness. Moreover, these models may generate the potential amount and rate of base soil erosion under various geo-hydraulic constraints or an estimate of the probability of filter failure brought about by clogging on the base soil – granular filter interface.

Real soils consist of particles of many sizes, and at their densest packing the voids between large particles contain smaller particles, and the voids between these contain yet smaller particles (Fig. 9.7a). The Fuller and Thompson [32] packing model is the idealised limit of this concept wherein the largest particles just touch each other, while there are enough intermediate size particles to occupy the voids between the largest without holding them apart, and smaller particles occupy the voids between intermediate sizes (Fig. 9.7c). In the loosest state, it is possible for groups of real soil particles to form "arch" structures, which can be sustained if left undisturbed (Fig. 9.7b). This unstable structure may collapse under the influence of a sudden shock, vibration, or inundation.

When a collection of equal spheres is in face to face contact, their extreme states of packing can be represented diagrammatically in two dimensions, as shown in Figure 9.8. They may be densely packed to attain a porosity of 26% as shown in Figure 9.8(a), or loosely packed with a porosity of 48%, as shown in Figure 9.8(b).

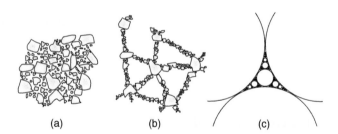

(a)　　　　(b)　　　　(c)

Figure 9.7 States of packing of soil particles: (a) densely packed, (b) loosely packed, and (c) idealised "Fuller and Thompson" packing (modified after Head, [33]).

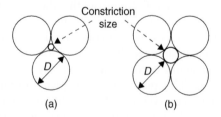

(a)　　　　(b)

Figure 9.8 Particle packing arrangement for (a) the densest, and (b) the loosest state (modified after Indraratna and Locke, [34]).

The largest particle diameter d, which can fit between spheres of diameter D, becomes the controlling constriction size of the pack. While these models give some idea of the geometry and controlling constriction of filters, these spherical particles and regular packing are too far from reality to be used for design purposes. Partially offset by the more irregular shape of real sand grains, Head [33] expected that the extreme limits of porosity values of many natural sands do not differ greatly from the theoretical values of equal spheres.

9.3.1 Geometric and probabilistic modelling

Recent mathematical approaches include geometric-probabilistic methods of modelling base soil – filter combinations. These methods consider the fact that soil masses are made up of a random distribution of an array of particle sizes, and recognises the geometric requirement that a base soil particle must be smaller than the pore constriction (the smallest opening between pore voids) through which it should pass. This approach tends to represent a combined probabilistic comparison of the base particle size and filter constriction size, hence, a constriction size distribution (CSD).

Silveira [35] was the first to adopt this approach to examine the migration of base soil particles into filters using a theoretical packing model. With further advancement, Silveira et al. [36] defined the constriction size D_{cD} (Fig. 9.9a) as the diameter of the largest circle that can fit within three tangential filter particles, as described by Equation (9.7). Humes [37] assumed that in a filter of maximum density only the densest arrangements exist. This equation can be solved for D_{cD} by an iterative process for a given set of values of particle sizes P_1, P_2 and P_3.

$$\left(\frac{2}{P_1}\right)^2 + \left(\frac{2}{P_2}\right)^2 + \left(\frac{2}{P_3}\right)^2 + \left(\frac{2}{D_{cD}}\right)^2$$

$$= \frac{1}{2}\left[\left(\frac{2}{P_1}\right) + \left(\frac{2}{P_2}\right) + \left(\frac{2}{P_3}\right) + \left(\frac{2}{D_{cD}}\right)\right]^2 \tag{9.7}$$

Since filters are not expected to sustain their maximum density during seepage, Silveira et al. [36] presented an alternative void model for the loosest state of a soil where four particles combine to form a void, as shown in Figure 9.9(b). Unlike the dense model,

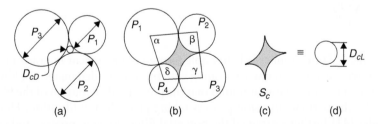

Figure 9.9 Filter particles in (a) dense, and (b) loose packing arrangement, (c) constriction area formed by tangent particles, S_c, and (d) circle of equivalent area (modified after Silveira et al., [36]).

Silveira et al. [36] noted that any analytical solution for the constriction size in this case is difficult without any reasonable simplification to the problem. Referring to Figure 9.9(c) to Figure 9.9(d), the constriction size, D_{cL}, is the diameter of equivalent circle with the same area as the enclosed area, S_c, formed by four tangent particles. For any set of four particles of sizes P_1, P_2, P_3, and P_4, the constriction area, S_c, can be determined by the following:

$$S_c = \frac{1}{8}[(P_1 + P_2)(P_1 + P_4)\sin\alpha + (P_2 + P_3)(P_2 + P_4)\sin\gamma$$
$$- (\alpha P_1^2 + \beta P_2^2 + \gamma P_3^2 + \delta P_4^2)] \tag{9.8}$$

The angles β, γ and δ can be related to α by plane geometry. For a particular angle α, when the value of S_c is maximum, then the constriction size in the loosest arrangement based on equivalent diameter D_{cL} is given by:

$$D_{cL} = \sqrt{\frac{4S_{c\,max}}{\pi}} \tag{9.9}$$

The dense constriction model of Silveira [35] had been shown to be an acceptable approximation for uniform filters where filter PSDs either by mass or by number of particles are used. However, De Mello [38] showed the limitations of the PSD by mass to model constrictions of GW filters. Large particles, with high individual mass but low in number, are over represented in the model and produce a high number of large pores. It was shown that as C_u increases; the number of small particles filling the voids between the larger particles would increase, leading to smaller constriction sizes. Federico and Musso [39] overcame this problem by converting the PSD by mass to PSD by number of particles. Raut and Indraratna [40] showed that if a filter material is composed of n discretised diameters $P_1, P_2, P_3, \ldots, P_n$ (Fig. 9.10) and their mass probabilities of occurrence $P_{M1}, P_{M2}, P_{M3}, \ldots, P_{Mn}$ respectively, then their probabilities by number can be obtained by multiplying the mass probabilities by their corresponding coefficients $P_{N1}, P_{N2}, P_{N3}, \ldots, P_{Nn}$ given by the generalised equation below:

$$P_{Ni} = \frac{\frac{P_{Mi}}{P_i^3}}{\sum\limits_{i=1}^{n} \frac{P_{Mi}}{P_i^3}} \tag{9.10}$$

Although the PSD by number is better able to predict movement in graded materials, limitations were encountered when dealing with broadly graded materials. Eliminating the fundamental restrictive assumption that the pore size distribution of the filter is not modified by the particle diffusion process taking place during filtration, Humes [37] and Schuler [41] further examined the potential filter clogging process and suggested an improvement by adopting a PSD by the surface area method. Accordingly, particle probabilities of occurrence by the surface area can be obtained by multiplying the mass

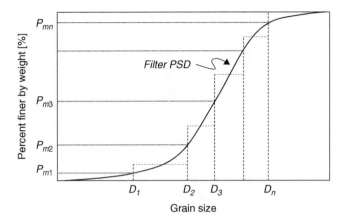

Figure 9.10 Discretised filter PSD (modified after Federico and Musso, [39]).

probabilities by their corresponding coefficients $P_{SA1}, P_{SA2}, P_{SA3}, \ldots, P_{SAn}$ given by the following generalised equation [37]:

$$P_{SAi} = \frac{\frac{P_{Mi}}{P_i}}{\sum\limits_{i=1}^{n} \frac{P_{Mi}}{P_i}} \qquad (9.11)$$

This is considered more representative of the possible particles which may form a constriction, since although there will be a small number of larger particles, they have a great number of contacts with other particles, due to their large surface area. Figure 9.11(a) shows a GU filter having a C_u of 1.4. It is clear that any PSD whether it is by mass, number, or surface area, results in nearly the same CSDs for a uniformly graded (GU) filter.

For a non-uniform filter, the resulting CSDs for the same PSD are very different. Figure 9.11(b) shows the CSDs by number, surface area, and mass for a well graded (GW) filter as simulated by Raut and Indraratna [40] using a numerical solution. The CSD by mass overestimates the bigger constrictions. The CSD by number of particles, on the other hand, overestimates smaller constrictions and underestimates the larger constrictions. The option involving the CSD by surface area estimates the CSD well by eliminating the misrepresentations caused by mass and number considerations. It was shown that the CSD with percent passing by surface area of the particles is a better option to quantify the filter characteristics, particularly for non-uniform filters.

Physical geometric modelling in combination with the probabilistic analysis that developed into a means of measuring the CSD, had been attempted by various researchers. A summary of these attempts at mathematical modelling in chronological order conducted over the years is shown in Table 9.2, which provides some background to the progression of the constriction based retention criterion for granular filter material.

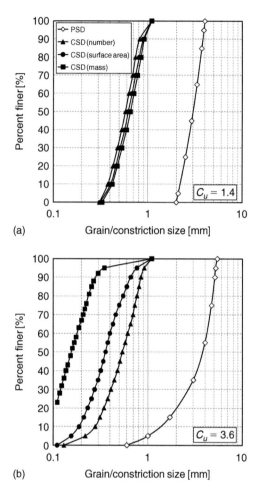

Figure 9.11 CSDs by number, surface area, and mass for a GU filter (modified after Raut and Indraratna, [40]).

9.3.2 Particle infiltration models

Honjo and Veneziano [21] presented a model based on the conservation of mass in the solid and liquid (slurry of soil and water) phases that can describe the absorption and release of soil particles with time in different elements of the base and filter. The model was used to demonstrate the self healing of the base soil as coarser particles collect at a screen with a systematic pattern of apertures. In addition, internal stability was also investigated using this model.

Kenney et al. [20] used a multi-layered one dimensional constriction model to analytically investigate the size of controlling constriction in a filter that is defined as the size of the largest base soil particle that can potentially penetrate the filter. Although this model is a good approximation of uniform filters and provides a sound

Table 9.2 Chronological progression in mathematical filtration models (modified after Indraratna, et al., [3]).

Year	Highlight	Author
1907	Theoretical grading and the densest possible state of packing of a collection of uniform spheres	Fuller and Thompson [32]
1965	Simple Filter Void Model: base soil particles encounter constrictions at uniform spacing in the direction of flow	Silveira [35]
1975	Constriction sizes for the densest state and loosest state of packing of a soil are defined	Silveira et al. [36]
1977	Limitations of PSD by mass to model constrictions of well graded filters are shown	De Mello [38]
1979	A model of a flow path in the form of a pore channel with irregular width in the direction of flow is developed	Wittmann [42]
1985	Multi-layered Void Network Model: estimates the number of confrontations with random constrictions until a base particle is retained; developed the PSD by numbers to model constrictions of GW filters	Kenney et al. [20]
1989	Particle Transport Model: based on conservation of mass in the solid and liquid (slurry of soil and water) phases	Honjo and Veneziano [21]
1993	Experimental results on CSD of filter according to geometric-probabilistic filtration theory are presented	Soria et al. [43]
	Enhancement of the 1975 model is presented	Silveira [44]
	Three Dimensional Pore Network Model: spheres as pores interconnected by pipes as pore constrictions	Witt [45]
1996	Cubic Pore Network Model: a regular cubic network of pores interconnected by six constrictions similar to Witt (1993) model	Schuler [41]
	The densest packing state in certain locations even in a medium dense soil is found to exist	Giroud [46]
1997	Pore Channel Model: improves the particle transport model showing smallest of the pore constrictions within the pore channel governs the size of a base particle that can pass through the pore channel	Indraratna and Vafai [47]
2000	The Indraratna and Vafai [42] model is improved by incorporating the cubic pore network model	Indraratna and Locke [34]
2004	CSD Model: numerical evaluation of the effectiveness of non-uniform granular filter	Raut and Indraratna [40]
2006	D_{c95} Model: the dominant constriction size (D_{c95}) is used to delineate effective from ineffective granular filters	Indraratna and Raut [48]
2007	D_{c35} Model: a retention criterion based on the controlling constriction size (D_{c35})	Indraratna et al. [49]
2010	Assessing the potential of internal erosion and suffusion of granular soils.	Nguyen et al. [50]

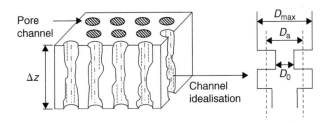

Figure 9.12 Void channel model (modified after Indraratna and Vafai, [47]).

understanding of the fundamental filtration mechanisms, it considers the flow channels to be independent.

Indraratna and Vafai [47] integrated the Honjo and Veneziano [21] model into their pore channel model to provide a geometric constraint to movement. They also considered the hydraulic forces required to mobilise the particles. If the minimum hydraulic force is exceeded then particle movement is modelled by the conservation of mass and momentum to produce a particle transport model rather than an analysis of the probability of particle movement. Indraratna and Locke [34] extended this work further to consider a more accurate three dimensional cubic void network model which could accommodate broadly graded filter materials with an allowance for energy loss due to particle transport, and the filtration of cohesive soils. They used the pore model shown in Figure 9.12 to evaluate the number of elements at the base soil – granular filter interface where particle movement is modelled by a finite difference procedure and the elements considered were thicker than $300D_5$. The minimum single controlling constriction size (D_0) was implemented. The model quantified the gradual change in PSD, hydraulic conductivity, and porosity of the materials with time, which means it described what occurred at the base soil – granular filter interface for the entire range of particles.

9.4 CONSTRICTION SIZE DISTRIBUTION MODEL

The void constrictions within the filter, not the filter particles, affect the filtration mechanism. Base particles are trapped by the smallest part of a connection between two voids, the size of which depends on the size and packing geometry of the filter particles. Locke et al. [2] highlighted the inadequacies of the PSD based retention criteria when describing filter effectiveness. With the introduction of geometric-probabilistic models and development of particle infiltration models, the appropriateness of using CSD in filter design is emphasised more.

9.4.1 Filter compaction

Schuler [41] examined the CSD of a soil at varying relative densities and reported that all the CSD curves have the same shape. However, Giroud [46] suggested that at certain locations within a medium to dense granular material, a number of particles would group together to form a maximum density arrangement. These two observations implied that within a granular filter, the smallest pore constrictions would be the same size regardless of its density and the distribution of coarser pore constrictions would vary, having the same shape as the minimum and maximum CSD curve. Real filters are unlikely to exist either in the densest or loosest states, but rather at an intermediate density. Hence, a more representative pore model should also consider the relative density of a filter.

Based on these findings, Indraratna and Locke [34] assumed that the coarser pore constrictions between the dense and loose constriction models expand in linear proportion to a decrease in relative density. In addition, the smallest constrictions are the same size as the smallest constrictions of the dense packing arrangement. This allowed for a simple formulation for the actual CSD based on (a) the dense CSD, (b) the loose CSD and (c) the filter relative density (R_d) defined in Equation (9.12):

$$R_d = \frac{e_{max} - e}{e_{max} - e_{min}} \tag{9.12}$$

where e_{max} and e_{min} are the maximum and minimum void ratios respectively, and e is the actual void ratio of the filter. The actual CSD is calculated using Equation (9.13). The dense and loose CSDs are divided into n equal discrete portions. The integer i represents these discrete portions of the CSD such that $P_{ci} = i/n$ is the fraction of constrictions finer than constriction diameter D_c representing the median diameter of the ith portion of the CSD, hence:

$$D_{ci} = D_{cDi} + P_{ci}(1 - R_d)(D_{cLi} - D_{cDi}) \tag{9.13}$$

In Equation (9.13), $i = 0, 1, 2, \ldots, n$, and D_{cDi} and D_{cLi} are the $100P_{ci}$ coarsest constrictions from the densest and loosest CSDs, respectively. In order to explain the application of Equation (9.13), for $i = 0$ the finest constriction diameter D_{c0} is the finest diameter of constrictions from the dense CSD. If $n = 10$, then $i = 1$ corresponds to the constriction diameter with 10% (i.e., 1/10) of constrictions finer. A typical behaviour of the CSDs of the same filter material but with varying relative densities is shown in Figure 9.13.

9.4.2 Filter thickness

The pore channel model in Figure 9.12 was based on the least single controlling constriction size (D_0) and the thickness of the filter element considered was greater than $300D_5$. Further refinements by Indraratna et al. [49] produced a model that suggested a minimum filter thickness based on the mean value of controlling constriction size (D_m). According to Figure 9.14, the number of layers (n_l) becomes high when the probability of forward movement of base particles approaches unity and the particles were 35% or finer. At a 95% confidence interval the rapidly increasing nature of the n_l-curve for the percentage of finer less than 35%, clearly indicates that any further increase in thickness beyond $225D_m$ does not contribute to base soil retention significantly.

Given that the computation of D_m is based on surface area principle, it varied from D_5 to D_{15} in most practical dam filters. In this respect a filter thickness of $225D_m$ is in agreement with the laboratory observations of $300(D_5$ to $D_{10})$, as suggested by Witt [45] and $200D_5$ by Kenney et al. [20]. For typical filter gradations (e.g., ICOLD, [31]),

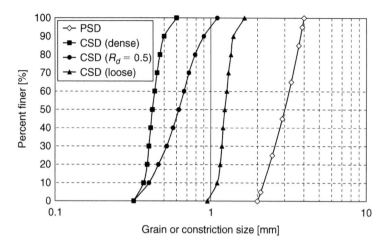

Figure 9.13 Influence of relative density on CSD based on numerical solution (modified after Raut and Indraratna, [40]).

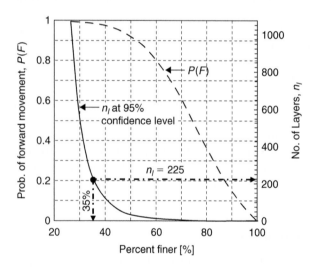

Figure 9.14 Influence of relative density on CSD based on numerical solution (modified after Raut and Indraratna, [40]).

all these values vary from 40–60 mm and may be used as preliminary guidance in the design of filters. In practice, the thickness of dam filters is usually much greater than the above mentioned values. For both construction feasibility and structural stability, the actual thickness of dam filters often exceeds 500 mm [31].

9.4.3 Dominant filter constriction size

The existence of a dominant filter constriction size was indirectly introduced into the formation of self-filtration layers in internally stable broadly graded soils [20, 28].

Probabilistic studies conducted by Locke et al. [2] found there was a 95% chance that a base particle larger than D_{c95} could not penetrate a single layer of the filter, and therefore would not influence self-filtration. Indraratna and Raut [48] clearly demonstrated that dominant constriction sizes of various filter types occur at 95% finer. The proposed approach of using the largest dominant constriction size D_{c95} for disregarding coarser particles that do not influence filtration is more comprehensive than Terzaghi's method of using particle size ratios, especially with GW soils. This modification of the base soil PSD also explains why the coarser particle fraction could be ignored in filter designs that involve GW and internally unstable gap graded base soils [28]. In other words, the PSD of the self-filtration layer is formed by filter particles and base particles finer than the constriction size D_{c95}.

9.4.4 Controlling filter constriction size

In order to improve the multi-layered one dimensional constriction model of Kenney et al. [20], which assumed the flow channels to be independent, Indraratna and Locke [34] considered a three dimensional pore network model. Possible sideways exits for the base soil particles were incorporated into a mathematical investigation of the size of controlling constriction in a filter. This new model increased the value of probability of forward movement corresponding to the percentage value of larger constrictions. Further analysis by Locke et al. [2] led to a constriction model that presented an exceedingly high probability of forward movement with a confidence interval of 95%. The probability approaches unity at a constriction size finer than 35% (i.e. D_{c35}). A particle of base soil smaller than D_{c35} would not be retained by a granular filter unless the constrictions become progressively finer due to self-filtration. Based on this, Indraratna et al. [49] proposed that the controlling constriction in a granular filter can be given by the specific constriction size D_{c35}.

Comparative calculations on the controlling constrictions D_{c35} by Indraratna et al. [49] indicated a close agreement with the findings of other authors [20, 23, 27, 45]. Deviations of the computational findings from those of Witt's [45] experimental measurements were attributed to the effect of R_d on the behaviour of the size of the constrictions. Witt's [45] approach for calculating constriction sizes of filter particles on silicon rubbers did not include the role of R_d. The effect that C_u had on controlling constriction sizes was clearly established which verified that the controlling constriction sizes in non-uniform filters were smaller than those in uniform filters for the same D_{15} and for a given level of compaction [23].

9.4.5 Base soil representative parameter

Modelling filters and base soils by the PSD based on surface area, Indraratna et al. [49] clearly demonstrated why the filter effectiveness tends to decrease as the base soil becomes increasingly non-uniform. Base soils having the same d_{85} by mass, but with increasing C_u values, they were shown to have a reducing effective amount of base soil particles larger than d_{85}. The proposed d_{85} by surface area (d_{85sa}) as the base soil representative parameter offers an advantage of taking into consideration the PSD and C_u in a single value. Particularly for non-uniform soils and in agreement with data taken from various past studies [21, 27, 30], d_{85sa} should satisfy the condition that at least 15% of the base soil particles are retained whereas the use of conventional d_{85} does not.

9.5 CONSTRICTION BASED CRITERIA FOR ASSESSING FILTER EFFECTIVENESS

Filter effectiveness has been evaluated by guidelines based on the grain size ratio of the base soil – granular filter combination. Considering that the size of the voids within a filter rather than the actual particles effect filtration, it is more appropriate to develop filter design criteria in terms of constriction sizes.

9.5.1 D_{c95} model

A new procedure incorporating the CSD of the filters (Fig. 9.15) was developed by Raut and Indraratna [40] to assess the effectiveness of the same test data used by Indraratna and Vafai [47]. Here the largest particle of the base soil was smaller than the smallest constriction of filter F2 that meant that the filter could not stabilise the base soil. The constrictions of filter F1 are larger than the finer particles of the base soil, whereas the coarser base particles are large enough to initiate self-filtration, which means the filter can retain the base soil. Furthermore, it seems possible that filter F2 could retain base soil that is equivalent in size to filter F1.

As an extension to their work in 2004, Indraratna and Raut [48] proposed constriction size D_{c95} as a cut off value where base particles larger than the constriction size do not influence the process of self-filtration because they do not penetrate the filter. The base soil PSD must also be modified accordingly. When eroded base particles are transported to the filter, only coarser particles larger than the controlling constriction size are captured initially. These finer constrictions progressively retain finer base particles to form a self-filtration layer. The representative diameter of the modified base soil PSD is then reduced to d_{85sa}.

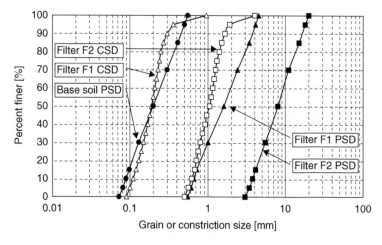

Figure 9.15 CSDs of effective (F1) and ineffective (F2) filters (modified after Raut and Indraratna, [40]).

Figure 9.16(a) and Figure 9.16(b) are examples of determining filter effectiveness using the criterion D_{c95} for filters F1 and F2, and the corresponding modified base soil PSD. The values of C_u for filters F1 and F2 are 1.20 and 5.23, respectively. Each filter's constriction size D_{c95}, the PSD of the self-filtration layer, and the PSD of the base soil were determined and examined. The analyses confirmed laboratory observations that filter F1 was ineffective while F2 was effective. These results were subsequently verified through Kenney and Lau's [51] H/F method for calculating the filter's internal stability wherein F is the mass per cent passing diameter D, and H is the mass per cent between diameters D and 4D. A ratio greater than 1 suggests a stable grading provided that $F \leq 30\%$ for uniform coarser part $(C_u < 3)$, and $F \leq 20\%$ for widely graded coarser part $(C_u > 3)$. Furthermore, the absence or presence of a gap in the self-filtration layer PSD plot indicates effectiveness or ineffectiveness, respectively.

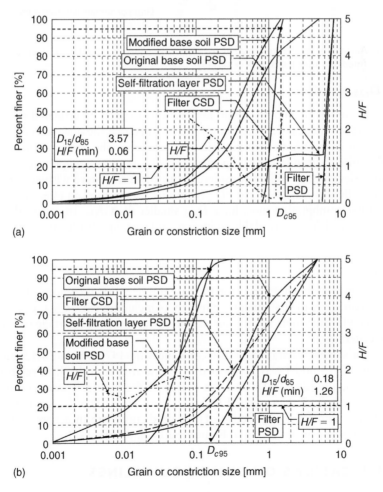

Figure 9.16 Analyses of (a) an ineffective uniform filter F1, and (b) an effective well graded filter F2 with a well graded base soil (modified after Indraratna and Raut, [48]).

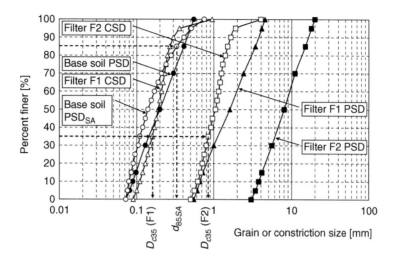

Figure 9.17 Analysis of moderately graded base soil and effective filter F1 and ineffective filter F2 (modified after Indraratna et al., [49]).

9.5.2 D_{c35} model

Several past studies, including Honjo and Veneziano [21], investigated the filtration process using mechanical sieves as filters, and revealed that the sieve can only retain the base soils if at least 15% of their particles are larger than the sieve aperture. Although a granular filter of randomly compacted particles is more complex than a regular mechanical sieve, it can still be considered equivalent to a sieve with apertures equal to the controlling constriction size (D_{c35}). In this perspective, an effective base soil – granular filter combination must have D_{c35} smaller than d_{85sa} to ensure that at least 15% of base particles are available to initiate and sustain self-filtration, hence:

$$\frac{D_{c35}}{d_{85sa}} < 1 \tag{9.14}$$

The above constriction based criterion for base soil retention is comprehensive because it considers an array of fundamental parameters including PSD, CSD, C_u, and R_d, in comparison with the single filter grain size of D_{15} and the base particle size d_{85} in the Terzaghi criterion.

Using the same test data from Indraratna and Vafai [47] in Figure 9.15, an analysis based on the D_{c35} model was performed by Indraratna et al. [49]. In relation to the current model, the filter CSDs and the PSDsa of the base soil were computed and plotted in Figure 9.17. Here $D_{c35}(F1) < d_{85sa}$, and $D_{c35}(F2) > d_{85sa}$, which classifies F1 as effective and F2 as ineffective. These predictions are in accordance with the experimental observations reported by Indraratna and Vafai [47].

9.6 IMPLICATIONS ON DESIGN GUIDELINES

The proposed constriction based D_{c95} and D_{c35} criteria cannot be directly compared with the two well known existing design guidelines applied in professional practices

Table 9.3 Comparison of capacities between particle based and constriction based filter criteria (Indraratna et al., [3]).

Criteria capabilities	Terzaghi [6]	NRCS [52]	D_{c95} Model [48]	D_{c35} Model [49]
Regrading required	✓	✓	X	X
Inherent internal stability analysis	X	X	✓	X
Enhanced design certainty due to self-filtration PSD	X	X	✓	X
Clear distinction between effective and ineffective filter	X	X	✓	✓
Porosity, R_d, and C_u considered	X	X	✓	✓
Analytical principles applied	X	X	✓	✓

namely the NRCS [52] and Lafleur [31] methods. The existing design guidelines provide varying filter boundaries depending on the percentage of fines in the base soils. However, as the existing methods are based on experimental studies on cohensionless soils, a comparison can be made by first applying the necessary regraded criterion $D_{15}/d_{85R} \leq 4$ [49].

The lack of reliability and adequacy of $D_{15}/d_{85} < 5$ [6] as a criterion for effectiveness of filter was emphasised when used on tests results involving base soils with increasing C_u. The proposed constriction based criteria clearly establish an effective zone away from some filters involving retention ratios D_{15}/d_{85} well below 4–5 that failed to retain the GW base soils but still plotted in the effective zone [48, 49]. Furthermore, the proposed models did not require the base soil to be regraded. Table 9.3 highlights the advantages of the constriction based models compared to the empirically developed particle based criteria.

The proposed D_{c95} model had an inherent capability of satisfying internal stability requirements. A prior analysis to examine the internal stability of the base soil was unnecessary because it was taken care of through the H/F technique of Kenney and Lau's [51]. In plotting the self-filtration PSDs, the 'gap' in all ineffective base soil – granular filter combinations were clearly established, providing more certainty to the design.

The use of CSD and PSD by surface area in the D_{c35} model eliminated the limitations of the particle size based retention criterion. Indraratna et al. [49] clearly demonstrated that for highly GW cohesionless tills where the conventional (by mass) d_{85} size was usually much larger than d_{85sa}, a cluster of coarse and uniformly ineffective filters fell into the predicted effective zone that required the introduction of additional constraints [6] to ensure effective filtration. According to the proposed model, none of the filters proven experimentally to be ineffective fell into the predicted effective zone.

The existing granular filter design criteria [52] proposed considerable improvements over the original Terzaghi's criterion [6]. Further improvements were demonstrated by Indraratna and Raut [48] and Indraratna et al. [49] in their proposed constriction size approach, which have equally acceptable methods for distinguishing between effective and ineffective filters. The integration of filter compaction, porosity and C_u, together with the incorporation of analytical principles capturing the surface area and constriction size concepts, have essentially made the models more comprehensive, quantifiable, and realistic.

Overall, a number of empirical and analytical models of the filtration phenomenon in granular materials have been developed for embankment dams [2, 47, 49], but the loading system in a rail track environment is cyclic unlike the steady seepage force that usually occurs in them. There is a need to assess the impact of cyclic loading in order to improve our understanding of the mechanisms of filtration, interface behaviour, and time dependent changes to the filtration that occurs within subballast as a filter medium. These advances may potentially improve rail performance and safety, extend system life cycles and reduce maintenance costs.

9.7   STEADY STATE SEEPAGE HYDRAULICS OF POROUS MEDIA

9.7.1   Development of Kozeny-Carman equation – a rationale

Steady state seepage hydraulics through porous media is governed by the three dimensional equations of continuity and Navier-Stokes. Using the Hagen-Poiseuille solution of the Navier-Stokes general equation, in addition to the application of Darcy's law and some relevant geometric assumptions, Kozeny [53] presented a relationship between the hydraulic conductivity (k) and porosity n (or voids ratio e) of a porous medium. Considering the tortuosity (τ) and shape of the channels within which a fluid particle has to travel through, Carman [54] modified the equation and came up with the more general and well known Kozeny-Carman (KC) formula:

$$k[\text{ms}^{-1}] = \frac{1}{72 \cdot \tau} \frac{\gamma}{\mu} \frac{d_e^2}{\alpha} \frac{n^3}{(1-n)^2} \tag{9.15}$$

where γ = unit weight of the permeant (N/m^3), μ = dynamic viscosity of the permeant (Pa-s), d_e = diameter of the spherical solid particle (m), and α = shape coefficient.

One of the other commonly used versions of the KC formula is written in terms of the specific surface of the particle (S_0). This is derived from the fundamental definition that the S_0 of a spherical solid particle (also applicable for cubical solid grain) is equal to the ratio between its surface area and its volume, thus giving the following equation:

$$d_e = \frac{6}{S_0} \tag{9.16}$$

where S_0 has units of 1/m. Since the surface area/volume ratio is influenced by the shape of individual soil grains, the original formula is supplemented by the use of α, whereby α is equal to 6 for spheres. If the version of the KC formula similar to Equation (9.15) is used, the value of α becomes 1. The introduction of this coefficient accounts for the difference in angularity between the sphere and the actual natural materials and its range of empirically derived values differ from one researcher to another.

Recent advances in steady state hydraulics have shown that the KC formula can provide satisfactory k estimates of fully saturated homogenised soils. In comparison to an extensive collection of experimental data, Chapuis and Aubertin [55] validated the adequacy of the KC formula for an even wider variety of materials, including both granular and clayey soils, provided that S_0 is calculated properly. For non-plastic soils S_0 can be simply estimated from the conventional particle size distribution by mass (PSD_m) [56, 57] and a number of improvements have been proposed ever since [58, 59].

Since its inception, the KC formulation has been popularised through classical soil mechanics and hydrogeology textbooks [60, 61, 62]. A summary of further modifications on the KC formula in its application to different porous media is provided by Xu and Yu [63]. Engineering practitioners however, are less adaptable at determining S_0 and they still rely on traditional empirical plots for estimating k. One possible way to encourage the use of the KC formula is by using its form in terms of d_e.

Although his work was a key corner stone for all investigations related to the determination of the characteristic diameter for a heterodisperse sample, Kozeny [53] originally developed the model for materials with uniform sized particles. Heterodisperse samples are composed of granular materials of wide distribution of sizes (well graded). The particle size that passes 10% by mass of the total sample (d_{10}) has been used and accepted as a good representation of d_e for uniform materials. For non-uniform materials, large particles with a high individual mass but low in number are over represented because it is unlikely that these few large particles would meet together to form a large pore [2]. Consequently, the errors introduced by using this method diminish the reliability of this predictive formula for non-uniform materials. In research involving filtration analysis, Humes [37] suggested that although there are only a small number of large particles in a non-uniform material, they impose significant contact with other particles due to their larger surface area.

9.7.2 Formulation for the effective diameter

The Kozeny-Carman equation was developed after considering a porous material as an assembly of capillary tubes for which the equation of Navier-Stokes can be used. Based on the assumption that the ratio of the surface area of the capillary tube to its inner volume should be equal to the ratio of the grain surface area to the pore volume, Kozeny [53] recommended d_e as the effective diameter representing the characteristic diameter of a heterodisperse sample undergoing seepage hydraulics. According to the original definition, this is the diameter of a sphere whose homodisperse sample has the same surface area/volume ratio as the heterodisperse spherical sample in question. Homodisperse samples are composed of granular materials of uniform distribution of sizes (uniformly graded).

The d_e discussed in the preceding paragraph is only related to a single grain. In nature, no layer is built up from particles of identical size and shape. To accommodate a more general application, the technique of calculating the effective parameter was further developed by way of discretising a given PSD_m curve into a number of segments. If a non-uniform soil material is composed of j discretised diameters $d_1, d_2, d_3, \ldots, d_j$ and their corresponding mass percent finer $p_1, p_2, p_3, \ldots, p_j$, then the d_e can be calculated as follows:

$$d_e = \frac{100\%}{\sum\limits_{i=1}^{j} (p_i/d_{avei})} \tag{9.17}$$

where d_{avei} is the geometric average of two adjacent diameters. This technique, together with Equation (9.16), has been integrated into the KC equation and a structured demonstration on the computational procedure discussed by Head [33].

During seepage, the resistance of channels formed by pores between grains connected almost continuously and distributed at random in the flow space, must be overcome by the forces accelerating and maintaining movement. The resistance of the network depends mostly on the size and shape of the pores forming the channels. These geometrical parameters of the network depend on the size and shape of the grains, the degree of sorting of grain sizes (in terms of C_u), and the porosity (n) [58]. Moreover, the adhesive forces are affected by the mineralogical and chemical character of the grains. To investigate seepage through grains of non-spherical shapes, a particular value α has to be chosen that properly suits the process being investigated.

A number of methods are available to use α for each pre-determined interval of the PSD_m. However, the determination of α for every fraction of a PSD_m requires mineralogical and microscopic examination. For a practical application of this method, highly empirical data based on statistical categorisation according to their origin (alluvial or aeolian), mineralogical composition (quartz, feldspar, mica, clay mineral), and size of soil grains, are available. For sand and gravel mix with a d_e not more than 3 mm, a wide range of α have been suggested by different authors [58, 64] and each set of values is correlated to its corresponding version of the KC equation.

9.8 SUBBALLAST FILTRATION BEHAVIOUR UNDER CYCLIC CONDITIONS

In rail track environments, the loading system is cyclic unlike the monotonic seepage force that usually occurs in embankment dams. The mechanisms of filtration, interface behaviour, and time-dependent changes of the drainage and filtration properties occurring within the filter medium require further research to improve the design guidelines. A novel cyclic process simulation filtration apparatus was designed and commissioned at the University of Wollongong, and a standard testing procedure was established. The test apparatus was designed to simulate heavy haul train operations. Key parameters that influence the change in porosity and pore-water pressure within the subballast layer under cyclic conditions in rail track environments were identified. In general, the objective of the investigation was to monitor the performance of a granular filter which was previously identified as satisfactory based on existing available filtration criteria.

9.8.1 Laboratory simulations

In current industry practice the inclusion of a layer of subballast involves typical road base material with particles ranging from 0.075 to 20 mm. A naturally well-graded,

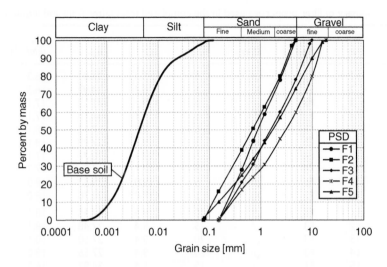

Figure 9.18 The particle size distribution of the filters and base soil used in the tests (after Trani and Indraratna, [67]).

commercially and locally available crushed basaltic rock road base was used as a granular filter. It was carefully sieved into a range of sizes, then washed, oven-dried, and mixed into a predetermined particle size distribution (PSD). The PSD of the filters F1 to F5 is shown in Figure 9.18.

To minimise base particle flocculation a low plasticity (LL = 48%, PI = 29%), highly dispersive and erodible silty clay (ASTM D4647, [65]) was used as the base soil. The slurry was formed by mixing 1500 g of dried fine base soil powder with 8 litres of water and then introduced into the subballast from the bottom of the permeameter using a computer controlled pump, to simulate clay pumping. A constant water pressure of 15 kPa was calibrated to adhere to a typical in situ excess pore water pressure associated with liquefaction [66]. The PSD of the base soil obtained by using a Malvern particle size analyser is also shown in Figure 9.18.

In preparing the specimen, a 10 kg plate of 225 mm diameter is placed on top of a 30 mm layer of granular material before the shaking table is switched on for approximately 30 seconds. This method is preferred over the compaction method in order to prevent unwanted breakage of the particles. This is done five more times until 150 mm of thickness is achieved. The calculated relative densities are above 97%. The summary of the initial filter material properties is shown in Table 9.4.

The standard constant head permeameter (Fig. 9.19) had to be modified in order to carry out the simultaneous action of dynamic train loading and clay pumping. The change in vertical hydraulic gradient was monitored with pressure transducers. Soil moisture sensors based on the concept of Amplitude Domain Reflectometry (ADR) were calibrated to measure the real-time changes to filter porosity. All these devices were connected to a data acquisition system. The effluent flow rate was determined at regular intervals and samples were taken to measure turbidity. The thickness of the specimen reflects the typical depth of subballast used on the actual rail track [7] while a diameter of 240 mm was chosen to minimise the effect of higher vertical seepage along

Table 9.4 Summary of filter properties used in the completed tests (after Trani and Indraratna, [67]).

	Test No.	Filter Type	With Slurry	Freq. (Hz)	C_u	n_0	γ_{dry} (kN/m^3)	R_d (%)
Phase I	1	F1	✓	5	5.4	34.94	16.3	98.1
	2	F1	✗	5	5.4	34.92	16.3	98.4
	3	F2	✓	5	9.1	31.22	17.1	97.7
	4	F2	✗	5	9.1	31.19	17.1	97.8
	5	F3	✓	5	9.4	27.52	18.1	98.7
	6	F3	✗	5	9.4	27.48	18.1	98.1
	7	F4	✓	5	17.1	28.71	17.1	98.7
	8	F4	✗	5	17.1	28.81	17.7	98.1
	9	F5	✓	5	18.0	28.23	18.0	98.7
	10	F5	✗	5	18.0	28.30	18.0	98.1
Phase II	11	F3	✓	10	9.4	27.45	18.1	98.4
	12	F3	✓	15	9.4	27.48	18.1	98.2
	13	F3	✓	20	9.4	27.50	18.1	98.4
	14	F3	✓	25	9.4	27.56	18.0	98.3
	15	F1	✓	15	5.4	34.87	16.2	98.8
	16	F1	✓	25	5.4	35.03	16.2	97.2

Notes: f = frequency; C_u = uniformity coefficient; n_0 = initial porosity; γ_{dry} = dry unit weight.

the side of the cell. The key features of the permeameter, the calibration of the ADR probes (Fig. 9.20), and the procedure for preparing the test specimen for the novel filtration apparatus are discussed in detail by Trani and Indraratna [68].

The cyclic wheel load simulating a typical heavy haul train was replicated in the modified permeameter, by imposing a uniform cyclic stress via a dynamic load actuator over a specified number of cycles, and at a desired frequency. Every specimen was subjected to a minimum stress of 30 kPa and a maximum stress of 70 kPa which is comparable to the vertical stress measurements induced by heavy haul freight trains recorded at the Bulli (NSW, Australia) experimental track [69].

The laboratory investigation was organised in 2 phases (Fig. 9.21). In the first half of phase 1, non-slurry pseudo-static filtration tests were conducted to investigate the internal stability of the chosen filters [51]. Pseudo-static tests are cyclic tests run at a frequency of 5 Hz. These tests served as a control for the corresponding slurry filtration tests. Effluent turbidity readings were used to indicate internal stability. All filter types, including those that exhibited washout and poor drainage capacity, were subjected to slurry filtration tests during the second half of Phase 1. All these tests were terminated after 100,000 cycles.

Filters that were considered acceptable after Phase 1 were subjected to Phase 2 slurry filtration tests. These tests were conducted while the loading frequency was increased to a predetermined level. All the specimens were fresh and the tests were terminated after 100,000 cycles. Filters that showed acceptable filtration and drainage potential under increased loading frequency were subjected to long-term filtration tests of up to 1 million cycles or until the filter failed under the drainage capacity criterion. The one-dimensional saturated vertical permeability of the filter, which indicates drainage capacity, was calculated using Darcy's law. To ensure a steady-state laminar

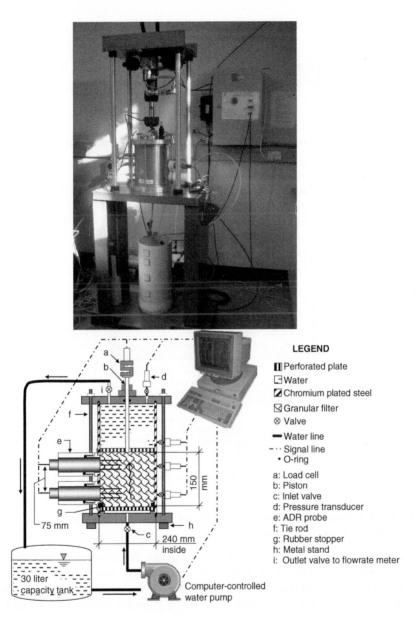

LEGEND

▮ Perforated plate
⊟ Water
▨ Chromium plated steel
⊠ Granular filter
⊗ Valve
➖ Water line
––·· Signal line
 • O-ring

a: Load cell
b: Piston
c: Inlet valve
d: Pressure transducer
e: ADR probe
f: Tie rod
g: Rubber stopper
h: Metal stand
i: Outlet valve to flowrate meter

150 mm

75 mm

240 mm
inside

30 liter
capacity tank

Computer-controlled
water pump

Figure 9.19 Actual set up (top) and schematic (bottom) of the cyclic loading permeameter (after Trani and Indraratna, [67]).

flow (the average Reynolds number Re, was 0.117), the pressure difference across the sample and the effluent flow rate used in the calculations were those recorded when the cyclic load was stopped, while keeping a surcharge of 15 kPa. Post-test analyses for every test where trapped fines were collected during wet and dry sieving were carried out.

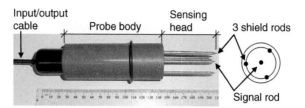

Figure 9.20 Actual Amplitude Domain Reflectometry (ADR) probe (after Trani, [15]).

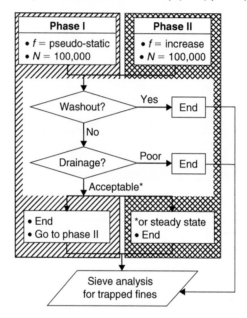

Figure 9.21 Experimental program (after Trani and Indraratna, [67]).

9.8.2 Deformation characteristics of subballast under cyclic loading

The accumulation of compressive and frictional plastic deformation is one of the major causes of geometric deterioration of railway substructures. In order to obtain a better understanding by way of a detailed analysis of track deterioration it is imperative to study the mechanical behaviour of the individual granular components during cyclic loading.

9.8.2.1 Pseudo-static loading

The effect of cycling the stress between two fixed limits ($\sigma_{min} = 30$ kPa, $\sigma_{max} = 70$ kPa) is shown in Figure 9.22. For all filter types, rapid strain development occurred during the first 7,500 cycles and eventually attained stability at about 20,000 cycles. This compression behaviour for all subballast types was uniform with respect to the number of load cycles irrespective of its grading or the range of particle sizes it is composed of.

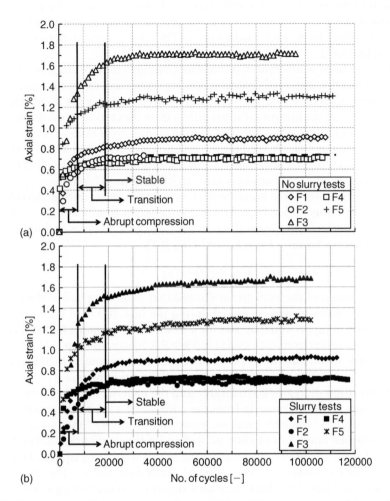

Figure 9.22 Development of strain under cyclic loading for all filter types during (a) non-slurry test, and (b) slurry tests (after Trani and Indraratna, [67]).

The introduction of base soil during slurry tests does not alter the strain development behaviour of the filter.

As shown in Table 9.4, all specimens were prepared to attain a dense state. Since there was no lateral strain during the tests, the axial strain was exactly equal to the volumetric strain. In several cycles of loading, only a portion of the strain that occurred while loading was recovered during subsequent unloading. The strains that resulted from sliding between particles or from fracturing of particles were largely irreversible. The rebound upon unloading was caused by the elastic energy stored within the individual particles as the soil was loaded [70].

There are some reverse sliding between particles during unloading. The sequence of events during cyclic loading can be explained by using results from a theoretical study of an ideal packing of elastic spheres [71]. In a one-dimensional array of elastic spheres, the normal forces at the contact points compress the spheres, but sliding occurs so that the resultant relative motion is purely vertical. Upon unloading, the particles regain their original shape and sliding occurs in the reverse direction. Some small amount of energy is absorbed during each loading cycle. The same general pattern of events must occur in actual soils.

During confined compression, particle motions are on the average in one direction only. Thus when the tangential contact forces are summed over the contact points lying on some surface, there should be a net tangential force (i.e., a net shear stress on the surface). In general, the horizontal stress differs from the vertical stress during confined compression, and its ratio is defined as K_0 (the lateral stress ratio at rest).

When a granular soil is loaded for the first time, the frictional forces at the contact points act in such a direction that $K_0 < 1$. During unloading, the direction of the frictional forces at contact points between particles begins to reverse during unloading. For a given vertical stress, the horizontal stress becomes larger during unloading when compared to the original loading. At some later stages of unloading, the horizontal stress may even exceed the vertical stress. During cyclic loading and unloading, the lateral stress ratio alternates from K_0 and $1/K_0$ [61].

9.8.2.2 Immediate response to cyclic loading

The cyclic evolution of the axial strain showed that more than half of the compaction of the granular material was generated during the first 400 load cycles (Fig. 9.23a). Within this period (approximately 80 s for a 5 Hz load frequency), higher levels of permanent strain resulted from each cycle until a stable hysteresis loop [61] was obtained, generating little or no additional permanent strain for a cycle of loading. This characteristic of a particulate system is known as the conditioning phase [72] wherein the elastic deformation decreases considerably and the material becomes stiffer.

The strains resulted primarily from the collapse of a relatively unstable arrangement of particles. As the stress was increased, the relatively loose array of granular particles collapsed into a more tightly packed and stiffer configuration. Finally, a stage was reached at which the already dense arrangements were being squeezed more tightly together as contact points crush, thus allowed a little more sliding. This phase is named as cyclic densification [73] where the elastic deformation did not significantly change anymore.

Figure 9.23(b) illustrates a marked increase in secant constrained modulus at about 400 cycles. Prior to the 400-cycle point, the lower magnitude of constrained modulus is related to the rapid axial deformation of the filters. As the filter stiffness increased as it is loaded and reloaded, the subsequent recorded deformation tapered off and stabilised. Also, for a given relative density (almost 100%, see Table 9.4), the starting modulus of the angular filter decreased as the particle size and grading led to a smaller void ratio. However, the effect of composition was bound to disappear during subsequent cycles of repeated loading [61].

9.8.3 Strain-porosity relationship of subballast under cyclic loading

9.8.3.1 Pseudo-static loading

Repeated loading on a confined assembly compresses the granular mass. The compression of the granular filter was observed to go together with the reduction of the voids of

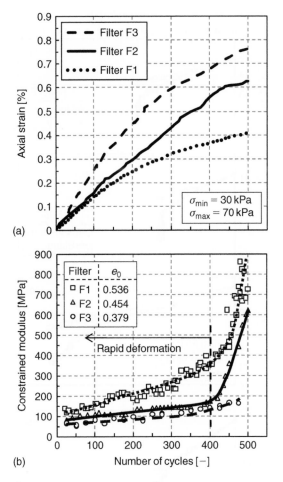

Figure 9.23 Immediate displacement reaction of granular filters subjected to cyclic loading (after Trani and Indraratna, [67]).

the filter medium skeleton. The reduction of the inter-particle voids, which effectively reduces the porosity, could be caused by repositioning of the particles. This could also be due to the filling of the voids by the relatively smaller particles present in the matrix or the fines generated by the attrition or breakage. The amount of fines coming from the degradation of filter grains with time, which has a potential to become part of the filter skeleton or may fill the voids, is of insignificant level. The average mass percentage of fines less than or equal to $150\,\mu m$ produced after test is less than 5%. This is mainly explained by the existence of optimum internal contact stress distribution and increased inter-particle contact areas.

Figure 9.24 shows the strain-porosity characteristics of each of the filter types used in the experimental program. In general, as the axial strain developed due to cyclic loading, the recorded filter porosity reduced. As shown earlier in Figure 9.22, the maximum amount of accumulated plastic strain was different for each filter type,

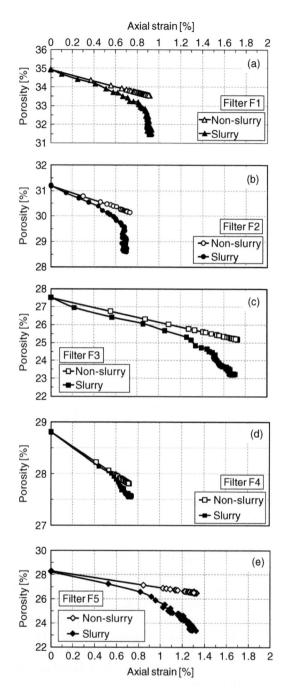

Figure 9.24 Effect of base soil intrusion to the strain-porosity relationship of the filters during cyclic loading (after Trani and Indraratna, [67]).

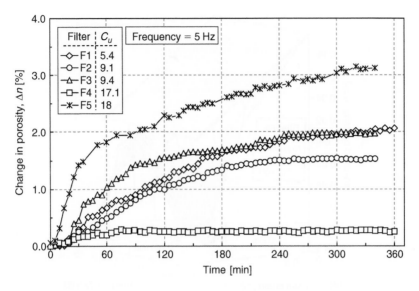

Figure 9.25 Measured change in porosity for filters FI to F5 (after Trani and Indraratna, [67]).

hence, the separation of the plots. During non-slurry tests, the reduction of the porosity of the filters is linearly related to their respective strain development and this creates the compression plane. Upon reaching the stable level of strain, no further reduction of porosity could be observed.

During slurry test, however, the filter porosities were gradually reduced. This was an indication that the filter interstices were being filled with base soil particles as the number of cyclic loading increased. The recorded reduction in porosity during slurry tests, therefore, is the combined effect of compression and reduced voids sizes due to trapped base soil particles within the filter voids.

The difference of measured porosity between the non-slurry (control) and slurry (actual) filtration tests is shown in Figure 9.25. Filters with lower values of Cu exhibited a more consistent capacity of trapping fines. For highly well-graded filters (F4 and F5), extreme behaviour is revealed. A correlation between the filter's Cu and its efficiency in trapping fines is premature at this stage.

9.8.3.2   Increased loading frequency

Figure 9.26 shows a comparison of the porosity-strain relationship between a well-graded filter (F3) and a uniformly graded filter (F1). The compression plane is the plot of the measured change in porosity during non-slurry tests. Based on Figure 9.26(a), an apparent threshold frequency of 10 Hz is shown wherein a minimum change in porosity reading was recorded. Comparing this to the 5 Hz test, the 10 Hz test took 170 min less to complete 102,000 cycles. Assuming that the capacity of the filter to capture fines was not affected by the change in frequency, this test completion time difference explains the difference in the porosity measurements. The longer the filtration tests are conducted, the more base particle fines are captured within the filter voids. However,

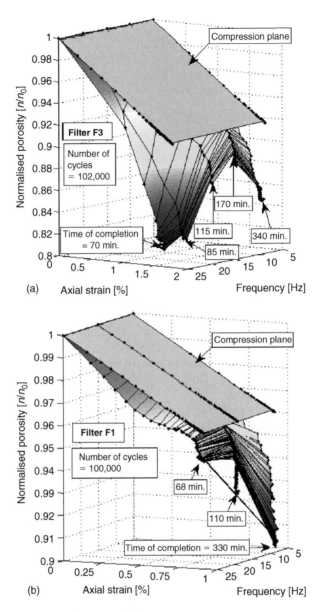

(a) Axial strain [%] Frequency [Hz]

(b) Axial strain [%] Frequency [Hz]

Figure 9.26 Porosity-strain relationship for a (a) well-graded filter F3, and (b) uniformly graded filter F1 (after Trani and Indraratna, [67]).

increasing the frequency to 15, 20, and 25 Hz (thus correspondingly reducing the test completion time) showed a rapid decrease in porosity readings. The filter is affected with variation in frequency and is unpredictable over time.

The porosity-strain behaviour of the uniformly graded filter (F1) is more predictable over time (Fig. 9.26b). The difference in test completion time played a role in the amount of reduced porosity. Unlike filter F3, the variation in frequency did not alter the capacity of the filter to capture base soil particles.

The deformability of the pore medium itself affected the filter condition due to the changes in porosity and hydraulic conductivity. Soil particle accumulations inside the filter layer may lead to stable or unstable filter capacity. However, the apparent equilibrium is endangered by the changing of load conditions, mainly caused by the loading frequency of applied stresses and the development of high hydraulic gradients in and underneath the filter.

9.8.4   Seepage hydraulics of subballast under cyclic loading

9.8.4.1   Turbidity measurements and trapped fines

Measuring the turbidity of the effluent during filtration tests is a useful indicative tool of the level of washing out within the filter. The washed out particles could be coming from an internally unstable filter or the slurry particles escaping through the filter voids. For the turbidity measurements of all filter types (Fig. 9.27b), filters F2 and F5 showed high turbidity readings that signal filter ineffectiveness at pseudo-static level. Further tests of increased loading frequencies for filters F1 and F3 showed an increase in turbidity reading. However, a high turbidity reading may not necessarily mean excessive washout similar to clay pumping and hydraulic erosion scenarios. A post-test sieve analysis would determine how effective the filter in trapping base soil particles.

Post-test wet and dry sieve analyses at every 30 mm layer were conducted for each of the tests. Figure 9.27(a) shows the amount of trapped fines collected through the profile of each of the filters. As expected, high amount of base soil fines were collected at the interface bottom layer (layer 1). Apart from F4, all other filter types exhibited the capacity to capture fines during pseudo-static filtration tests. Filter F4, on the other, failed to demonstrate its capacity to capture fines within its voids as clogging occurred at the filter-slurry interface. No further tests were performed on filter F4.

The collective amount of captured fines for all layers of filter F3 is in the average of 80% (Fig. 9.28a). The captured fines of 50 g at the top level (level 5), on the other hand, showed the proximity of the fines to being expelled at the top surface. This illustrates the possibility that a substantial percentage of the fines collected during the post-test wet and dry sieving were the transient fines captured during the termination of the test. The agitation generated by the increased loading frequency caused these transient fines to get washed out as shown by the high turbidity measurements in Figure 9.28(b).

Filter F1 exhibits a relatively high collective amount of accumulated fines of the first 2 layers (Fig. 9.29a). This is an indication that the slurry base soil particles were contained within the bottom half of the filter. Furthermore, the gradual decrease of the profile of accumulated fines as the loading frequency increased showed a time-dependent filtration behaviour. All tests in Phase 2 were conducted to a maximum of 100,000 cycles and a shorter period of time was required to complete a test conducted at a higher frequency. The corresponding turbidity measurements taken on the effluent of the slurry filtration tests on filter F1 is shown in Figure 9.29(b).

Looking at the first three layers from the bottom, the average accumulated fines in F3 exceeded 35% when compared to that of F1. Comparing the average accumulated

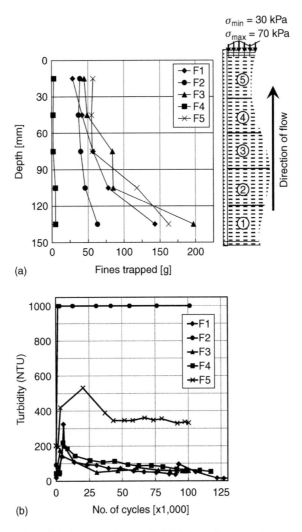

(a)

(b)

Figure 9.27 For all filters: (a) trapped fines collected through post-test wet and dry sieving, and (b) effluent turbidity measurements during slurry tests (after Trani and Indraratna, [67]).

fines from the upper layers 4 and 5, the collective amount in F3 has increased to about twice as much as that of F1. This amount of fines in the upper half of F3 represents the base particles that escaped through the constrictions at the lower layer that also have the potential of getting flushed toward the filter surface.

9.8.4.2 Short-term drainage performance

The water must be permitted to drain freely out of the deforming soil and filter sample so that the reduction of the pore volume is exactly equal to the volume of pore fluid expelled perpendicular to the subballast layer. Adequately designed filters should resist

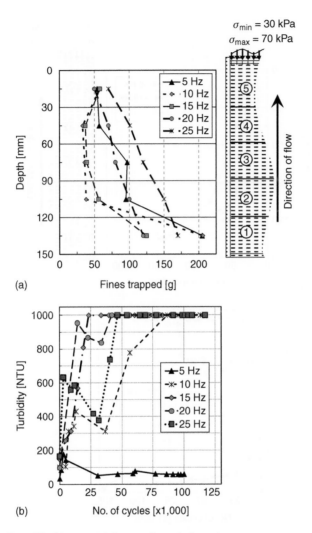

Figure 9.28 For filter F3: (a) trapped fines collected through post-test wet and dry sieving, and (b) effluent turbidity measurements during slurry tests (after Trani and Indraratna, [67]).

the imposed load conditions in an acceptable way, whether this flow is directed in opposite directions or caused by steady or changing hydraulic gradients, or cycling load conditions occurring with and without rapid load changes and dynamic influences.

Looking into the seepage characteristics of the filters, Figure 9.30(a) shows the behaviour of measured filter hydraulic conductivity (k) during slurry filtration tests with pseudo-static loading. The values of k for filters F1, F3 and F4, which lie above the 10^{-5} m/s threshold [33, 61], indicate that both filter types possess good drainage capacity. In contrast, filters F2 and F5 are categorised as poor drainage granular layers. The presence of at least 15% very fine sand in well-graded filters F2 and F5 ($C_u \geq 9$) effectively reduced the values of k of the whole filter matrix. In Figure 9.30(b), a

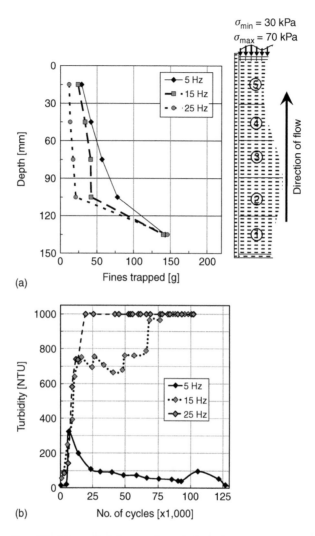

Figure 9.29 For filter F1: (a) trapped fines collected through post-test wet and dry sieving, and (b) effluent turbidity measurements during slurry tests (after Trani and Indraratna, [67]).

decrease in values of k is observed for all filters subjected to slurry filtration tests. The subsequent values of k for filters F2 and F5 deteriorated close to the k-value of a loosely compacted base soil.

9.9 TIME DEPENDENT GEO-HYDRAULIC FILTRATION MODEL FOR PARTICLE MIGRATION UNDER CYCLIC LOADING

In this section, three steps of mathematical description and the physical basis of the filter mechanism under cyclic loading regime are discussed. The first step is to investigate

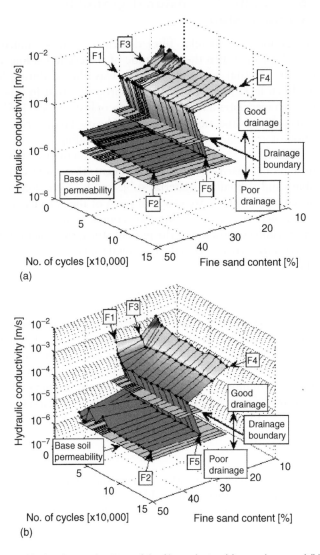

Figure 9.30 Measured hydraulic conductivity of the filters during (a) non-slurry, and (b) slurry filtration test under cyclic loading (after Trani and Indraratna, [67]).

the one dimensional cyclic compression behaviour of the subballast and its effect on the reduction of its controlling constriction size relative to the base soil representative diameter [49]. The coupling effect of the consolidation behaviour, which is developed in the framework of post shakedown plastic analysis, is then investigated with respect to base soil particle migration mechanism through the network of filter voids. In addition, a temporal porosity reduction function is proposed and the Kozeny-Carman formula is extended to provide a practical tool in predicting the longevity of the drainage layer.

9.9.1 Time based one dimensional granular filter compression

The evolution of permanent granular filter deformation was studied over a large number of load cycles (N). When the amplitude of the cyclic loading was above the shakedown level, the internal material structure was altered during loading which caused the shakedown level to evolve [74]. It is proposed that in this study a stress domain of Drucker-Prager potential applied in a viscoplastic model [75] in the form of a post shakedown cyclic densification regime would be used to describe the progressive plastic deformation of granular material under cyclic loading.

Suiker and de Borst [73] proposed a detailed mathematical development of a cyclic densification model based on triaxial experiments that showed the plastic deformation of a ballast and subballast material subjected to cyclic loading. This model described two mechanisms which are essential parts of the granular material densification process, frictional sliding and volumetric compaction. Due to the existing one dimensional compression constraints of the present study, the function for irreversible plastic strain (ε_p) is set to correspond to the frictional shakedown evolution framework and is proposed as:

$$\varepsilon_p = \varepsilon_f(1 - e^{-tf/k_s}) \tag{9.18}$$

where, ε_f = the shakedown plastic strain obtained from one dimensional cyclic consolidation test on a fully saturated specimen, t = time (sec), f = frequency (Hz), and k_s = scaling factor equal to $N_{max}/10$, where N_{max} is the maximum number of cycles used in the model. Figure 9.31 shows a comparison between the proposed plastic strain evolution model over a number of cycles and their corresponding experimental data.

From the one dimensional compression principle, the plastic axial strain is given as:

$$\varepsilon_p = \frac{\Delta e}{1 + e_0} \tag{9.19}$$

where Δe is the change in voids ratio and e_0 is the initial voids ratio of the filter matrix.

Using the voids ratio-porosity relationship ($n = e/(1 + e)$), the change in porosity of a porous medium caused by axial compression during a single time step (Δn_c) is represented by:

$$\Delta n_c = \frac{\varepsilon_p \left(1 + \frac{n_0}{1 - n_0}\right)}{1 + \varepsilon_p \left(1 + \frac{n_0}{1 - n_0}\right)} \tag{9.20}$$

$$\Delta n_c = \frac{\varepsilon_p}{1 - n_0 + \varepsilon_p} \tag{9.21}$$

where n_0 is the initial filter porosity. Substituting Equation (9.18) into Equation (9.21) results into:

$$\Delta n_c = \frac{\varepsilon_f(1 - e^{-tf/k_s})}{1 - n_0 + \varepsilon_f(1 - e^{-tf/k_s})} \tag{9.22}$$

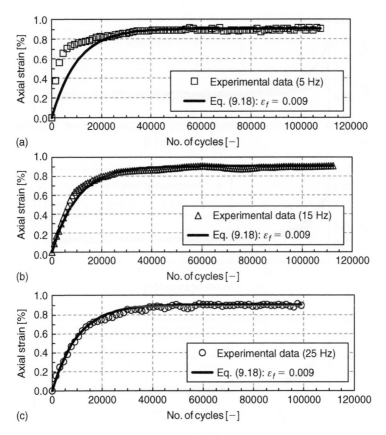

Figure 9.31 Comparison between the experimental data and proposed plastic strain model at (a) 5 Hz, (b) 15 Hz, and (c) 25 Hz (after Trani, [15]).

The proposed function takes into account the reduction of porosity with time that depends on the densification energy (natural or imposed stress state) through the parameter ε_f. The prediction of actual material porosity is comparable with the experimental observations as shown in Figure 9.32.

9.9.2 Accumulation factor

The development of the accumulation factor (F_a) is based on the assumption that the dominant constriction size (D_{c35}) of the filter is smaller than or equal to the representative diameter (d_{85sa}) of the base soil [76]. Depending on the manner by which the filter is prepared, the initial size of D_{c35} can be controlled by the level of compaction through the relative density (R_d) of the filter. However, none of the initial D_{c35} of the filter used in this study satisfied the effective filter criteria even though the values of R_d were already close to unity. Despite the initial conditions, the actual pseudo-static tests produced two successful filters in F1 and F3. The densification energy generated from

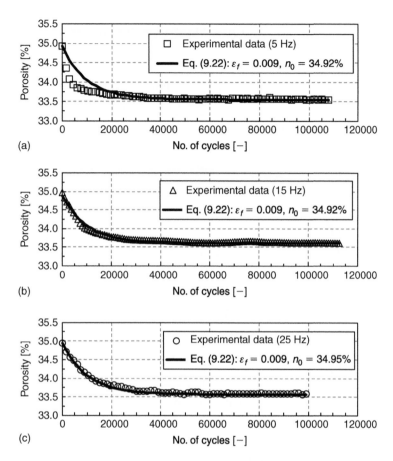

Figure 9.32 Comparison between the experimental data and proposed porosity predictive model for granular soils subjected to cyclic loading with a frequency of (a) 5 Hz, (b) 15 Hz, and (c) 25 Hz (after Trani, [15]).

cyclic loading led to granular filter permanent deformation over time. The eventual irrecoverable plastic strain affected the geometry of the constrictions in a way that the apparent D_{c35} was as close to the size where it satisfied the filtration criterion.

Shown in Figure 9.33 is the progressive reduction of the constriction size profile of the filter F1 with time in comparison with the d_{85} and d_{50} of the base soil. Each data point of the filter constriction curve represents the geometric – weighted harmonic mean of the CSD of the combined mass of the filter and the fines enmeshed in the filter matrix obtained through sieve analysis. The sieve analysis was conducted for each of the five filter layers of equal thickness. The CSD by mass is an acceptable representation of the constrictions of the new base soil-filter PSD by mass since the filter is considered uniform and the base soil mass is sufficiently small relative to the original filter matrix [2]. The formation of the self-filtration layer at the bottom is also shown by the drastic reduction of constrictions within a few load cycles (represented

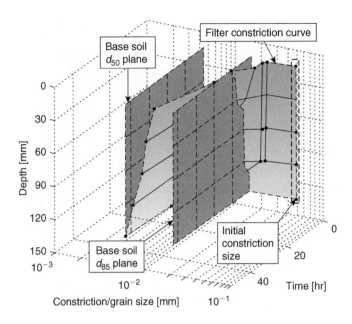

Figure 9.33 Reduction of filter constriction size due to accumulation of base particles (after Trani, [15]).

by time axis). The gradual reduction of the constriction size profile over time indicates stability of the accumulated fines within the filter voids. This stability of the new base soil-filter formation consequently created finer constriction sizes much smaller than the d_{85} of the base soil.

The base soil particles with size d_{50}, which was still marginally smaller than the estimated filter constriction size, have the capacity to migrate upwards to the next filter layer. However, these particles only represent 50% of the total original soil mass. Compounded by the gradual formation of finer constrictions within the lower filter layers further limited the mass of the base particles from being transported into the next upper layer. This successive reduction of accumulation of fines along the profile of the filter is controlled by a depth dependent F_a parameter which can be described by a rate law relationship:

$$F_a = F_1 e^{F_2 z} \tag{9.23}$$

In the above equation, F_1 and F_2 are empirical indices related to slurry concentration and slurry loading rate, respectively. This proposed function creates an apparent threshold amount of fines that could occupy the voids spaces in between the filter grains at a given depth.

The parameter $m_{a\,max}$ is defined as the apparent maximum amount of subgrade fines that could occupy a part of the volume of voids that remained after compaction. The equivalent volume taken up by the trapped and accumulated fines is much less than the theoretical volume of voids (V_v) and it is given as:

$$m_{a\,max} = F_a \frac{V_0 \rho_a}{1 + \varepsilon_f} \tag{9.24}$$

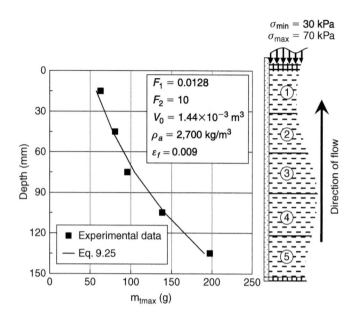

Figure 9.34 Comparison between the experimental and predicted amount of accumulated fines within filter FI (after Trani, [15]).

where F_a is the dimensionless accumulation factor, V_0 is the bulk volume of soil specimen [m³], ρ_a = solid density of accumulated fines [kg/m³].

By substituting Equation (9.23) into Equation (9.24), the maximum amount of fines that can be accumulated with respect to the thickness profile of a filter could be predicted by:

$$m_{a\,max} = F_1 e^{F_2 z} \frac{V_0 \rho_a}{1 + \varepsilon_f} \tag{9.25}$$

Figure 9.34 shows a good agreement between the amount of fines collected through post test sieve analysis and the predicted values. Note also that F_a can be used as a predictive tool of filter porosity deterioration. The value of F_a of 0.049 at layer 1 is comparable to the amount of porosity reduction during filtration tests by Locke et al. [2].

9.9.3 Mathematical description of porosity reduction due to accumulated fines

In an ideal coarse packing, the filter matrix is assumed to be supported by the skeleton created by the contacts among the filter grains. The porosity of the filter matrix is traditionally defined as the ratio of the volume of voids and the bulk volume of soil specimen ($n = V_v/V_0$). The plastic deformation due to compression ($V_0/(1 + \varepsilon_p)$) impacts the filtering capacity of the filter by effectively reducing the size of its constrictions.

Due to the cyclic loading action of the passing train, base soil particles are pumped upwards into the subballast filter from the fully saturated subgrade. Fines are trapped

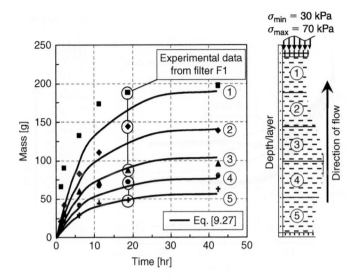

Figure 9.35 Comparison between the experimental and predicted amount of accumulated fines within filter F1 over time (after Trani, [15]).

by the filter constrictions and are deposited within the pore network. With the simultaneous action of one dimensional compression and pumping of subgrade fines, the volume of voids is reduced by the volume of accumulated fines (V_a) trapped in the original voids while the bulk volume of the filter reduces with time:

$$\Delta n_a = m_a \frac{1 + \varepsilon_p}{V_0 \rho_a} \tag{9.26}$$

where, m_a = mass of accumulated fines within the filter voids (kg). The amount of fines trapped by the constrictions of an effective filter is proposed to follow the given relationship:

$$m_a = m_{a\,max}(1 - e^{-tf/k_s}) \tag{9.27}$$

The comparison between the collective results of post test sieve analyses for all tests performed on filter F1 versus the predictions provided by Equation (9.27) is shown in Figure 9.35. Each line and the corresponding experimental data points represent a layer of the filter profile. As expected, more fines were captured and collected at the bottom section of the profile (layer 1) while the least amount was collected at the topmost section of the filter (layer 5). Subsequent back substitution of Equation (9.27) into Equation (9.26) yields:

$$\Delta n_a = \frac{m_{a\,max}}{V_0 \rho_a}(1 + \varepsilon_p)(1 - e^{-tf/k_s}) \tag{9.28}$$

Combining Equations (9.25) and (9.28) simplifies into:

$$\Delta n_a = F_1 e^{F_2 z} \frac{1 + \varepsilon_p}{1 + \varepsilon_f}(1 - e^{-tf/k_s}) \tag{9.29}$$

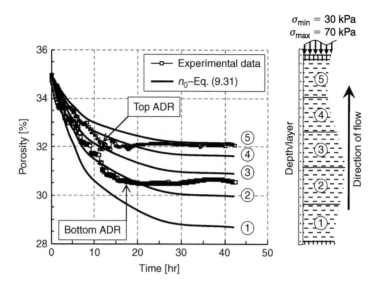

Figure 9.36 Comparison between the experimental and predicted porosity of filter FI (after Trani, [15]).

By using Equation (9.18) and the more compact Equation (9.23) into Equation (9.29), a time dependent porosity reduction function due to accumulated base soil fines is derived as follows:

$$\Delta n_a = F_a \frac{1 + \varepsilon_f (1 - e^{-tf/k_s})}{1 + \varepsilon_f}(1 - e^{-tf/k_s}) \tag{9.30}$$

The sum of Equations (9.22) and (9.30) is the time dependent total porosity reduction of the filter matrix as a collective effect of one dimensional compression and the accumulation of fines within the filter voids (Eq. (9.31)).

$$\Delta n_T = \frac{\varepsilon_f (1 - e^{-tf/k_s})}{1 - n_0 + \varepsilon_f (1 - e^{-tf/k_s})} + F_a \frac{1 + \varepsilon_f (1 - e^{-tf/k_s})}{1 + \varepsilon_f}(1 - e^{-tf/k_s}) \tag{9.31}$$

Figure 9.36 shows the predictive values of filter porosity caused by compression and accumulation of base particles in comparison with the ADR measurement. Note that the top ADR was located at approximately within level 4 of the filter while the bottom ADR was at level 2.

9.9.4 Time based hydraulic conductivity model

The Kozeny-Carman equation forms the basis of the derivation of the formulation employed to predict the deterioration in hydraulic conductivity of the granular filter specimens. The initial hydraulic conductivity of the granular filters (k_0) can be

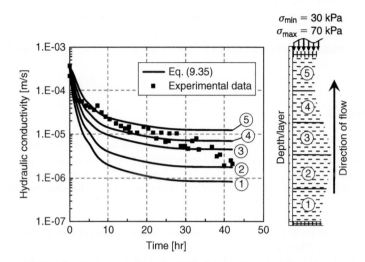

Figure 9.37 Comparison between the experimental and predicted hydraulic conductivity of filter FI (after Trani, [15]).

estimated as follows (similar to Eq. (9.15)):

$$k_0 [ms^{-1}] = \frac{1}{72\tau} \frac{\gamma}{\mu} \frac{d_{e.0}^2}{\alpha} \frac{n_0^3}{(1-n_0)^2} \tag{9.32}$$

where τ = tortuosity, γ = unit weight of the permeant [N/m³], μ = dynamic viscosity of the permeant [Pa-s], $d_{e.0}$ = initial effective diameter of the granular filter, and α = shape coefficient. Considering that the filter porosity decreases as the filter layer is being compressed and the clogging material accumulates with time (Eq. (9.31)), the reduced hydraulic conductivity with time (k_t) can be obtained from:

$$k_t = \frac{1}{72\tau} \frac{\gamma}{\mu} \frac{d_{e.t}^2}{\alpha} \frac{(n_0 - \Delta n_T)^3}{[1 - (n_0 - \Delta n_T)]^2} \tag{9.33}$$

In the above, $d_{e.t}$ = effective diameter of the granular filter at any time t. Rearranging Equation (9.32), the constants can be expressed as:

$$\frac{1}{72\tau\alpha} \frac{\gamma}{\mu} = \frac{(1-n_0)^2}{n_0^3} \frac{k_0}{d_{e.0}^2} \tag{9.34}$$

Substituting Equation (9.34) into Equation (9.33), a decreased hydraulic conductivity as a result of time based compression and clogging can be represented as a function of the initial hydraulic conductivity, the change in porosity, and the change in effective matrix diameter. The resulting equation is as follows:

$$k_t = k_0 \frac{(1-n_0)^2}{n_0^3} \left[\frac{d_{e.t}^2}{d_{e.0}^2} \frac{(n_0 - \Delta n_T)^3}{(1 - (n_0 - \Delta n_T))^2} \right] \tag{9.35}$$

In this expression, the effective diameters $d_{e.0}$ and $d_{e.t}$ are the geometric-weighted harmonic mean of their respective PSDs. Figure 9.37 illustrates the comparison between the measured and predicted hydraulic conductivity profile of filter F1 during a long term test. Using Equation (9.35), the values of k_t in each of the 5 layers of the filter is estimated. With reference to the position of the pressure transducers on the filtration cell, the experimental measurements of k represents approximately the middle layer of the filter.

REFERENCES

1. Alobaidi, I.M. and Hoare, D.J.: Mechanism of pumping at the subgrade-subbase interface of highway pavements. *Geosynthetics International*, 1999, Vol. 6, No. 4, pp. 241–259.
2. Locke, M., Indraratna, B. and Adikari, G.: Time-dependent particle transport through granular filters. *Journal of Geotechnical and Geoenvironmental Engineering*, ASCE, 2001, Vol. 127, No. 6, pp. 521–529.
3. Indraratna, B., Trani, L.D.O. and Khabbaz, H.: A critical review on granular dam filter behavior – from particle sizes to constriction – based design criteria. *Geomechanics and Geoengineering*, 2008, Vol. 3, No. 4, pp. 279–290.
4. Trani, L.D.O. and Indraratna, B.: Assessment of subballast filtration under cyclic loading. *Journal of Geotechnical and Geoenvironmental Engineering*, 2010, Vol. 136, No. 11, pp. 1519–1528.
5. Bertram, G.E.: *An Experimental Investigation of Protective Filters*, Report, Harvard Graduate School of Engineering, 1940, Vol. 6, No. 267.
6. USACE: *Investigation of filter requirements for underdrains*. 1953, Tech. Memo. No. 3–360, United States Army Corps of Engineers, U.S. Waterways Experiment Station, Vicksburg, Mississippi.
7. Selig, E.T. and Waters, J.M.: Track technology and substructure management. 1994, Thomas Telford, London.
8. Byrne, B.J.: *Evaluation of the Ability of Geotextiles to Prevent Pumping of Fines into Ballast*. M.Sc. project report No. AAR89-367P, 1989, Department of Civil Engineering, University of Massachusetts.
9. ASTM Annual Book of Standards, D1241: *Standard specification for soil-aggregate subbase, base, and surface flow courses*. Vol. 04.08, Philadelphia.
10. RIC 2001, T.S. 3402: *Specification for supply of aggregate for ballast*. Rail Infrastructure Corporation, NSW, Australia.
11. Haque, A., Bouazza, A. and Kodikara, J.: Filtration behaviour of cohesionless soils under dynamic loading. *Proceedings of the 9th ANZ Conference in Geomechanics*, 2004, Auckland, Vol. 2, pp. 867–873.
12. Standards Australia, Part 7: *Railway ballast*, AS 2758.7. Aggregates and rock for engineering purposes. 1996, Sydney.
13. Salim, W.: *Deformation and degradation aspects of ballast and constitutive modelling under cyclic loading*. PhD thesis, 2004, University of Wollongong.
14. RIC: T.S. 3402 – Specification for supply of aggregate for ballast. Rail Infrastructure Corporation, 2001, NSW, Australia.
15. Trani, L.D.O.: *Application of constriction size based filtration criteria for railway subballast under cyclic conditions*. PhD thesis, 2009, University of Wollongong.
16. Radampola, S.S.: *Evaluation and modeling performance of capping layer in rail track substructure*. PhD thesis, 2006, Central Queensland University.
17. Queensland Rail: Civil Engineering Standard Specification – *Earthworks*. QR 1998 – Civil Engineering Section, Part No. 6 (Revision C).
18. Vaughan, G. and Soares, F.: Design of filters for clay cores of dams. *Journal of Geotechnical Engineering*, ASCE, 1982, Vol. 108, pp. 18–31.

19. Kwang, T.: *Improvement of dam filter criterion for cohesionless base soil.* MEng thesis, 1990, Asian Institute of Technology, Bangkok.
20. Kenney, T.C., Chahal, R., Chiu, E., Ofoegbu, G.I., Omange, G.N. and Ume, C.A.: Controlling constriction sizes of granular filters. *Canadian Geotechnical Journal*, 1985, Vol. 22, No. 1, pp. 32–43.
21. Honjo, Y. and Veneziano, D.: Improved filter criterion for cohesionless soils. *Journal of Geotechnical Engineering Division*, ASCE, 1989, Vol. 115, No. 1, pp. 75–83.
22. Sherard, J. and Dunnigan, L.: Filters and leakage control in embankment dams. *Proceedings of the Symposium on Seepage and Leakage from Dams and Impoundments*, Volpe, RL and Kelly, WE (eds), 1985, ASCE, pp. 1–30.
23. Sherard, J., Dunnigan, L. and Talbot, J.: Basic properties of sand and gravel filters. *Journal of Geotechnical Engineering Division*, ASCE, 1984, Vol. 110, No. 6, pp. 684–700.
24. Sherard, J., Dunnigan, L. and Talbot, J.: Filters for silts and clays. *Journal of Geotechnical Engineering Division*, ASCE, 1984, Vol. 110, No. 6, pp. 701–718.
25. Sherard, J. and Dunnigan, L.: Critical filters for impervious soils. *Journal of the Geotechnical Engineering Division*, ASCE, 1989, Vol. 115, No. 7, pp. 927–947.
26. Indraratna, B. and Locke, M.: Design methods for granular filters – critical review. *Proceedings of the Institution of Civil Engineers – Geotechnical Engineering*, 1999, Vol. 137, pp. 137–147.
27. Foster, M. and Fell, R.: Assessing embankment dam filters that do not satisfy design criteria. *Journal of Geotechnical and Geoenvironmental Engineering*, ASCE, 2001, Vol. 127, No. 5, pp. 398–407.
28. Lafleur, J., Mlynarek, J. and Rollin, A.L.: Filtration of broadly graded cohesionless soils. *Journal of Geotechnical Engineering*, ASCE, 1989, Vol. 115, No. 12, pp. 1747–1768.
29. Locke, M. and Indraratna, B.: Filtration of broadly graded soils: the reduced PSD method. *Géotechnique*, 2002, Vol. 52, No. 4, pp. 285–287.
30. Lafleur, J.: Filter testing of broadly graded cohesionless soils. *Canadian Geotechnical Journal*, 1984, Vol. 21, No. 4, pp. 634–643.
31. ICOLD: Embankment Dams – Filters and Drains, *International Commission on Large Dams*, 1994, France.
32. Fuller, W.B. and Thompson, W.E.: The laws of proportioning concrete. *Transactions on American Society of Civil Engineers*, 1907, Vol. 59.
33. Head, K.H: *Manual of soil laboratory testing, permeability, shear strength and compressibility tests.* 1982, Vol. 2, Pentech Press, London.
34. Indraratna, B. and Locke, M.: Analytical modelling and experimental verification of granular filter behaviour. Keynote paper, *Filters and drainage in Geotechnical and Geoenvironmental Engineering*, 2000, Wolski, W. and Mlynarek, J. (eds), AA Balkema, Rotterdam, pp. 3–26.
35. Silveira, A.: An analysis of the problem of washing through in protective filters. *Proceedings of the 6th International Conference on Soil Mechanics and Foundation Engineering*, Toronto, Canada, 1965, Vol. 2, pp. 551–555.
36. Silveira, A., Peixoto, D.L.T. and Nogueira, J.B.: On void size distribution of granular materials. *Proceedings of the 5th Pan-American Conference of Soil Mechanics and Foundations Engineering*, 1975, pp. 161–176.
37. Humes, C.: A new approach to compute the void-size distribution curves of protective filters. *Proceedings of Geofilters '96*, Lafleur, J. and Rolin, A.L. (eds), Bitech Publications, Montreal, 1996, pp. 57–66.
38. De Mello, V.: Reflections on design decisions of practical significance to embankment dams. *Géotechnique*, 1977, Vol. 27, No. 3, pp. 279–355.
39. Federico, F. and Musso, A.: *Some advances in the geometric-probabilistic method for filter design.* in Filters in Geotechnical and Hydraulic Engineering, Brauns, J., Heibaum, M. and Schuler, U. (eds), Balkema, Rotterdam, 1993, pp. 75–82.

40. Raut, A.K. and Indraratna, B.: Constriction size distribution of a non-uniform granular filter. *Proceedings of the 15th South East Asian Geotechnical Conference*, Bangkok, Thailand, 2004, pp. 409–414.

41. Schuler, U.: Scattering of the composition of soils: an aspect for the stability of granular filters. *Proceedings of Geofilters '96*, Lafleur, J. and Rolin, A.L. (eds), Bitech Publications, Montreal, 1996, pp. 21–34.

42. Wittman, L.: The process of soil-filtration – its Physics and the approach in engineering practice. *Proceedings of the 7th European Conference of Soil Mechanics and Foundation Engineering*, Brighton, UK, 1979, Vol. 1, pp. 303–310.

43. Soria, M., Aramaki, R. and Viviani, E.: *Experimental determination of void size curves. in Filters in Geotechnical and Hydraulic Engineering*, Brauns, J., Heibaum, M. and Schuler, U. (eds), Balkema, Rotterdam, 1993, pp. 43–48.

44. Silveira, A.: *A method for determining the void size distribution curve for filter materials in Filters in Geotechnical and Hydraulic Engineering*, Brauns, J., Heibaum, M. and Schuler, U. (eds), Balkema, Rotterdam, 1993, pp. 71–74.

45. Witt, K.: Reliability study of granular filters. in Filters in Geotechnical and Hydraulic Engineering, Brauns, J., Heibaum, M. and Schuler, U. (eds), *Balkema, Rotterdam*, 1993, pp. 35–42.

46. Giroud, J.: Granular filters and geotextile filters. Proceedings of Geofilters '96, Lafleur, J. and Rolin, A.L. (eds), *Bitech Publications*, Montreal, 1996, pp. 565–680.

47. Indraratna, B. and Vafai, F.: Analytical model for particle migration within base soil – filter system. *Journal of Geotechnical and Geoenvironmental Engineering*, ASCE, 1997, Vol. 123, No. 2, pp. 100–109.

48. Indraratna, B. and Raut, A.K.: Enhanced criterion for base soil retention in embankment dam filters. *Journal of Geotechnical and Geoenvironmental Engineering*, ASCE, 2006, Vol. 132, No. 12, pp. 1621–1627.

49. Indraratna, B., Raut, A.K. and Khabbaz, H.: Constriction-based retention criterion for granular filter design. *Journal of Geotechnical and Geoenvironmental Engineering*, ASCE, 2007, Vol. 133, No. 3, pp. 266–276.

50. Nguyen, V.T., Indraratna, B. and Rujikiatkamjorn, C.: Assessing the potential of internal erosion and suffusion of granular soils. *Journal of Geotechnical and Geoenvironmental Engineering*, ASCE, 2010, (Accepted, September 2010).

51. Kenney, T.C. and Lau, D.: Internal stability of granular filters. *Canadian Geotechnical Journal*, 1985, Vol. 22, No. 2, 215–225.

52. NRCS: Gradation design of sand and gravel filters. Part 633 National Engineering Handbook, 1984, *Natural Resources Conservation Services*, United States Department of Agriculture, Washington, DC.

53. Kozeny, J.: Ueber kapillare leitung des wassers im boden. Sitzungsberichte der Akademie der Wissenschaften, *Wien*, 1927, Vol. 136, No. 2a, pp. 271–306.

54. Carman, P.C.: Determination of the specific surface of powders I. *Transactions on Journal of the Society of Chemical Industries*, 1938, Vol. 57, pp. 225–234.

55. Chapuis, R.P. and Aubertin, M.: On the use of the Kozeny-Carman equation to predict the hydraulic conductivity of soils. *Canadian Geotechnical Journal*, 2003, Vol. 40, No. 3, pp. 616–628.

56. Dullien, FAL, Porous Media: *Fluid Transport and Pore Structure*, 1979, Academic Press, New York.

57. Lowell, S. and Shields, J.E.: *Powder Surface Area and Porosity*, Chapman and Hall, 1991, London.

58. Kovacs, G.: *Seepage Hydraulics*, Elsevier Publishers, 1981, New York.

59. Chapuis, R.P. and Légaré, P.P.: *A simple method for determining the surface area of fine aggregates and fillers in bituminous mixtures in Effects of Aggregate and Mineral Fillers on Asphalt Mixture Performance*, ASTM STP 1147, 1992, Philadelphia, pp. 177–186.

60. Taylor, D.W.: *Fundamentals of Soil Mechanics*, John Wiley, 1948, New York.
61. Lambe, T.W. and Whitman, R.V.: *Soil Mechanics*, Wiley, 1969, New York.
62. Freeze, R.A. and Cherry, J.A.: *Groundwater.* Prentice Hall, 1979, New Jersey.
63. Xu, P. and Yu, B.: Developing a new form of permeability and Kozeny–Carman constant for homogeneous porous media by means of fractal geometry, *Advances in Water Resources*, 2008, Vol. 31, pp. 74–81.
64. Loudon, A.G.: The computation of permeability from simple soil tests. *Géotechnique*, 1952, Vol. 3, pp. 165–183.
65. ASTM, ASTM Annual Book of Standards, D4647: *Standard test method for identification and classification of dispersive clay soils by the pinhole test.* 2006, Vol. 04.08, Philadelphia.
66. Chang, W.J., Rathje, E.M., Stokoe II, K.H. and Hazirbaba, K.: In situ pore pressure generation behavior of liquefiable sand. *Journal of Geotechnical and Geoenvironmental Engineering*, ASCE, 2007, Vol. 133, No. 8, pp. 921–931.
67. Trani, L.D.O. and Indraratna, B.: Experimental investigations into subballast filtration behavior under cyclic conditions. *Australian Geomechanics Society Journal*, 2009.
68. Trani, L.D.O. and Indraratna, B.: The use of impedance probe for estimation of porosity changes in saturated granular filters under cyclic loading: calibration and application. *Journal of Geotechnical and Geoenvironmental Engineering*, ASCE, 2010, Vol. 136, No. 10, pp. 1469–1474.
69. Christie, D.: *Bulli Field Trial, Vertical and Lateral Pressure Measurement*, Presentation, Rail CRC Seminar, 2007, University of Wollongong, delivered 29 June.
70. Whitman, R.V.: *Stress-strain-time Behaviour of Soil in One-dimensional Compression*, Report R63-25, 1963, by MIT Department of Civil Engineering to U.S. Army of Engineer Waterways Experiment Station.
71. Miller, E.T.: *Stresses And Strains in an Array of Elastic Spheres*, Report R53-39, 1963, by MIT Department of Civil Engineering to U.S. Army Engineers Waterways Experiment Station.
72. Galjaard, P.J., Paute, J.L. and Dawson, A.R.: Recommendations for repeated load triaxial test equipment for unbound granular materials. in Flexible Pavements, *Gomes Correia*, A (ed), AA Balkema, Rotterdam, 1996, pp. 23–34.
73. Suiker, A.S.J. and de Borst, R.: A numerical model for the cyclic deterioration of railway tracks. *International Journal of Numerical Methods in Engineering*, 2003, Vol. 57, pp. 441–470.
74. Melan, E.: Theorie statisch unbestimmer systeme aus ideal-plastischen baustoff. *Sitzungsberichte der Akademie der Wissenschaften*, Wien, 1936, pp. 145–195.
75. Perzyna, P.: Fundamental problems in viscoplasticity. *Advance in Applied Mechanics*, 1966, Vol. 9, pp. 243–377.
76. Raut, A.K. and Indraratna, B.: Further advancement in filtration criteria through constriction-based techniques. *Journal of Geotechnical and Geoenvironmental Engineering*, ASCE, 2008, Vol. 134, No. 6, pp. 883–887.

Chapter 10

Field Instrumentation for Track Performance Verification

For designing new track structures and for reducing track maintenance costs, an understanding of the complex mechanisms of track deterioration is necessary. Most of the design methods prevalent in practice are based on conservative estimates of settlements and stress-transfer between the track layers. Due to complexities in the behavior of the composite track system consisting of rail, sleeper, ballast, sub-ballast and subgrade subjected to repeated rail traffic loading, the track design techniques are still far from advanced. In order to gain more insight into the stress-strain mechanism of the track substructure, a field trial was conducted on a section of instrumented railway track in the town of Bulli. The benefits of a geocomposite layer installed at the ballast-capping interface and the relative performances between moderately-graded recycled ballast and traditionally very uniform fresh ballast were also examined during this study. The design specifications for the instrumented track were provided by University of Wollongong and the field trial was sponsored by RailCorp, Sydney. The details of new equipment, field installation and monitoring procedures alongwith records of measurements are described in the following sections.

10.1 SITE GEOLOGY AND TRACK CONSTRUCTION

10.1.1 Site investigation

The site investigation was carried out to investigate the condition of subgrade and comprised of 8 test pits and 8 Cone Penetrometer tests. Test pits were excavated using Bobcat backhoe excavator to a maximum depth of 860 mm below the sleeper and the subgrade encountered was silty clay with shale cobbles and gravels. Longitudinal section of the track showing subsurface profile is shown in Figure 10.1.

Cone Penetrometer testing (sometimes referred as a Dutch Cone) was carried out using Electrical Friction Cone Penetrometer (EFCP). The high values of cone resistance (q_c) and friction ratio (R_f) obtained in EFCP tests as evident in Figures 10.2(a and b) revealed that the subgrade soil was stiff overconsolidated and of sufficient strength to support the train loads [2].

Bedrock was found at a depth of 2.3 m below the excavation level at centre of Section 4 and based on other EFCP test results, it was anticipated that its depth gradually increased towards Section 1. The bedrock was highly weathered sandstone having weak to medium strength [1].

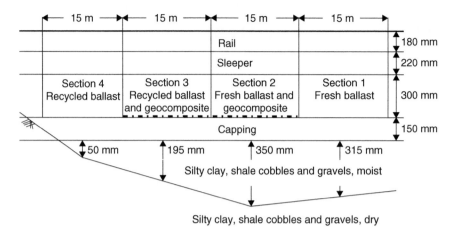

Figure 10.1   Longitudinal section of instrumented track at Bulli (adapted from Choudhury, [1]).

10.1.2 Track construction

Track reconditioning was required due to the inhomogeneity of the soil conditions along the track. This warranted a minimum 450 mm depth of excavation below the sleeper and proof rolling at the exposed surface, and involved excavation near Section 4 (Fig. 10.1). The 150 mm thick sub-ballast layer was placed in compliance with Australian standards [3] with cross fall of $1V:30H$. Then a 300 mm thick ballast layer was placed on the top of capping layer.

The track was constructed between two turnouts at Bulli along the New South Coast. The total length of the instrumented track section was 60 m and was divided into four sections, each of 15 m length. Fresh and recycled ballast were used at sections 1 and 4, respectively without inclusion of a geocomposite layer, while sections 2 and 3 were built by placing a geocomposite layer at the base of the fresh and recycled ballast, respectively. A layer of bi-axial geogrid was placed over the non-woven polypropylene geotextile to form the geocomposite as shown in Figure 10.3. The settlement pegs and displacement transducers were installed at the centre of each section whereas pressure cells were installed at locations 1C and 1D in Section 1 as shown in Figure 10.4(a). Figures 10.4(b) and 10.4(c) show the schematic diagram of a ballasted track bed with and without the inclusion of a geocomposite layer. Concrete sleepers were used in the test track.

The overall track bed thickness was 450 mm including a ballast layer of 300 mm and a capping layer of 150 mm in thickness. The particle size, gradation, and other index properties of fresh ballast used at the Bulli site were in accordance with the Technical Specification of RailCorp, Sydney [5] which represents sharp angular coarse aggregates of crushed volcanic basalt (latite). Recycled ballast was collected from spoil stockpiles of a recycled plant commissioned by RailCorp at their Chullora yard near Sydney. The finest fraction (less than 9.5 mm) was removed by screening (i.e. $d_{min} = 9.5$ mm; see Table 10.1). The capping material was comprised of sand-gravel

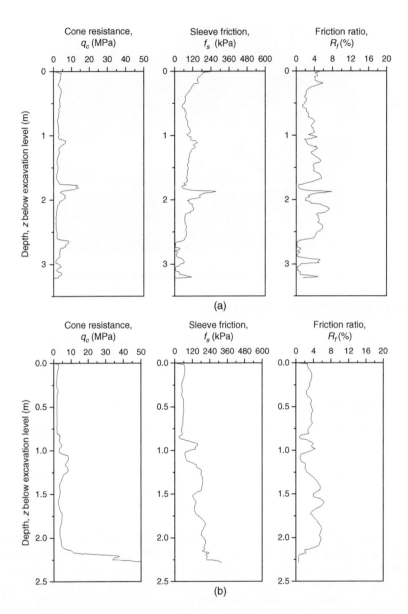

Figure 10.2 (a) EFCP test record at centre of Section 2 (adapted from Choudhury, [1]) and (b) EFCP test record at centre of Section 4 (adapted from Choudhury, [1]).

mixture. The particle size distribution of fresh ballast, recycled ballast and the capping (sub-ballast) materials are shown in Figure 10.5. Table 10.1 shows the grain size characteristics of fresh ballast, recycled ballast and the capping materials used in the Bulli instrumented track.

Figure 10.3 Placement of geocomposite over capping layer.

10.2 FIELD INSTRUMENTATION

To accurately measure cyclic stresses and deformations in the track, robust and high precision instruments were used at the site. The details of these instruments are given below.

10.2.1 Pressure cells

Two important prerequisites should be maintained when attempting to measure stresses in soils [6]:

(i)   Inclusion of the measuring device must not alter the actual stress field in the soil.
(ii) The measuring device must respond to the applied stress conditions in a fashion identical to the material in which it is embedded.

The measurement of stresses inside a deforming soil mass, ballast mass in particular, is therefore a challenging task. In the present study, the vertical and horizontal stresses developed in the track bed under repeated wheel loads were measured by pressure cells. The pressure cells were rapid-response hydraulic earth pressure cells with grooved thick active faces based on semi-conductor type transducers. Several factors, including the aspect ratio and size of cell, placement effects, corrosion and temperature affect measurements [7, 8, 10–13]. In accordance, relatively thin but robust pressure cells made of stainless steel (thickness 12 mm, diameter 230 mm) were adopted.

The pressure cells were installed by excavating beneath the sleeper up to the bottom of the capping layer and then backfilled at the appropriate levels, with care taken to avoid any damage during placement and subsequent material compaction. The cells were designed for minimum sensitivity to temperature (temperature range of −20°C to +80°C). In house calibration was carried out by the manufacturer, and the cell output at zero pressure was recorded before installation and load application.

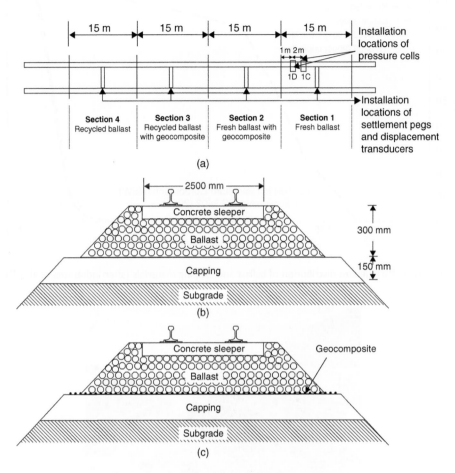

Figure 10.4 (a) Details of instrumented track at Bulli (b) section of ballasted track bed (c) section of ballasted track bed with geocomposite layer at the ballast-capping interface (after Indraratna et al., [4]).

Table 10.1 Grain size characteristics of ballast and capping materials (after Indraratna et al., [4]).

Material	d_{max} (mm)	d_{min} (mm)	d_{10} (mm)	d_{30} (mm)	d_{50} (mm)	d_{60} (mm)	C_u	C_c
Fresh Ballast	75.0	19.0	24.1	29.1	35.0	36.1	1.5	1.0
Recycled Ballast	75.0	9.5	23.1	31.5	38.0	41.5	1.8	1.0
Capping	19.0	0.05	0.07	0.17	0.26	0.35	5.0	1.2

The pressure cells were placed in a staggered pattern as shown in Figure 10.6. While vertical stresses were measured at three different levels i.e. sleeper-ballast, ballast-capping and capping-subgrade interfaces, horizontal stresses were measured only at two levels, i.e. sleeper-ballast and ballast-capping interfaces mainly due to budget limitations. Pressure cells were installed under the rail and at the bottom edge of

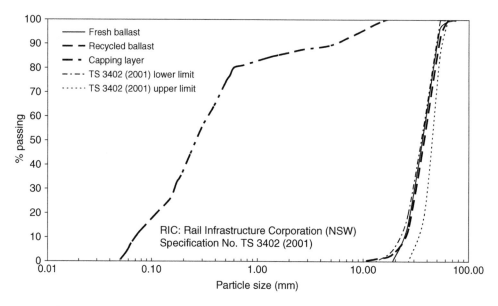

Figure 10.5 Particle size distribution of ballast and capping materials (after Indraratna et al., [4]).

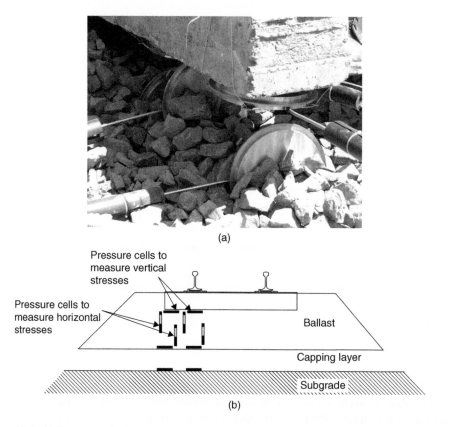

Figure 10.6 (a) Pressure cells for measuring stresses in the track bed (b) schematic diagram showing installation of vertical and horizontal pressure cells (adapted from Indraratna et al., [4]).

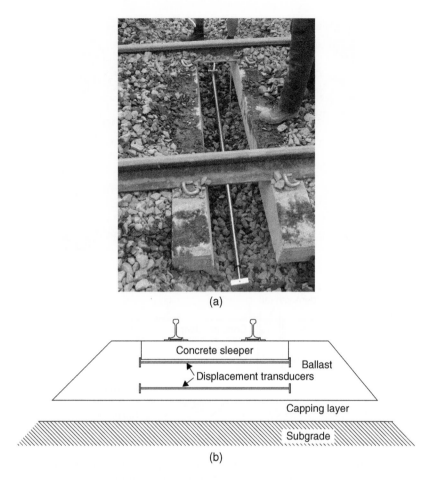

(a)

(b)

Figure 10.7 (a) Displacement transducers for measuring lateral deformations in the track bed, (b) schematic diagram showing installation of vertical and horizontal pressure cells (adapted from Indraratna et al., [4]).

sleeper near each interface. A total of 20 pressure cells were installed to record the vertical and horizontal stresses.

10.2.2 Displacement transducers

To measure vertical and horizontal deformations of ballast, settlement pegs and displacement transducers were installed in different track sections. The use of displacement transducers is an established practice for measuring vertical displacements [9]. In this field trial, special purpose displacement transducers were used to measure the transient horizontal track movements. These potentiometric transducers were protected inside 2.5 m long stainless steel housing, which consisted of two tubes that can slide over each other with 100 mm × 100 mm end caps as anchors while providing protection from moisture ingress and damage under harsh track conditions (Fig. 10.7a).

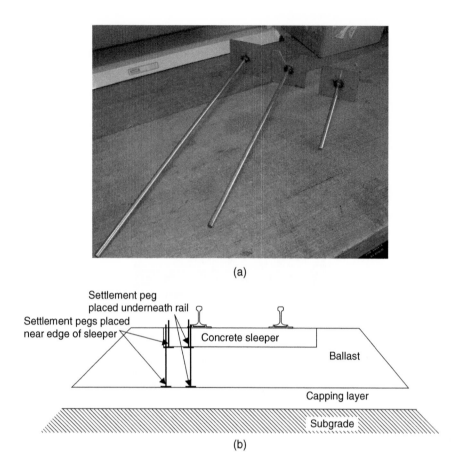

(a)

(b)

Figure 10.8 (a) Settlement pegs for measuring vertical deformations in the track bed, (b) schematic diagram showing installation of vertical settlement pegs (adapted from Indraratna et al.,[4]).

The typical arrangement of displacement transducers is shown in Figure 10.7(b). Displacement transducers were installed both at the sleeper-ballast and ballast-capping interfaces to measure the horizontal track deformations. Data loggers were connected to displacement transducers to obtain a continuous record of permanent track deformations.

10.2.3 Settlement pegs

Track deformation is considered to be a primary indicator for predicting track strength, life, and quality. Excessive deflection causes accelerated movements and breakage of ballast. To measure vertical and horizontal deformations of ballast, settlement pegs and displacement transducers were installed in different track sections. The settlement pegs consisted of 100 mm × 100 mm × 6 mm stainless steel base plates attached to 10 mm diameter stainless steel rods with length matching for burial in track layers (Fig. 10.8a). The typical arrangement of settlement pegs is shown in Figure 10.8(b).

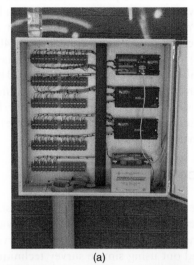

(a)

(b)

Figure 10.9   (a) Control box equipped with data acquisition system, (b) flexible conduits connected to data acquisition system.

The settlement pegs were installed at sleeper-ballast and ballast-capping interface at all sections. To measure the settlement of subgrade soil, settlement pegs were also installed at the capping-subgrade interface in Section 1. The settlement pegs were also placed under the rail and beneath the edge of sleeper to study the variation of deformation along the track section.

10.2.4   Data acquisition system

Electric cables were run through flexible conduits along the ballast shoulder and under the track at the central location, and connected to an automated data logger in a control box mounted on a signal box adjacent to the track (Figs. 10.9a and b).

To record the maximum values of pressure transmitted from the sleeper through the ballast, pressure cells were connected directly to the data logger and triggering was carried out manually for each train. A maximum of eight cells could be connected to the data logger which could operate at a frequency of 40 Hz. While these results appear to be successful, it is clear that the maximum value of pressure transmitted from the sleeper was not always recorded. At a speed of 60 km/hr, a wheel will travel 0.4 m in 1/40th of a second, thus it could not be ascertained that the wheel would be over the instrumented sleeper at the time of recording. Therefore, the maximum values recorded for each train were taken as the best estimate of the maximum dynamic pressure from the wheel load.

10.3 DATA COLLECTION

The settlement pegs were surveyed immediately after installation and again after 2 days, then at weekly intervals for 3 weeks, monthly intervals for the next 3 months, 3 monthly intervals for the next 9 months and a final survey after 17 months. The measurements were carried out using simple survey techniques recording the change in the reduced level of the surface of each layer with time. The recording of horizontal deformations from data loggers was initially conducted on an hourly basis and later transferred to a daily record in the monitoring history. The data was downloaded from the data logger manually on a daily basis.

10.4 RESULTS AND DISCUSSION

The vertical and horizontal deformations were measured against time in the field. In order to establish a suitable correlation with other research methodologies, an appropriate scale of 'number of load cycles' is selected in addition to the 'time' scale. A relation between million gross tons (MGT) of rail traffic annually and number of cycles (N) could be used to determine number of load cycles [14]

$$C_m = \frac{10^6}{(A_t \times N_a)} \tag{10.1}$$

where, C_m = number of load cycles/MGT, A_t = axle load in tons, N_a = number of axles/load cycle.

Considering the annual traffic tonnage of 60 MGT and four axles per load cycle, an axle load of 25 tons gives 600,000 load cycles per MGT. Therefore results are plotted against both the time and number of load cycles as discussed below.

10.4.1 Vertical deformation of ballast both under rail and edge of sleeper

The vertical deformation of ballast layer both under the rail position (S_{vr}) and edge of sleeper position (S_{vs}) are obtained by deducting the vertical displacements of sleeper-ballast and ballast-capping interfaces. The vertical strains (ε_{vr}, ε_{vs}) of the ballast layer are obtained by dividing the vertical deformations (S_{vr}, S_{vs}) of the ballast layer by the initial layer thickness. The vertical deformations (S_{vr}, S_{vs}) and vertical strains (ε_{vr}, ε_{vs})

thus obtained are plotted against the time (t) and number of load cycles (N) as shown in Figures 10.10(a) and 10.10(b). It is observed that the vertical deformations (S_{vr}, S_{vs}) of ballast layer are highly non-linear under cyclic loading and are similar to observations reported in previous studies [15–17]. A rapid increase in vertical deformations

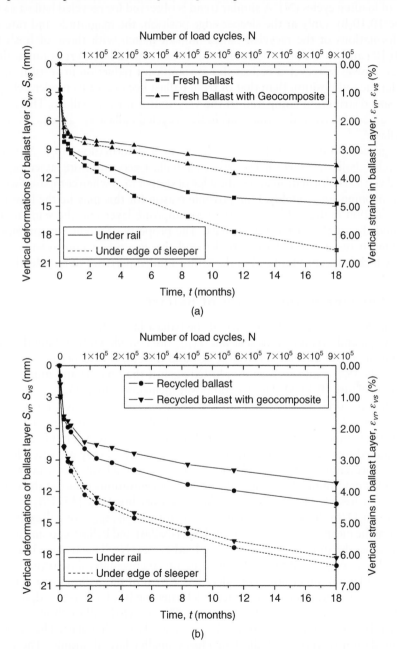

Figure 10.10 Vertical deformations (S_{vr}, S_{vs}) and vertical strains (ε_{vr}, ε_{vs}) measured in (a) fresh ballast (with and without geocomposite), (b) recycled ballast (with and without geocomposite), respectively (after Indraratna et al., [4]).

(S_{vr}, S_{vs}) is observed during first 120,000 load cycles, beyond which deformations (S_{vr}, S_{vs}) show marginal increase.

It is evident from Figure 10.10(a) that fresh ballast exhibits greater vertical deformation under the edge of sleeper (S_{vs}) compared to that under the rail (S_{vr}) for increasing number of loading cycles (N). A similar trend is observed for recycled ballast as shown in Figure 10.10(b). Only at the sleeper edge position, the magnitude and rate of vertical deformations of the recycled ballast almost match with those of fresh ballast (Figs. 10.10a and b). This can be attributed to the reduced lateral restraint at the edge of sleeper. However, recycled ballast shows significant reduction in vertical deformation under the rail position than that of fresh ballast because of its moderately-graded particle size distribution compared to the very uniform fresh ballast. Therefore, the average values of vertical deformations in the recycled ballast are always less than the fresh ballast.

The geocomposite layer decreases the vertical deformations (S_{vr}, S_{vs}) of fresh and recycled ballast. Nevertheless recycled ballast-geocomposite assembly shows increased vertical deformation under the edge of sleeper (S_{vs}) when compared with fresh ballast-geocomposite assembly. One possible reason for this may be the lower global interface friction mobilised between the geocomposite layer and the semi-angular or semi-rounded particles of recycled ballast. The property of angularity enables better interlocking between the ballast particles and the geogrid to improve the interface friction, which is less pronounced in the semi-rounded particles of recycled ballast.

10.4.2 Average deformation of ballast

To investigate the overall performance of the ballast layer, the average vertical deformation $(S_v)_{avg}$ and average vertical strain $(\varepsilon_1)_{avg}$ are considered by taking the mean of measurements taken under the rail $(S_{vr}, \varepsilon_{vr})$ and the edge of sleeper $(S_{vr}, \varepsilon_{vr})$ at each interface. The $(S_v)_{avg}$ and $(\varepsilon_1)_{avg}$, are plotted against the time (t) and number of load cycles (N) in Figure 10.11(a). The geocomposite inclusion reduces $(S_v)_{avg}$ and $(\varepsilon_1)_{avg}$ for both fresh and recycled ballast at a large number of cycles (N). Also, Figure 10.11(a) shows that the $(S_v)_{avg}$ and $(\varepsilon_1)_{avg}$ in the recycled ballast are less than the fresh ballast. The better performance of selected recycled ballast (if placed as a moderately-graded or well-graded mix) can also benefit from less breakage as they are often less angular thereby preventing corner breakage due to high contact stresses. Under a typical railway track environment, considerable stress concentrations occur at the corners of sharp angular fresh ballast particles, leading to corner breakage [17–19].

Figure 10.11(b) shows the average lateral deformation $(S_h)_{avg}$ of ballast (i.e., determined from the mean of measurements at sleeper-ballast and ballast-capping interfaces) plotted against the time (t) and number of load cycles (N). The average lateral strain of ballast layer $(\varepsilon_3)_{avg}$ is obtained by dividing the average lateral deformation $(S_h)_{avg}$ by the initial lateral dimension (considered as 2.5 m) of the ballast layer. The ballast layer exhibits an increase in average lateral deformation [i.e. lateral spread, represented by negative $(S_h)_{avg}$ and $(\varepsilon_3)_{avg}$] in all sections. The recycled ballast show significantly lower lateral deformation $(S_h)_{avg}$ and $(\varepsilon_3)_{avg}$ compared to fresh ballast. The moderately-graded gradation of recycled ballast produces smaller lateral strains. The inclusion of geocomposite in fresh ballast decreases $(S_h)_{avg}$ and $(\varepsilon_3)_{avg}$ significantly, however inclusion of the same in the recycled ballast shows a negligible effect on $(S_h)_{avg}$ and

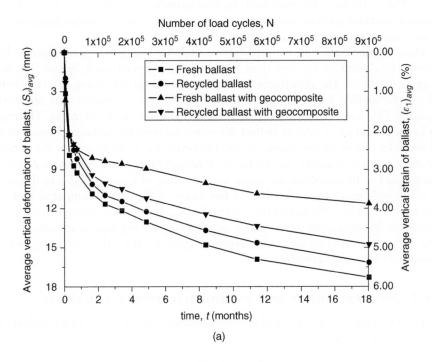

(a)

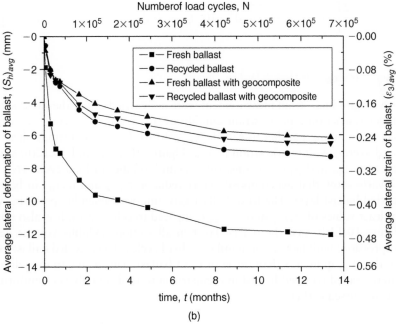

(b)

Figure 10.11 (a) Average vertical deformation $(S_v)_{avg}$ and average vertical strain $(\varepsilon_1)_{avg}$ (b) average lateral deformation $(S_h)_{avg}$ and average lateral strain $(\varepsilon_3)_{avg}$, of the ballast layer (after Indraratna et al., [4]).

$(\varepsilon_3)_{avg}$. This is due to highly frictional, angular particles of fresh ballast which develop increased interface friction with the geocomposite layer in the lateral direction, thus resisting lateral movement to a greater extent.

More significantly, the recycled ballast stabilised with the geocomposite layer exhibits $(S_h)_{avg}$ and $(\varepsilon_3)_{avg}$ less than those of unreinforced fresh ballast (i.e. without geosynthetics). The effectiveness of geocomposite in stabilising recycled ballast under cyclic loading has also been confirmed by laboratory triaxial tests reported in chapter 5. This has a significant bearing on the maintenance of rail tracks. The reduction in the lateral movement of ballast decreases the need for additional layers of crib and shoulder ballast during maintenance. However, questions related to the potential reduction in track drainage due to use of a considerably more well-graded recycled ballast needs to be addressed for much higher values of C_u ($C_u > 2.5$). In this study, the moderately-graded recycled ballast has value of C_u of 1.8 compared to 1.5 of more uniform fresh ballast, and this increase of C_u is not large enough to cause segregation during transport or to reduce permeability to any significant extent. Also, in the absence of fouling (screening removed particles finer than 9.5 mm), reduced permeability was not a concern.

10.4.3 Average shear and volumetric strain of ballast

The average shear strain $(\varepsilon_s)_{avg}$ and average volumetric strain $(\varepsilon_v)_{avg}$ of the ballast layer can be determined by [20]:

$$(\varepsilon_s)_{avg} = \frac{\sqrt{2}}{3}\left[\sqrt{((\varepsilon_1)_{avg} - (\varepsilon_2)_{avg})^2 + ((\varepsilon_2)_{avg} - (\varepsilon_3)_{avg})^2 + ((\varepsilon_3)_{avg} - (\varepsilon_1)_{avg})^2}\right]$$

(10.2)

$$(\varepsilon_v)_{avg} = (\varepsilon_1)_{avg} + (\varepsilon_2)_{avg} + (\varepsilon_3)_{avg}$$

(10.3)

Since longitudinal strain measurement were not carried out at the site due to time and budget restrictions, plane strain conditions are assumed (average intermediate principal strain acting parallel to rail, $(\varepsilon_2)_{avg} = 0$) to determine average shear strain $(\varepsilon_s)_{avg}$ and average volumetric strain $(\varepsilon_v)_{avg}$. Figures 10.12(a) and 10.12(b) show the variation of $(\varepsilon_s)_{avg}$ and $(\varepsilon_v)_{avg}$ against time (t) and number of load cycles (N). These results clearly show that geocomposite layer reduces $(\varepsilon_s)_{avg}$ and $(\varepsilon_v)_{avg}$ in both fresh and recycled ballast layer. The fresh ballast-geocomposite assembly performs well in terms of least values of $(\varepsilon_s)_{avg}$ and $(\varepsilon_v)_{avg}$ compared to other cases. It is also observed from Figure 10.12(b) that the ballast layer in all sections exhibits volume decrease (i.e. compression) with increase in number of load cycles. The recycled ballast exhibits $(\varepsilon_s)_{avg}$ and $(\varepsilon_v)_{avg}$ quite less than those of fresh ballast. This is due to the selection of moderately-graded recycled ballast in comparison to traditionally very uniform fresh ballast as discussed earlier.

10.4.4 In-situ stresses across different layers

Figure 10.13(a) shows that the maximum vertical cyclic stresses $(\sigma_{vr}, \sigma_{vs})$ and maximum horizontal cyclic stress $(\sigma_{br}, \sigma_{bs})$ recorded in the Section 1 due to the passage of train

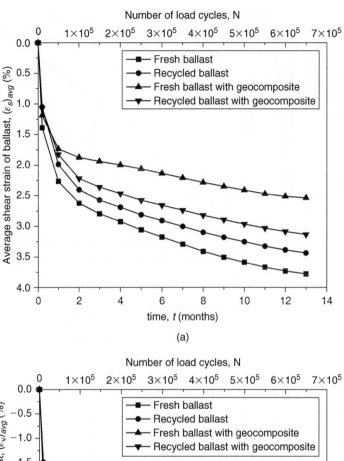

(a)

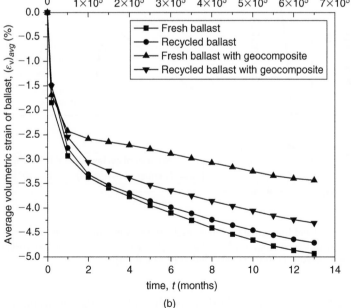

(b)

Figure 10.12 (a) Average shear strain $(\varepsilon_s)_{avg}$ and (b) average volumetric strain $(\varepsilon_v)_{avg}$, of the ballast layer (after Indraratna et al., [4]).

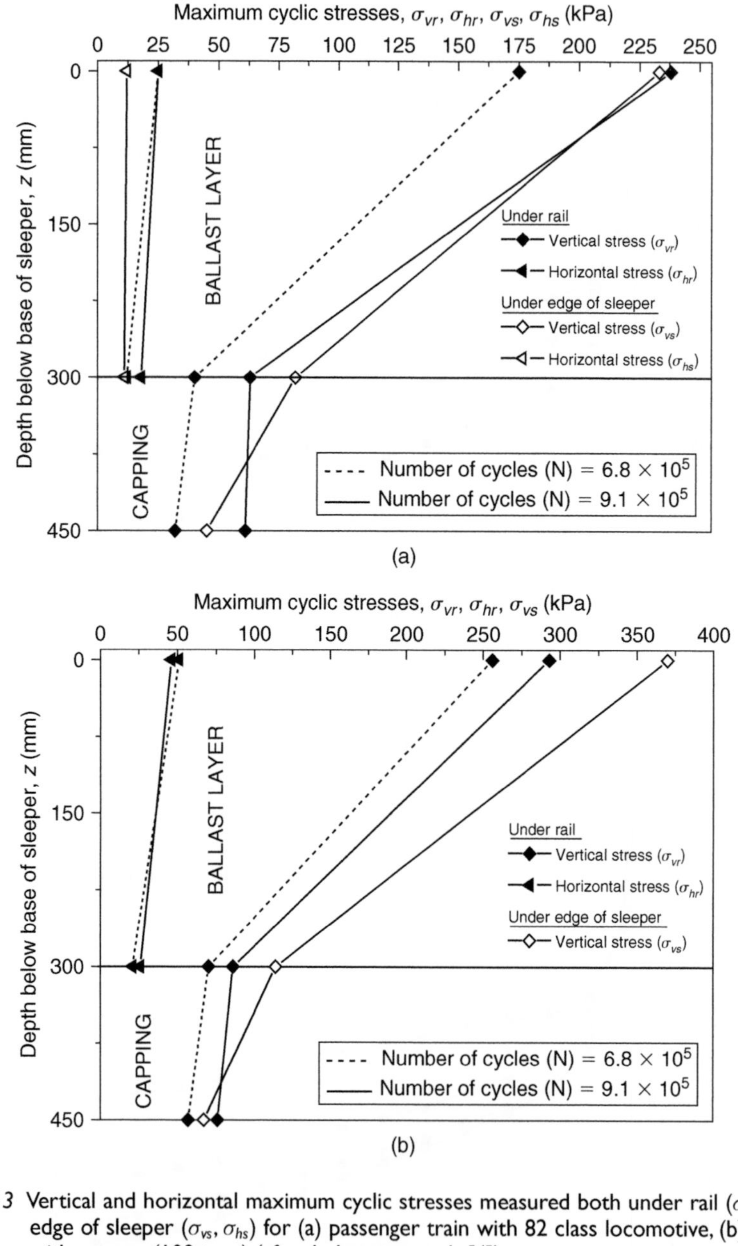

Figure 10.13 Vertical and horizontal maximum cyclic stresses measured both under rail (σ_{vr}, σ_{hr}) and edge of sleeper (σ_{vs}, σ_{hs}) for (a) passenger train with 82 class locomotive, (b) coal train with wagons (100 tons) (after Indraratna et al., [4]).

at 60 km/h (20.5 tons axle load) both under the rail and edge of sleeper position. It is observed that σ_{vr} and σ_{vs} are much higher than σ_{hr} and σ_{hs}, thus producing large shear strains in the rail track. Under normal rail track environment, there is significant lateral movement observed in the ballast layer. It is the large vertical stress and relatively small lateral (confining) stress that cause large shear strains in the track. The corresponding ease for horizontal spreading of ballast in the absence of sufficient confinement leads to increased vertical compression of the layer, as also confirmed by Selig and Waters [14]. Also, σ_{vr}, σ_{hr}, σ_{vs} and σ_{hs} increase with increase in number of load cycles leading to further degradation of track bed. It is evident that σ_{vr} and σ_{vs} decrease significantly with depth, while σ_{hr} and σ_{hs} decrease only marginally with depth. If a greater internal confining pressure on track could be applied by placing a geosynthetic layer within the ballast bed itself, lateral strains of ballast would also decrease. The track substructure is essentially self-supporting with minimal lateral restraints and the effective confining pressure is a key parameter governing the design of railway tracks with implications on ballast movement and associated track maintenance [17]. The study reported in chapter 5 has also clearly highlighted the increase in the track confinement as a result of placing the geosynthetic layer.

Figure 10.13(b) shows the maximum cyclic stresses (σ_{vr}, σ_{hr}, σ_{vs} and σ_{hs}) recorded in Section 1 due to the passage of a coal train with 100T wagons (25 tons axle load), where the stresses are measured both under the rail and edge of sleeper. As expected, maximum cyclic stresses (σ_{vr}, σ_{hr}, σ_{vs} and σ_{hs}) measured in the ballast and capping layer are higher due to a coal freight train than those attributed to a passenger train. It is anticipated that the greater axle load of the coal train imposes higher σ_{vr}, σ_{hr}, σ_{vs} and σ_{hs} resulting in greater deformation and degradation of ballast, implying the need for earlier track maintenance.

10.4.5 Comparison of current results with previous literature

The maximum vertical cyclic stresses (σ_{vr}) measured beneath the rail in the Bulli track are compared with results of analytical models and field studies reported in the literature as shown in Figure 10.14. Rose et al. [21] conducted trials at Transportation Technology Centre Inc (TTCI), and also used the software KENTRACK to validate the field data [22, 23]. They used a slightly different track bed configuration viz. 304.8 mm ballast layer underlain by 101.6 mm hot mix asphalt (HMA) layer and a wheel load of 200 kN (40 tons axle load). A wheel load of 145 kN, a ballast depth of 380 mm and a sub-ballast depth of 150 mm were considered in MULTA (three-dimensional equations of linear elasticity for multilayered systems), PSA (Fourier series for linear elastic behavior of materials) and ILLI-TRACK (finite element method employing nonlinear elastic material behaviour), as further elaborated by Adegoke et al., [24]. In addition, GEOTRACK (modified version of MULTA) was used with a wheel load of 146 kN and a ballast depth of 300 mm [14], while the Bulli field trial was based on 125 kN wheel load (i.e. 25 tons axle load).

While the authors recognise the limitations of a direct comparison, due to these variations in input parameters, an acceptable match could be found with the results of this study, the field data and analytical predictions. In chapter 12, records of field measurements are compared with the predictions of finite element analysis employing PLAXIS. The track instrumentation scheme employed in the present study leads to significant understanding of the stress-transfer and strain accumulation mechanisms. Field results demonstrate the potential benefits of using geocomposite (for

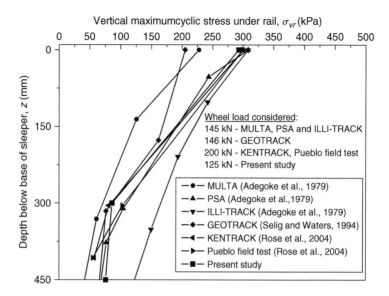

Figure 10.14 Comparison of vertical maximum cyclic stresses (σ_{vr}) measured under the rail at Bulli with analytical predictions (after Indraratna et al., [4]).

stabilising fresh and recycled ballast) in railway track with obvious implication on reduced maintenance costs.

REFERENCES

1. Choudhury, J.: Geotechnical Investigation Report for Proposed Bulli Track Upgrading between 311 & 312 Turnouts: Dn track 71.660~71.810 km, Up track 71.700~71.780 km. Memorandum, Engineering Standards & Services Division, Geotechnical Services, NSW, Australia, 2006.
2. Robertson, P.K.: Soil classification using the cone penetration test. *Canadian Geotechnical Journal*, Vol. 27, 1990, pp. 151–158.
3. T.S. 3422: *Standard for Formation Capping Material.* Rail Infrastructure Corporation of NSW, Sydney, Australia, 2001.
4. Indraratna, B., Nimbalkar, S., Christie, D, Rujikiatkamjorn, C. and Vinod, J.S.: Field assessment of the performance of a ballasted rail track with and without geosynthetics. *Journal of Geotechnical and Geoenvironmental Engineering*, ASCE, Vol. 136, No. 7, 2010, pp. 907–917.
5. T.S. 3402: *Specification for Supply of Aggregates for Ballast.* Rail Infrastructure Corporation of NSW, Sydney, Australia, 2001.
6. Talesnick, M.: Measuring soil contact pressure on a solid boundary and quantifying soil arching, *Geotechnical Testing Journal*, Vol. 28, No. 2, pp. 1–9.
7. Clayton, C.R.I., and Bica, A.V.S.: The design of diaphragm-type boundary total stress cells. *Geotechnique*, Vol. 43, No. 4, 1993, pp. 523–535.
8. Dunnicliff, J. Geotechnical Instrumentation for Monitoring Field Performance, John Wiley and Sons, New York, 1988.

9. Grabe P.J., and Clayton C.R.I.: Permanent deformation of railway foundations under heavy axle loading. *Proceedings of the Conference of the International Heavy Haul Association*, Dallas, May 2003, pp. 3.25–3.33

10. Richards, D.J., Clark, J., Powrie, W., and Heymann, G.: Performance of push-in pressure cells in overconsolidated clay, Proceedings of the Institution of Civil Engineers, *Geotechnical Engineering*, Vol. 160, No. GEI, 2007, pp. 31–41.

11. Selig, E.T.: A review of stress and strain measurement in soil. *Proceedings of the Symposium on Soil Structure Interaction*, University of Arizona, Tucson, 1964, pp. 172–186.

12. Selig, E.T.: Soil Stress Gage Calibration. *Geotechnical Testing Journal*, Vol. 3, No. 4, 1980, pp. 153–158.

13. Weiler, W.A. and Kulhawy, F.H.: Factors affecting stress cell measurements in soil. *J. of the Geotechnical Engineering Division*, ASCE, Vol. 108, No. GT12, 1982, pp. 1529–1548.

14. Selig, E.T., and Waters, J.M.: *Track Geotechnology and Substructure Management*, Thomas Telford, London. Reprint 2007.

15. Ionescu, D., Indraratna, B., and Christie, H.D.: *Behaviour of railway ballast under dynamic loads*. Proc. 13th Southeast Asian Geotechnical Conference, Taipei, 1998, pp. 69–74.

16. Jeffs, T., and Marich, S.: Ballast characteristics in the laboratory. *Conf. on Railway Engineering*, Perth, 1987, pp. 141–147.

17. Lackenby, J., Indraratna, B., and McDowel, G.: The role of confining pressure on cyclic triaxial behaviour of ballast. *Geotechnique*, Vol. 57, No. 6, 2007, pp. 527–536.

18. Housain, Z., Indraratna, B., Darve, F., and Thakur, P.: DEM analysis of angular ballast breakage under cyclic loading, *Geomechanics and Geoengineering: International Journal*, Vol. 2, No. 3, 2007, pp. 175–182.

19. Indraratna, B., Lackenby, J., and Christie, D.: Effect of confining pressure on the degradation of ballast under cyclic loading. *Geotechnique*, Vol. 55, No. 4, 2005, pp. 325–328.

20. Timoshenko, S.P., and Goodier, J.N.: *Theory of Elasticity*, McGraw Hill, New York, 1970.

21. Rose, J., Su, B., and Twehues, F.: Comparisons of railroad track and substructure computer model predictive stress values and in-situ stress measurements. *Proc. Annual conf. & Expo. American Railway Engineering and Maintenance-of-Way Association*, 2004, September, 17 pages.

22. Huang, Y.H., Lin, C., Deng, X., and Rose, J.: KENTRACK, a computer program for hot-mix asphalt and conventional ballast railway trackbeds. Asphalt Institute (Publication RR-84-1) and National Asphalt Pavement Association (Publication QIP-105), 164 pages, 1984.

23. Rose, J., Su, B., and Long, B.: KENTRACK: A railway trackbed structural design and analysis program. *Proc. Annual conf. & Expo. American Railway Engineering and Maintenance-of-Way Association*, 2003, October, 25 pages.

24. Adegoke, C.W., Chang, C.S., and Selig, E.T.: Study of analytical models for track support systems. *Transportation Research Record*, Vol. 733, 1979, pp. 12–20.

Chapter 11

DEM Modelling of Ballast Densification and Breakage

Granular materials consist of grains in contacts and the surrounding voids. The micromechnical behaviour of granular materials is, therefore, inherently discontinuous and heterogeneous. The macroscopic (overall or averaged) behaviour of granular materials is determined not only by how discrete grains are arranged in space, but also by the interactions between them. A constitutive model for granular media based on the continuum approach usually includes a number of material constants, which sometimes have no clear physical meaning. The ambiguous characters of the material constants based on continuum approaches may have their origin in the implicit expressions of the geometry of a packed assembly of particles. Thus, we could expect to analyse granular materials in a more rational manner if we were to make use of discrete element approaches in which the particle arrangement would be modeled explicitly.

In the particulate mechanics approach, the granular medium is treated as an assemblage of particles where the fundamental physical process needs to be captured at the particulate level. Such studies include different methods such as analytical, physical and numerical, considering assemblies of discs, spheres, ellipses or oval shaped rods. Analytical models are limited to regular arrays (simple cubic array, body centered cubic array, face centered cubic and cubical tetrahedral) of spheres and discs of uniform size applicable to simple loading conditions only [1–8]. Nonlinear and hysteresis stress-strain behaviour was found even in these regular arrays.

In physical tests, aluminium rods, steel balls, glass beads, photoelastic discs and plaster of Paris models have been tried to model the behaviour of granular materials. Photoelastic discs were proposed by Dantu [9] and Wakabayashi [10]. Measured force distributions were analyzed and reported by De Josselin de Jong and Verruijt [11] in an assembly of cylinders made of photoelastic material packed randomly in a 2-D simple shear apparatus. Oda and Konoshi [12] made a detailed study of the change in fabric due to shear. Aluminium rods and oval shaped discs have also been used in physical modeling of granular media [13, 14]. These physical tests provide accurate measurement of the displacement and contact forces. They are time consuming and expensive, and this test data is insufficient for an accurate micromechanical analysis.

Serrano and Rodriguez-Ortiz [15] developed a numerical model based on finite element scheme for assemblies of discs and spheres. They computed the increments of contact forces by incremental displacement of the particle centers by Hertzian contact compliance's for normal forces. Tangential forces were calculated as per theories of Mindlin and Deresiewicz [16] and Nayak [17]. In this scheme, the stiffness matrix needs to be updated every time a new contact is developed or lost. This involves a major computational effort.

Difficulties in the latter approaches lie in the process of simulating real granular materials, where an infinite number of particles with various shapes are assembled. Thus, when using discrete approaches, it is inevitable to model granular material as idealized particle assemblies. Even if we can handle only idealized models, discrete element approaches enable us to investigate the micromechanics of granular materials in a way that can not be achieved by any of the above approaches.

This chapter presents the description of micromechanical based Discrete Element Modeling (DEM) that has been implemented in computer software PFC^{2D} [18] along with a detailed investigation on the modelling of irregular ballast particle and densification and breakage of ballast during cyclic loading.

11.1 DISCRETE ELEMENT METHOD AND PFC^{2D}

Discrete element method (DEM) employs an explicit finite difference scheme and can handle particles of different shapes and sizes. The discrete numerical simulation predicts the overall behaviour of the assembly due to the cumulative effect of the particle to particle interactions in the assembly. This method is strictly based on the modeling of the granular media at the grain scale level. The advantage of this method is that it has the flexibility in facilitating the isolation of the micro-mechanical effects, and thereby influences of the loading configuration, particle parameters such as grain size distribution, shape, roughness and physical properties in relation to the mechanical behaviour of the assembly. The discrete numerical simulation is powerful in developing an insight into the micromechanical behaviour of the granular assembly to facilitate the formulation of a micromechanical based constitutive model. The numerical simulation scheme is related to the plane assemblies of discs and polygon shaped particles and assemblies of spheres. The discrete element concept has been used by various researchers to study the constitutive behaviour of granular materials by developing quasi-static models to capture the constitutive response [19–29].

Initially the Distinct Element Method (DEM) was first developed by Cundall [30] for rock mass problems, followed by a model of two dimensional assembles of circular discs [31, 32, 33, 34]. A FORTRAN – code (BALL) implements the DEM and is used extensively as an aid to develop a general constitutive model for granular materials based on micromechanical considerations [33, 35, 36]. Modified versions of BALL have been used by various researchers [37–45]. Later Cundall and Strack [34] and Strack and Cundall [46] developed a three dimensional version TRUBAL for modeling the mechanical behaviour of three dimensional assemblies of spheres. The structure of TRUBAL resembles that of BALL, and since then it has been used extensively [26, 47–52].

Particle Flow Code in 2-Dimension, PFC^{2D} [18] can simulate the movement and interaction of stressed assemblies of rigid circular particles using DEM. The distinct particles displace independently from one another and interact only at contacts or interfaces between the particles. The particles are assumed to be rigid and have negligible contact areas (contact occurs at a point). The behaviour at the contacts uses the soft contact approach, whereby the rigid particles are allowed to overlap one another

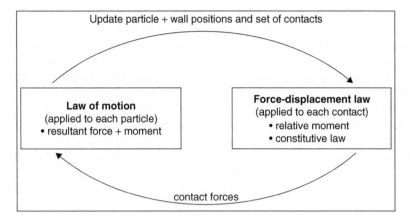

Figure 11.1   Calculation Cycle in PFC^{2D}/PFC 3D (Itasca, [18]).

at the contact points. The critical time step calculated for the time stepping algorithm in PFC^{2D} is not equal to the minimum Eigen-period of the total system because of the impractically lengthy computational time. PFC^{2D} uses a simplified procedure such that a critical time step is calculated for each particle and for each degree of freedom assuming that all degrees of freedom are uncoupled. The final critical time step is the minimum of all the calculated critical time steps. The actual time step used in any calculation cycle is then taken as a fraction of this estimated critical value. PFC^{2D} enables the investigation of features that are not easily measured in laboratory tests, such as the coordination numbers, interparticle contact forces and the distribution of normal contact vectors. Furthermore, it is possible to compose bonded particles into clusters and simulate fracture when the bonds break [18].

11.1.1   Calculation cycle

The calculation cycle in PFC^{2D} is a time stepping algorithm that consists of the repeated application of the law of motion to each particle, a force-displacement law to each contact, and a constant updating of wall positions. Contacts that exist between two balls or between a ball and a wall are formed and broken automatically during the course of a simulation. The calculation cycle is illustrated in Figure 11.1. At the start of each time step, the set of contacts is updated from the known particle and wall positions. The force-displacement law is then applied to each contact to update the contact forces based on the relative motion between the two entities at the contact and the contact constitutive model. Subsequently, the law of motion is applied to each particle to update its velocity and position based on the resultant force and moment arising from the contact forces and any body forces acting on the particle. Also, the wall positions are updated based on the specified wall velocities [18].

11.1.2   Contact constitutive model

The DEM keeps track of the motion of individual particles and updates any contact with neighbouring particles by using a constitutive contact law. The constitutive model acting at a particular contact consists of three parts: a stiffness model (consisting of a linear or a simplified Hertz-Mindlin Law contact model), a slip model, and a bonding model (consisting of a contact bond and/or a parallel bond model) [18].

The stiffness model relates the contact forces and relative displacements in the normal and shear directions via the force-displacement law. PFC^{2D} provides two types of contact stiffness model: a linear model and a simplified Hertz-Mindlin model. The linear contact model is defined by the normal and shear stiffnesses k_n and k_s (force/displacement) of the two contacting entities, which can be two balls or a ball and a wall. The normal stiffness is a secant stiffness, which relates the total normal force to the total normal displacement, while the shear stiffness is a tangent stiffness relating the increment of shear force to the increment of the shear displacement. The contact normal and shear stiffnesses K^n and K^s denoted by the upper case K are computed by assuming that the stiffnesses k_n and k_s of the two contacting entities act in series, therefore:

$$K^n = \frac{k_n^{[A]} k_n^{[B]}}{k_n^{[A]} + k_n^{[B]}} \tag{11.1}$$

$$K^s = \frac{k_s^{[A]} k_s^{[B]}}{k_s^{[A]} + k_s^{[B]}} \tag{11.2}$$

In the above equations, the superscripts [A] and [B] denote the two entities in contact. The simplified Hertz-Mindlin model is defined by the elastic properties of the two contacting balls: i.e. shear modulus G and Poisson's ratio v. When the Hertz-Mindlin model is activated in PFC^{2D}, the normal and shear stiffnesses are ignored and walls are assumed to be rigid. Hence, for ball to wall contacts, only the elastic properties of the ball are used and for the ball to ball contacts, the mean values of the elastic properties of the two contacting balls can be used. Tensile forces are not defined in Hertz-Mindlin model. Thus, the model is not compatible with any type of bonding model. It should also be noted that PFC^{2D} does not allow contact between a ball with the linear model or a ball with the Hertz model [18].

The slip model limits the shear force between two contacting entities. A ball and a wall can each be given a friction coefficient, and the friction coefficient at the contact, μ, is taken to be the smaller of the values of the two contacting entities. The slip model will be deactivated in the presence of a contact bond and will be automatically activated when the bond breaks. The maximum elastic shear force (F_{max}^s) that the contact can sustain before sliding occurs is given by:

$$F_{max}^s = \mu |F_i^n| \tag{11.3}$$

where, F_i^n is the normal force at the contact. If the shear force at the contact calculated by Eq. 11.3 exceeds this maximum elastic shear force, the magnitude of the shear force at the contact will be set equal to F_{max}^s. It should be noted that setting $\mu = 0$ means that the two contacting entities will slip at all times because elastic shear force cannot be sustained.

The bonding model in PFC^{2D} allows balls to be bonded together to form arbitrary shapes. There are two types of bonding model in PFC^{2D}: a contact-bond model and a

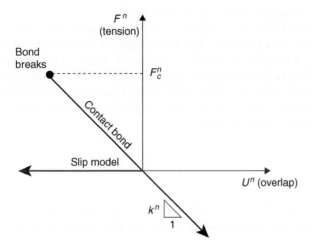

(a) Normal component of contact force

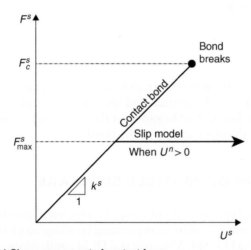

(b) Shear component of contact force

Figure 11.2 Constitutive behaviour for contact occurring at a point (Itasca, [18]).

parallel-bond model. The contact-bond model is a simple contact bond which can only transmit force and is defined by two parameters: the normal contact bond strength F_c^n and shear contact bond strength F_c^s. A contact bond can be envisaged as a point of glue with constant normal and shear stiffness at the contact point. The constitutive behaviour relating the normal and shear components of contact force and relative displacement for particle contact occurring at a point is shown in Figure 11.2. The contact bond will break if either the magnitude of the tensile normal contact force or the shear contact force exceeds the bond strength specified. Thus, the shear contact force is limited by the shear contact bond strength instead of the maximum elastic shear force given by Eq. 11.3. As a result, either the contact-bond model or the slip model is active at any given time at a contact [18].

A parallel bond (*PB*) approximates the physical behaviour of a cement-like substance joining the two particles. The constitutive behaviour of the parallel bond is similar to that of the contact bond, as shown in Figure 11.2. During loading, the parallel bonded particles develop force and moment within the bond due to a relative motion between the particles. The total force and moment associated with the parallel bond are denoted by $\overline{F}_i$ and $\overline{M}_i$ and can be resolved into normal and shear components with respect to the contact plane as:

$$\overline{F}_i = \overline{F}_i^n + \overline{F}_i^s \tag{11.4}$$

$$\overline{M}_i = \overline{M}_i^n + \overline{M}_i^s \tag{11.5}$$

where, $\overline{F}_i^n$, $\overline{F}_i^s$, $\overline{M}_i^n$, $\overline{M}_i^s$ are normal and shear component vectors respectively. The maximum tensile and shear stresses acting on the periphery of the bond are then given by:

$$\sigma_{\max} = \frac{-\overline{F}^n}{A'} + \frac{|\overline{M}|}{I}\overline{R} \tag{11.6}$$

$$\tau_{\max} = \frac{|\overline{F}^s|}{A'} \tag{11.7}$$

where, $\sigma_{\max}$ and $\tau_{\max}$ are the maximum tensile and shear stresses acting on the periphery of the bond, $\overline{F}^n$ and $\overline{F}^s$ are the normal and shear forces acting on the bond, $\overline{M}$ is the moment acting on the bond, $\overline{R}$ is the radius of the bond, and A' and I are the area and moment of inertia of the cross section of the bond.

11.2 MODELLING OF PARTICLE BREAKAGE

Very few studies have been carried out using the Discrete Element Method (DEM) to investigate the behaviour of ballast incorporating breakage mechanisms during monotonic and cyclic loading [53–56]. Most original DEM applications do not allow particle breakage [34].

However, various modelling techniques have been adopted by recent researchers to simulate particle breakage. McDowell and Harireche [57], Lu and McDowell [58] considered each particle as a cluster of bonded particles. The bonds which hold the particles in a cluster together can disintegrate during cyclic loading, and this represents breakage.

Another method of simulating particle breakage is to replace the particles fulfilling a predefined failure criterion with an equivalent set of smaller particles. This approach was adopted by Lobo-Guerrero and Vallejo [59], Hossain et al. [56]. Fig. 11.3 shows the agglomerate in single particle crushing test simulation [57]. They showed that it is possible to reproduce the average strength of agglomerates as a function of size and the correct statistical distribution of strengths for a given size, so that the strengths followed the Weibull distribution. McDowell and Harireche [57] then used these agglomerates

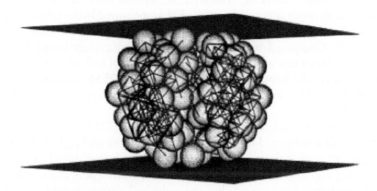

Figure 11.3 Final fracture of a typical 0.5 mm diameter agglomerate, showing intact contact bonds (McDowell and Harireche, [58]).

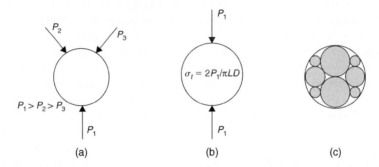

Figure 11.4 Idealisation of the induced tensile stress and arrangement of the produced fragments (Lobo-Guerrero and Vallejo, [59]).

to model one dimensional compression tests on silica sand. The results from these simulations showed that yielding coincided with the onset of bond fracture. This is shown to be consistent with the hypothesis by McDowell and Bolton [60] that yielding is due to the onset of particle breakage.

Lobo-Guerrero and Vallejo [59] developed a method to model particle crushing in two-dimensional simulations. In their method, they assumed that the breakage criterion applies only to a particle having a coordination number smaller than or equal to three. The real loading configuration (Fig. 11.4a) is equivalent to that obtained in a diametrical compression test (Fig. 11.4b). When the internal tensile stress of the disc is greater than its tensile strength, the disc is fractured into eight smaller discs, as shown in Fig. 11.4(c).

11.3   NUMERICAL SIMULATION OF MONOTONIC AND CYCLIC BEHAVIOUR OF BALLAST USING PFC^{2D}

Lim and McDowell [53] carried out a series of simulations on single particle crushing tests using agglomerates of bonded balls. In their simulations, they showed that the distribution of strengths correctly followed the Weibull distribution and that the size effect on average strength was consistent with that measured in the laboratory. Lim and McDowell [53] also simulated oedometer tests on crushable ballast using agglomerates of bonded balls. Compared to the experimental results, they found that the yield stress for the agglomerates was less than that for the real ballast. They indicated that the difference of results between laboratory tests and DEM simulations was due to the spherical shape of the agglomerates, which lead to columns of strong force in the simulated sample.

Box tests were simulated by Lim and McDowell [53] to study the mechanical behaviour of ballast subjected to traffic loading. Spheres and eight-ball clumps were used to represent each ballast particle to ascertain whether interlocking of ballast can be modelled and also whether the particle shape influences the resilient and permanent deformations. They found that the eight-ball clumps would provide particle interlocking and give more realistic mechanical behaviour under repeated loading. A similar conclusion was drawn by McDowell et al. [61] when they used both spheres and eight-ball cubic clumps in the simulation of large-scale triaxial tests. McDowell et al. [61] pointed out that dilation rather than contraction was observed at high confining pressures because breakage was not captured in their simulations, unlike the correct experimental observations reported by Indraratna et al. [62].

Lobo-Guerrero and Vallejo [59] studied the effect of crushing of railway ballast in a simulated track section using a circular disc to represent each single ballast particle. The simulated track sections were subjected to 200 load cycles via three simulates sleepers for particles. Numerical simulations were conducted on assembles of particles with and without crushing. Particle crushing was modeled based on the method developed by Lobo-Guerrero and Vallejo [59]. The permanent deformation was found to increase considerably when particle crushing was included. Moreover, particle crushing was concentrated underneath the sleepers (Fig. 11.5). The effect of particle shape was not considered in their simulations.

Hossain et al. [56] studied the effect of angular ballast breakage on the stress strain behaviour of railway ballast under different confining pressures using biaxial test simulations. Two dimensional angular shaped clumps were used to model particle interlocking (Fig. 11.6). Similar to the method introduced by Lobo-Guerrero and Vallejo [54], particle crushing was simulated by releasing discs from the clump when the internal tensile stress induced by contact forces was greater than or equal to 10 MPa. Hossain et al. [56] showed that particle breakage had significant influence on both the axial and the volumetric strain (Fig. 11.7).

Lu and McDowell [55] modelled ballast using clumps of spheres (Fig. 11.8). They formed the clumps from ten balls in a tetrahedral shape. They modelled interlocking and breaking of very small asperities using weak parallel bonds between clumps. Using those clumps, they simulated static and cyclic triaxial tests and compared the DEM results with existing experimental data. They also simulated tests using uncrushable clumps to highlight the important role of asperity abrasion. However, the number of cycles simulated was limited to 100 cycles. Moreover, the simulated ballast particles were of the same shape and sizes while ballast used in the experiment are of various shapes and sizes.

Indraratna et al. [62] introduced a novel approach to model an identical two dimensional (2D) projection of the ballast particles. The ballast particles were

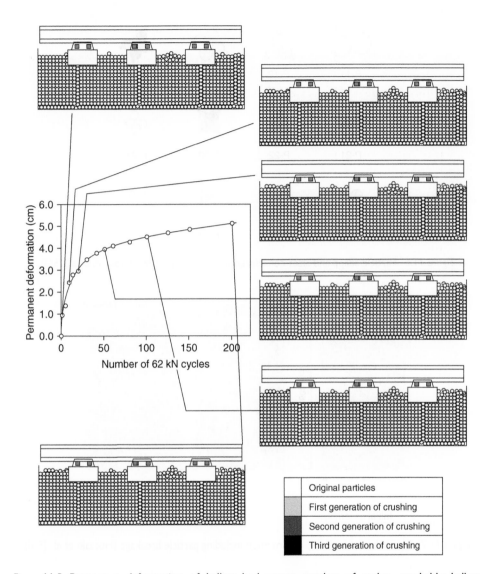

Figure 11.5 Permanent deformation of ballast bed versus number of cycles, crushable ballast (Lobo-Guerrero and Vallejo, [59]).

separated into five different sieve sizes, (i) passing 53 mm and retaining 45 mm, (ii) passing 45 mm and retaining 37.5 mm, (iii) passing 37.5 mm and retaining 31.5 mm, (iv) passing 31.5 mm and retaining 26.5 mm, and (v) passing 26.5 mm and retaining 19 mm.

Fifteen representative ballast particles (3 from each sieve size range) of different shapes (almost rectangular, circular and triangular) were selected. The images of each of the selected ballast particles were taken from the same elevation (Table 11.1). All of these images were then imported into AutoCAD in a single layer. The scale of the

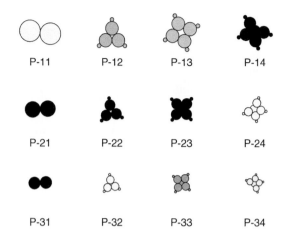

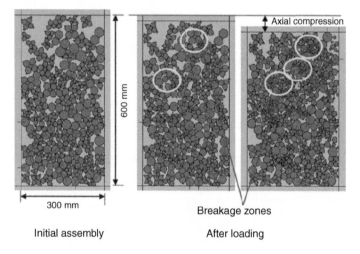

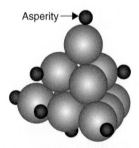

Figure 11.6 Particle shape and sizes considered for DEM simulations (Hossain et al., [56]).

Figure 11.7 A snapshot of assembly deformation including particle breakage (Hossain et al., [56]).

Figure 11.8 Ten-ball triangular clump with eight small balls (asperities) bonded as a ballast particle model (Lu and McDowell, [55]).

Table 11.1 Representative ballast particles for the DEM simulation (Indraratna et al., [62]).

Passing 53 mm and retaining on 45 mm sieve		
Ballast particles		
PFC particles	R1	R2	R3
Passing 45 mm and retaining on 37.5 mm sieve		
Ballast particles		
PFC particles	W1	W2	W3
Passing 37.5 mm and retaining on 31.5 mm sieve		
Ballast particles		
PFC particles	Y1	Y2	Y3
Passing 31.5 mm and retaining on 26.5 mm sieve		
Ballast particles		
PFC particles	G1	G2	G3
Passing 26.5 mm and retaining on 19 mm sieve		
Ballast particles		
PFC particles	N1	N2	N3

drawing was selected in such a way that the images represent the true size of the ballast. These images were then filled with tangential circles in another layer and every circle was assigned an identification number (ID). After this the ID, the radius and the central coordinates of each circular particle was extracted from AutoCAD, in order to

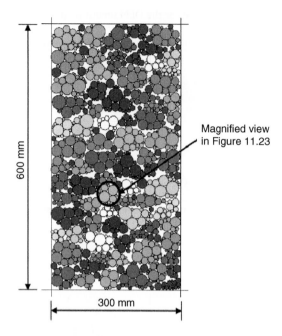

Figure 11.9 Initial assembly for the cyclic biaxial test simulations in PFC^{2D} (Indraratna et al., [62]).

generate 'Balls' in PFC^{2D}. An inventory of these particles was then created in AutoCAD by converting the group of circles representing a single ballast particle into a 'Block' i.e. a single object made from a combination of a number of objects. This procedure was used to model the irregular ballast particles. Table 11.1 shows the image of typical ballast particles created for the numerical simulation. These irregular particles were assigned names such as R1, R2, R3; W1, W2 etc. according to the specific colour scheme of the ballast.

11.3.1 Cyclic biaxial test simulations

A typical sample considered for the cyclic biaxial tests is shown in Figure 11.9. The properties assigned for the particles in this simulation are tabulated in Table 11.2. A linear contact model was used for the numerical simulation program. The biaxial sample generated was given a confining pressure of 60 kPa. Then the specimen was cycled to bring it into an isotropic stress state, until the ratio of mean unbalanced force to mean contact force, or the ratio of maximum unbalanced force to maximum contact force reached 0.005.

In order to prevent the particle breakage during compaction, the ballast particles were treated as clumps during the isotropic stress installation. After the isotropic stress state, the clumps were released and parallel bonds (PB) were installed to represent breakable particles. The two side walls that can be numerically controlled by a servo, were used to apply a constant confining pressure (σ'_3) of 60 kPa.

Table 11.2 Micromechanics Parameters used in the DEM Simulations (after Indraratna et al., [62]).

Micromechanics parameters	Values
Particle density (kg/m³)	2500
Radius of particles (m)	$16 \times 10^{-3} - 1.8 \times 10^{-3}$
Interparticle & wall friction	0.25
Particle normal & shear contact stiffness (N/m)	3×10^8
Side wall Stiffness (N/m)	3×10^7
Top & bottom wall stiffness (N/m)	3×10^8
Parallel bond radius multiplier	0.5
Parallel bond normal & shear stiffness (N/m)	6×10^{10}
Parallel bond normal & shear strength (N/m²)	5×10^6

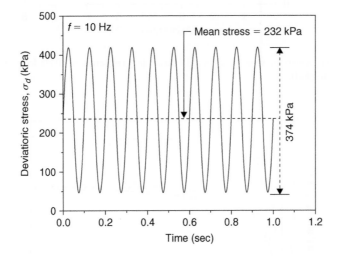

Figure 11.10 A typical deviatoric stress (σ_d) applied for 10 Hz frequency (Indraratna et al., [62]).

A subroutine was developed in PFC^{2D} to apply a stress-controlled cyclic biaxial test at the desired frequency (f) and amplitude of cyclic loading. Figure 11.10 shows a typical sinusoidal loading curve representing the applied cyclic deviatoric stress (q_{cyc}) with time (mean $= 232$ kPa; amplitude $= 374$ kPa) for a frequency of 10 Hz.

Figure 11.11 shows a typical response of the sample presented as an axial deviatoric stress (σ_d) versus axial strain (ε_a). It is evident that the response of the ballast during cyclic loading is similar to that obtained in the laboratory [63]. Cyclic tests at a frequency of 10 Hz, 40 Hz were simulated, and the corresponding data including axial strain (ε_a) number of cycles (N), bond breakage (B_r), and axial deviatoric stress (σ_d) were recorded for every cycle.

Figure 11.12 shows the comparison of ε_a obtained in the DEM with the experimental results carried out using large-scale cyclic triaxial apparatus [62]. The DEM results are in acceptable agreement with the laboratory data. After calibrating the model, it has been used to investigate the ballast behaviour under various combinations of

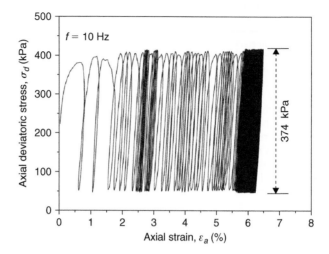

Figure 11.11 Axial deviatoric stress (σ_d) vs axial strain (ε_a) response at 10 Hz frequency (Indraratna et al., [62]).

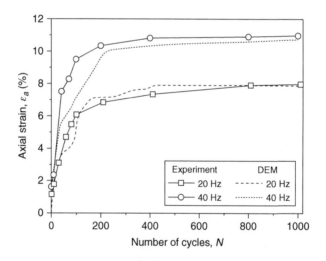

Figure 11.12 Calibration of cyclic biaxial test results (Thakur et al., [64]).

frequency and confining pressure followed by the study of micromechanical behaviour pertinent to breakage.

Figure 11.13 presents the variation of axial strain (ε_a) at various frequencies (f) with number of cycle (N) obtained from the DEM simulations in contrast to laboratory triaxial data. It is evident that the DEM simulation that has captured the ε_a response during cyclic loading is similar to the laboratory findings. This proves beyond doubt that the frequency of cyclic loading (f) has a significant influence on the ε_a.

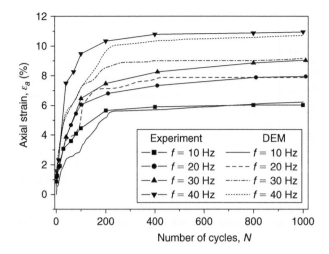

Figure 11.13 Comparison of axial strain (ε_a) observed in the Experiment and in the DEM (after Indraratna et al., [62]).

11.4 BREAKAGE BEHAVIOUR

Figure 11.14 shows the cumulative bond breakage (B_r), defined as a percentage of bonds broken compared to the total number of bonds at different f and N. The magnitude of B_r increases with the increase in f and N. Most of the bond breakages are observed during the initial cycles of loading (e.g. 200 cycles) causing a higher permanent ε_a (Fig. 11.14). Once the bond breakage ceases, only a marginal increase in ε_a occurs. This clearly highlights that particle degradation is one of the major source responsible for permanent deformation. Similar observations have been made by Harireche and McDowell (2003). Figure 11.15 illustrates the relationship between B_r and ε_a at various f (10 Hz–40 Hz). It is evident that there exist a linear relationship between B_r and ε_a for different values of f.

Figure 11.16 shows variation of bond breakage (B_r) with f after 1000 cycles. The variation of B_r with f is found to be very similar to the variation of BBI observed from the laboratory data (Fig. 4.6). Although these two indices are different, they both measure the intensity of particle breakage. As expected B_r increases with f until $f = 20$ Hz and insignificant increase in B_r between $20\,\text{Hz} \le f \le 30\,\text{Hz}$ followed by drastic increase of B_r for $f > 30\,\text{Hz}$

Figure 11.17 illustrates the vertical permanent deformation in terms of ε_a with N at various σ_3'. It has been observed that ε_a increases with N at all σ_3'. However, ε_a decreases as σ_3' increases. For instance, the maximum ε_a of 18% is observed at $\sigma_3' = 10$ kPa. Increasing σ_3' to 30 kPa reduces ε_a to 11%. A further increase of σ_3' to 60 kPa resulted in ε_a further decreasing to 8%. Increasing σ_3' from 60 kPa to 120 kPa did not show much influence on ε_a. Elevating σ_3' to 240 kPa has further reduced ε_a to 6% which is only 25% less than that observed at $\sigma_3' = 60$ kPa.

Figure 11.18 shows the response of ε_v with N at various σ_3'. At very low σ_3' (e.g. 10 kPa), the ballast compresses during initial cycles (e.g. first 200 cycles) and

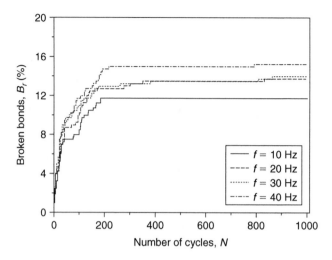

Figure 11.14 Effects of frequency (f) on bond breakage (B_r) with number of cycles (N) (after Indraratna et al., [62]).

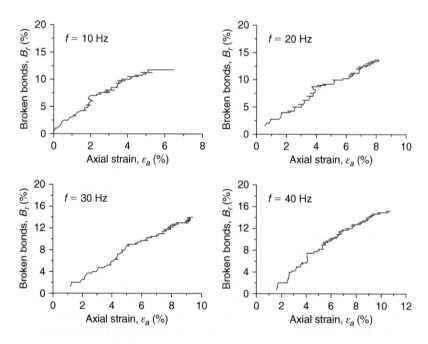

Figure 11.15 Trend of bond breakage (B_r) with axial strain (ε_a) at various frequencies (f) (after Indraratna et al., [62]).

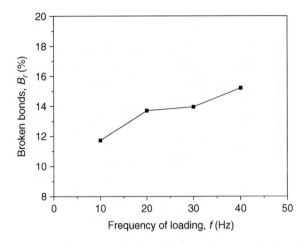

Figure 11.16 Breakage trend in DEM simulation with frequency (*f*) (after Indraratna et al., [62]).

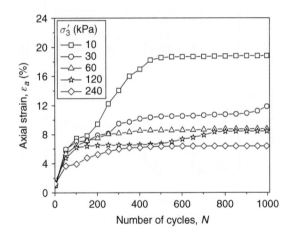

Figure 11.17 Variation of ε_a with *N* at different σ_3' after (Thakur et al., [65]).

then dilated causing higher permanent vertical deformation as shown in Figure 11.17. However, as σ_3' increases from 30 kPa to 240 kPa, the ballast compresses as *N* increases. Maximum compression observed at $\sigma_3' = 30$ & 60 kPa are around 3% and 4% respectively. Increasing σ_3' to 240 kPa results into a maximum volumetric compression of 4.5%

Figure 11.19 explains the particle breakage behaviour at various values of σ_3'. For $\sigma_3' < 30$ kPa, a very high B_r is observed. This is mainly caused by dilation of the assembly (Fig. 5.17). Indraratna et al. [66] categorized this zone as 'Dilatant, Unstable Degradation Zone' (DUDZ), and reported that degradation is attributed mainly to the shearing and attrition of angular projections due to excessive axial and radial strains in this zone. With further increase in σ_3', B_r is found to decrease, and it attains an

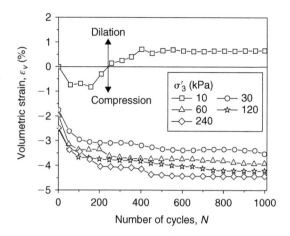

Figure 11.18 Variation of ε_v with N at different σ_3' after (Thakur et al., [65]).

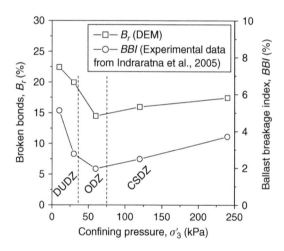

Figure 11.19 Particle breakage at various σ_3' and comparison of breakage trends observed in the DEM with the experiment after (Thakur et al., [65]).

optimum value in the range $30\,\text{kPa} < \sigma_3' < 75\,\text{kPa}$. This zone is named as the optimum degradation zone (ODZ). Within this zone of confining pressure, an optimum particle configuration (packing arrangement) is attained thereby significantly reducing the dilative behaviour of the assembly and ε_a decreases significantly. This shows that rail tracks can benefit through reduced maintenance costs by slightly increasing the lateral confining pressure (i.e., less settlement and degradation of ballast). For $\sigma_3' > 75\,\text{kPa}$, B_r starts increasing (Fig. 5.18), with a corresponding increase in ε_v and assigned a name CSDZ (Compressive, Stable, Degradation Zone) by Indraratna et al., [66]. The ε_a in this zone is not much reduced when compared to ODZ as optimum packing arrangement of the particles is already attained. Figure 11.19 also compares the bond breakage (B_r) with ballast breakage index (BBI) developed by Indraratna et al., [66]. Although these two

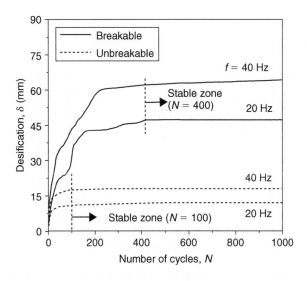

Figure 11.20 Comparison of cyclic densification (breakable and unbreakable particles) (Thakur et al., [67]).

indices are distinctly different, they both measure the intensity of particle breakage. It is interesting to see that DEM results have captured the same trends of breakage as those observed in the laboratory.

Figure 11.20 explains the role of particle breakage on cyclic densification of ballast with the number of cycles (N) at 20 and 40 Hz frequencies (f). It can be seen that the maximum densification observed for unbreakable particles is 12 and 17 mm at $f = 20$ and 40 Hz, respectively. In the case of breakable particles, the densification is around 45 and 67 mm, respectively, at the end of 1000 cycles (Fig. 11.20). Also, it is noted that the number of cycles required to reach stable permanent deformation depends on the breakability of the particles. For example, for unbreakable particles the stable zone is reached at around 100 cycles, whereas in case of breakable particles, is the stable zone occurs at around 400 cycles.

Figure 11.21 illustrates the effect of particle breakage, expressed in terms of broken bonds ($B_r\%$) on cyclic densification of ballast. Broken bonds has been defined as a percentage of bonds broken compared to the total number of bonds present in the initial assembly. It is noted that the shape of the densification curve and breakage curve is very similar. This signifies the fact that particle breakage has direct influence on the cyclic densification behaviour of ballast. Rapid deformation of the assembly in the initial cycles of loading for unbreakable particles (Fig. 11.21) is mainly governed by rolling and sliding of particles, however, in case of breakable particles, it is largely dominated by breakage. Once the particles are broken, they roll, slide and fill the nearby voids causing more densification. Consequently, more sample deformation is then observed. When particle breakage is ceased ($N > 400$), permanent deformation of the sample becomes almost stable. At $f = 20$ Hz, deformation is relatively constant after $N = 400$, however, it is still increasing for $f = 40$ Hz.

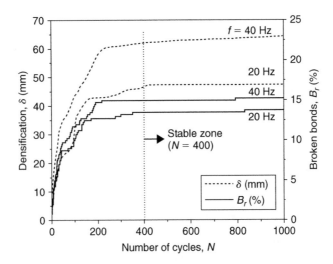

Figure 11.21 Effect of bond breakage on cyclic densification (after Thakur et al., [67]).

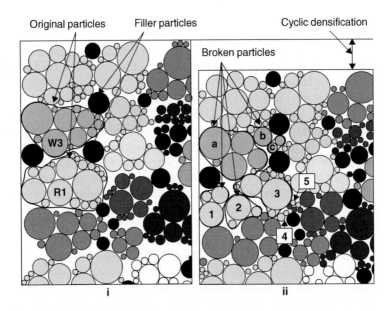

Figure 11.22 Portion of (i) Initial assembly and (ii) Assembly after 100 cycles at $f = 40$ Hz showing particle breakage and cyclic densification (Indraratna et al., [62]).

11.4.1 Micromechanical investigation of breakage

Figures 11.22(i) and (ii) illustrate a portion of initial assembly and assembly after 100 cycles of loading. These sets of data clearly highlight the particle breakage and the rearrangement of broken particles during cyclic loading. For example, as shown in Figure 11.21(ii), particle W3 (passing 45 mm and retaining on 37.5 mm sieve) has

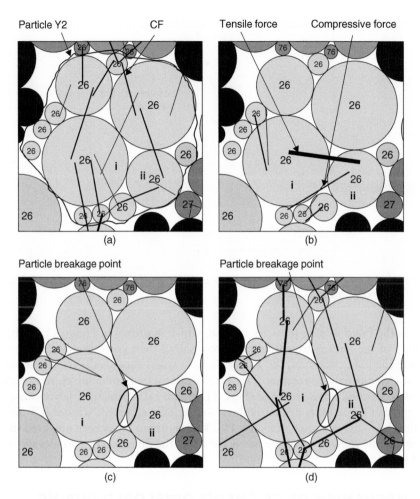

Figure 11.23 Details of particle condition (a) Contact force chains, and (b) Bond forces before bond breakage, (c) Bond forces and (d) Contact force chains after bond breakage at 40 Hz frequency during cyclic loading (Indraratna et al., [62]).

been broken into three pieces viz: a, b and c. Similarly, particle R1 (passing 53 mm and retaining on 45 mm sieve) has been broken into five pieces (Fig. 11.22(ii)). These broken particles eventually move to the void space in the assembly and cause permanent deformation. Thus breakage and subsequent rearrangement of broken particles contribute towards cyclic densification of the assembly as shown in Figure 11.22(ii).

Particle breakage is one of the major factors influencing the behaviour a rail track during cyclic loading by the passage of trains. An enlarged view of a typically irregular particle before and after breakage is shown in Figure 11.23.

The position of this particle can be identified from Figure 11.9. Figure 11.23 (a) shows the contact force (*CF*) chains for particle Y2 (passing 37.5 mm and retaining on 31.5 mm sieve) before breakage whereby the *CF* developed at this stage was compressive. The *CF* acting on particles Y2 (Fig. 11.23a) induce tensile and compressive bond

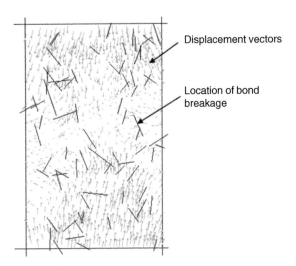

Displacement vectors

Location of bond breakage

Figure 11.24 Displacement vectors and location of bond breakage after 500 cycles at $f = 40\,\text{Hz}$ (Indraratna et al., [62]).

forces in the bond joining particles i and ii, which is shown in Figure 11.23 (b). The thickness of the lines representing the forces shows their corresponding magnitudes. As the cyclic load continues, the induced tensile stress exceeds the tensile strength of the particle causing it to fracture (Fig. 11.23c). Particles i and ii which were attached are now separated after breakage (Figs. 11.23c and d). Similar observations clearly highlight that breakage is primarily the result of induced tensile stresses. CF distribution after breakage is shown in Figure 11.23 (d). Further, it has been observed that the majority of the particle breakage has occurred in the direction of particles movement (Fig. 11.24).

11.5 MECHANISM OF CF CHAINS DEVELOPED DURING CYCLIC LOADING

Figure 11.25 explains the development of CF chains and associated bond breakage at different stages of cyclic loading. It can be seen that the major CF chains were developed in the major principal stress direction during 1st cycle of loading (Fig. 11.25a) attributed to good contacts between the particles. However, with increase in cyclic loading the contacts between the particles become weak due to bond breakage resulting in weaker CF chains in major principal stress direction (Fig. 11.25b). With further cycling the broken particles re-arrange, and become compacted halting further degradation, and this results in solid and more uniform CF chains in the major principal stress direction (Fig. 11.25c). This phenomenon clearly explains that the formation of CF chains in the assembly during cyclic loading is a dynamic process that is significantly influenced by the particle breakage.

As expected, displacement vectors of the ballast particles were convergent at the start of the cyclic loading (Fig. 11.26a). At this stage, very few particles were broken.

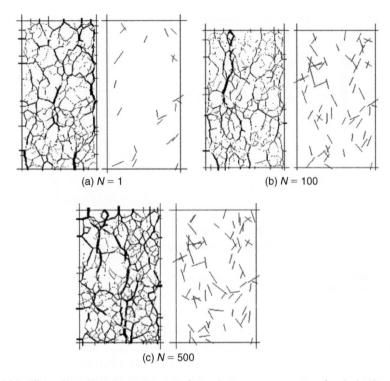

(a) $N = 1$ (b) $N = 100$

(c) $N = 500$

Figure 11.25   Effect of bond breakage on contact force chains at various stage of cyclic loading at $f = 40$ (after Indraratna et al., [62]).

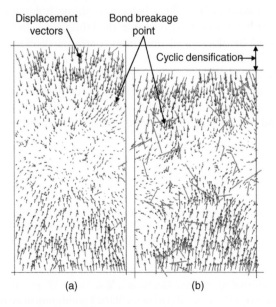

Figure 11.26   Displacement vectors and bond breakage (a) at 1st cycle of loading, (b) at 500 cycle of loading after (Thakur et al., [64]).

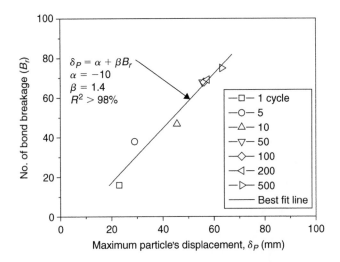

Figure 11.27 Relationship between particle breakage and maximum particles displacement (Thakur et al., [64]).

As the number of cyclic loading increased, more particles were subjected to breakage. Subsequently, the broken particles get the freedom to move around and densify into nearby voids (Fig. 11.26b). Furthermore, it is clearly observed that particles tend to move towards the major stress direction and away from minor stress direction. The displacement vectors highlight that the shearing of the particles are not along a particular plane as it is usually along an inclined plane in the case of fine grained soils under triaxial condition. The shear behaviour shows bulging of the sample. This bulging type of shear behaviour is also seen in the laboratory (Indraratna et al., [62]).

Number of bond breakage and maximum particles displacement vectors were recorded at various numbers of cycles to understand the effect of the bond breakage on the particle's displacement. These data plotted in Figure 11.27 clearly show that there exists a direct and linear relationship between particle breakage and particle displacement. This analysis confirms that when the bond breaks, particles get more freedom to slide and roll into the nearby voids causing increased overall displacement.

REFERENCES

1. Duffy, J. and Mendlin, R.D.: Stress-strain Relations and Vibrations of a Granular Medium, *Journal of Applied Mechanics Trans.* ASME, Dec., 1957. pp. 585–593.
2. Deresiewicz, H.: Mechanics of Granular Materials, *Advd. Applied Mech.* Vol. 5, 1958. pp. 233–306.
3. Duffy, J.: A differential stress-strain relations for the hexagonal closely packed array of Elastic spheres, *Journal of Applied Mechanics*, Vol. 26, 1959. pp. 88–94.
4. Rowe, P. W.: The Stress Dilatency Relation for Static Equilibrium of an Assembly of Particles in Contact, *Proc. Roy. Soc. of London*, A 269, 1962. pp. 500–527.

5. Hendron, A.J. Jr.: *The behavior of sand in one dimensional compression*, Ph.D. Dissertation, Department of Civil Engineering, University of Illinois, Urbana, IL. 1963.
6. Ko, H.Y. and Scott, R.F.: Deformation of sand in hydrostatic compression, *Journal of Soil Mechanics and Foundation Engineering Division*, ASCE Vol. 93, No. SM3, 1967. pp. 137–156.
7. Maklhouf, H. and Stewart, J.J.: Elastic Constants of Cubical-Tetrahedral and Tetragonal Spheroidal Arrays of Uniform Spheres, *Proc. Intl. Symposium on Wave Propagation and Dynamic Properties of Earth Materials*, Albuquerque, NM, Aug. 1967.
8. Petrikis, E. and Dobry, R.: *A Two Dimensional Numerical Micromechanical Model for a Granular Soil at Small Strains*, Report Ce-87-01, Rensselaer Polytechnic Institute, Troy, NY. 1987.
9. Dantu, P.: Contibution á l'étude mécanique et géometrique des milieux pulvérulents, *Proc. 4th Int. Conf. Soil Mech. Found. Eng., London*, 1, 1957. pp. 144–148.
10. Wakabayashi, T.: Photoelastic method for determination of stress in powdered Mass, *Proceedings of 7th Japan Nat. Congr. Appl. Mech.*, 1957. pp. 153–158.
11. De Josselin de Jong, G. and Verrujit, A.: Étude photo-élastique d' un empilement de disques, *Cahiers du Groupe Francais de Rhéologie*, Vol. 2, No. 1, 1969. pp. 73–86.
12. Oda, M. and Konoshi, J.: Microscopic deformation mechanism of granular material in simple shear, *Soils and Foundations*, Vol. 14, No. 4, 1974. pp. 25–38.
13. Matsuoka, E. and Geka, H.: A stress-strain models for granular materials considering mechanism of fabric changes, *Soils and Foundations*, Vol. 23, No. 2, 1983. pp. 83–97.
14. Oda, M., Nemat Nasser, S. and Konishi, J.: Stress induced anisotropy in granular masses, *Soils and Foundations*, Vol. 25, No. 3, 1985. pp. 85–97.
15. Serrano, A.A. and Rodrigues-Ortiz, J.M.: A contribution to the mechanics of Heterogeneous Granular Media, *Symp. On Plast. And Soil Mech.*, Cambridge, U.K. 1973.
16. Mindlin, R.D. and Deresiewicz, H.: Elastic spheres in contact under varying oblique forces, *Journal of Applied Mechanics*, ASME, Vol. 21, 1953. pp. 327–344.
17. Nayak, P.R. Surface roughness effects in rolling contacts. Trans. Am.Soc. Mech. Engrs. Ser. E, *J. Appl. Mech.* Vol. 39, No. 2, 1972. pp. 456–460.
18. Itasca.: *Particle Flow Code in Two Dimensions*. Itasca Consulting Group, Inc., Minnesota. 2003.
19. Bardet, J.P. and Proubet, J.: Adaptive dynamic relaxation for statics of granular materials, *Comput. and structures.*, Vol. 39, 1991 pp. 221–229.
20. Thornton, C. and Sun, G.: *Numerical Methods in Geotechnical Engineering*, in I.M. Smith (Ed.), Balkema, 1994. pp. 143–148.
21. Ting, J.M., Khwaja, M., Meachum, L.R. and Rowell, J.D.: An ellipse-based discrete element model for granular materials, *Int. J. Numer. Analyt. Methods Geomechanics*, Vol. 17, 1993. pp. 603–623.
22. Thornton, C. and Sun, G.: *Axisymmetric compression of 3D polydisperse systems of spheres*, In Powders and Grains, 93, (ed.C. Thoronton), Rotterdam: Balkema, 1993. pp. 129–134.
23. Borja, R.I. and Wren, J.R.: *Micromechanics of granular media, Part I: Generation of overall constitutive equation for assemblies of circular disks*, Comput. Methods Appl. Mech. Engrg., Vol. 127, 1995. pp. 13–36.
24. Wren, J.R. and Borja, R.I.: Micromechanics of granular media part II: Overall tangential moduli and localization model for periodic assemblies of circular disks, *Comput. Methods Appl. Mech.* Engrg., 141, 1997. pp. 221–246.
25. Thornton, C. and Antony, S.J.: 2000. Quasi-static shear deformation of a soft particle system, *Powder Technology*, 109, pp. 179–191.
26. Thornton, C.: Numerical simulations of deviatoric shear deformation of granular media, *Géotechnique*, Vol. 50, No. 1, 2000. pp. 43–53.

27. McDowell, G.R. and Harireche, O.: Discrete element modelling of soil particle fracture, *Géotechnique*, Vol. 52, No. 2, 2002a. pp. 131–135.

28. McDowell, G.R. and Harireche, O.: Discrete element modelling of yielding and normal compression of sand, *Géotechnique*, Vol. 52, No. 4, 2002b. pp. 299–304.

29. Mirghasemi, A.A., Rothenburg, L. and Matyas, E.L.: Influence of particle shape on engineering properties of two-dimensional polygon-shaped particles, *Géotechnique*, Vol. 52, No. 3, 2002. pp. 209–217.

30. Cundall, P.A.: Explicit finite difference methods in geomechanics, *Proc. 2nd Int. Conf. Numerical Methods in Geomechanics*, Blacksburg, Virginia, Desai, C. S. (Ed.), 1. 1976. pp. 132–150.

31. Strack. O.D.L. and Cundall, P.A.: *The Discrete Element Method as a tool for Research in Granular Media*, Part I, Report to National Science Foundation, Department of Civil and Mineral Engineering, University of Minnesota, Minneapolis, Minnesota, 97 pp. 1978.

32. Cundall, P.A. and Strack, O.D.L.: A Discrete Numerical Model for Granular Assemblies, *Géotechnique*, Vol. 29, No. 1, 1979a. pp. 47–65.

33. Cundall, P.A. and Strack, O.D.L.: The development of constitutive laws for soil using the discrete element method, *Proc. 3rd Int. conf.Numerical Methods in Geomechanics*, Aachen, W. Witke (Eds), 1, 1979b. pp. 289–298.

34. Cundall, P.A. and Strack, O.D.L.: The discrete element as a tool for research in granular media, Part II, *Report to National Science Foundation*, Department of Civil and Mineral Engineering, University of Minnesota, Minneapolis, Minnesota, 1979c. 204 pp.

35. Cundall, P.A and Strack, O.D.L.: *Modeling of microscopic mechanisms in granular materials: New Model* and Constitutive Relations, Jenkins, J. T. and Satake, M. (eds.), Elsevier, Amsterdam, 1983. pp. 137–149.

36. Cundall, P.A, Drescher, A and Strack, O.D.L.: Numerical experiments on granular assemblies: Measurement and Observations, *Proc. IUTAM Conf. On Deformation and Failure of Granular materials*, Delft, P. A. Vermeer and H. J Luger (eds.) Balkema, Rotterdam, 1982. 355–370.

37. Thornton, C. and Barnes, D.J.: Computer simulated deformation of compact granular assemblies, *Acta Mechanica*, 64, 1986. pp. 45–61.

38. Thornton, C.: Computer simulated experiments on particulate materials, *Tribology in Powder Technology*, B. J. Briscoe and M. J. Adams (eds), Adam Higher, Bristol, 1987. pp. 292–302.

39. Bathurst, R.J. and Rothenburg, L.: Note on a random isotropic granular material with negative poisson's ratio, *Int. J. Engng. Sci.*, Vol. 26, No. 4, 1988b. pp. 373–383.

40. Bathurst, R.J. and Rothenburg, L.: 1989, Investigation of micromechanical features of idealized granular assemblies using DEM, *Proc. 1st U.S. Conf. On Discrete Element Methods*, Golden, Colorado, 12 pp.

41. Bathurst, R.J. and Rothenburg, L.: Observations on stress-force-fabric relationships in idealized granular materials, *Mechanics of Materials*, Vol. 9, No. 1, 1990. pp. 65–80.

42. Rothenburg, L. and Bathurst, R.J.: Analytical study of induced anisotropy in idealized granular materials, *Géotechnique*, Vol. 39, No. 4, 1989. pp. 601–614.

43. Sitharam, T.G.: *Numerical simulation of hydraulic fracturing in granular media*, Ph.D. thesis, University of Waterloo, Waterloo, Ontario, Canada. 1991.

44. Mirghasemi, A.A., Rothenburg, L. and Matyas, E.L.: Numerical simulation of assemblies of two-dimensional polygon-shaped particles and effects of confining pressure on shear strength, *Soils and Foundations*, Vol. 37, No 3, 1997. pp. 43–52.

45. Sitharam, T.G.: Discrete element modeling of cyclic behaviour of granular materials, *Geotechnical and Geological Engineering*, Vol. 21. 2003. pp. 297–329.

46. Strack. O.D.L. and Cundall, P.A.: *Fundamental studies of Fabric in Granular Materials*, Interiuum Report to National Science Foundation, Department of Civil and Mineral Engineering, University of Minnesota, Minneapolis, Minnesota, 53 pp. 1984.

47. Tang-Tat Ng: *Numerical Simulation of Granular Soil Under Monotonic And Cyclic Loading*: A Particulate Mechanics Approach, Ph.D. thesis, submitted to Rensselaer Polytechnic Institute, Troy, New York. 1989.

48. Chantawarungal, K.: *Numerical simulations of three dimensional granular assemblies*, Ph.D. thesis, University of waterloo, waterloo, Ontario, Canada. 1993.

49. Thoronton, C., Ciomocos, M.T., Yin, K.K. and Adams, M.J.: *Fracture of particulate solids*, In Powders and Grains 97(eds Behringer, R.P. and Jenkis, J.T.), Rotterdam, Balkema 1997. pp. 131–134.

50. Itasca Consulting Group, Inc.: *Particle flow code in 3 dimensions*, (Software programme). 1999.

51. Dinesh, S.V.: *Discrete element simulation of static and cyclic behaviour of granular media*, Ph.D. thesis, submitted to Indian Institute of science, Bangalore, India. 2003.

52. Vinod, J.S.: *Liquefaction and dynamic properties of granular materials*: A DEM Approach, Ph.D. thesis, submitted to Indian Institute of science, Bangalore, India. 2006.

53. Lim, W.L., and McDowell, G.R.: Discrete Element Modelling of Railway Ballast. *Granular Matter*, Vol. 7, No. 1, 2005. pp. 19–29.

54. Lobo-Guerrero, S., and Vallejo, L.E.: Discrete Element Method Analysis of Railtrack Ballast Degradation during Cyclic Loading. *Granular Matter*, 8, 2006. pp. 195–204.

55. Lu, M., and Mcdowell, G.R.: Discrete Element Modelling of Railway Ballast under Monotonic and Cyclic Triaxial Loading. *Geotechnique*, Vol. 60, No. 6, 2010. pp. 459–467.

56. Hossain, Z., Indraratna, B., Darve, F., and Thakur, P.K.: DEM Analysis of Angular Ballast Breakage under Cyclic Loading. *Geomechanics and Geoengineering*, Vol. 2, No. 3, 2007. pp. 175–181.

57. McDowell, G.R., and Harireche, O.: Discrete Element Modelling of Soil Particle Fracture. *Geotechnique*, Vol. 52, No. 2. 2002. pp. 131.

58. Lu, M., and McDowell, G.R.: Discrete Element Modelling of Ballast Abrasion. *Geotechnique*, Vol. 56, No. 9, 2006. pp. 651–655.

59. Lobo-Guerrero, S., and Vallejo, L.E.: Discrete Element Method Analysis of Railtrack Ballast Degradation during Cyclic Loading. *Granular Matter*, 8, 2006. pp. 195–204.

60. McDowell, G.R., Bolton, M.D., and Robertson, D.: The Fractal Crushing of Granular Materials. *Journal of Mechanics and Physics of Solids*, Vol. 44, No. 12, 1996. pp. 2079–2102.

61. McDowell, G.R., Harireche, O., Konietzky, H., Brown, S.F., and Thom, N.H.: Discrete Element Modelling of Geogrid-reinforced Aggregates. *Proceedings of the Institution of Civil Engineers: Geotechnical Engineering*, Vol. 159 No. 1, 2006. pp. 35.

62. Indraratna, B., Thakur, P.K., and Vinod, J.S.: Experimental and Numerical Study of Railway Ballast Behaviour under Cyclic Loading. *International Journal of Geomechanics*, ASCE, Vol. 10, No. 4, 2010. pp. 136–144.

63. Selig, E.T. and Waters, J.M.: *Track Technology and Substructure Management*. Thomas Telford, London, 1994.

64. Thakur, P.K., Vinod, J.S and Indraratna, B.: The Role of Particle Breakage on the Shear Behaviour of Coarse Granular Materials: A Micromechanical Investigation. *International Symposium on Geomechanics and Geotechniques*: From Micro to Macro (IS-Shanghai-2010), 10 Oct 2010, China 2010a. (Accepted).

65. Thakur, P.K., Indraratna, B., and Vinod, J.S.: DEM Simulation of Effect of Confining Pressure on Ballast Breakage. *17th International Conference on Soil Mechanics and Geotechnical Engineering*, 5–9 October 2009, Alexandria, Egypt, 2009. pp. 602–605.

66. Indraratna, B., Lackenby, J., and Christie, D.: Effect of Confining Pressure on the Degradation of Ballast under Cyclic Loading. *Geotechnique*, Institution of Civil Engineers, UK, Vol. 55, No. 4, 2005. pp. 325–328.
67. Thakur, P.K., Vinod, J.S., and Indraratna, B.: Effect of Particle Breakage on Cyclic Densification of Ballast: A DEM Approach. WCCM/APCOM 2010, IOP Conf. Series: *Materials Science and Engineering Sydney*, Australia 10, 2010b. pp. 122–129.

Chapter 12

FEM Modelling of Tracks and Applications to Case Studies

In order to compete with other modes of transportation, rail industries face challenges to minimise track maintenance costs, and to find alternative materials and approaches to improve the track performance. The track should be designed to withstand large cyclic train loadings to provide protection to subgrade soils against both progressive shear failure and excessive plastic deformation. The design of track should also consider the deterioration of ballast due to breakage and subsequent implications on the track deformations. The potential use of geosynthetics in the improvement of track stability and reducing the maintenance cost is well established. The stabilisation of the track by means of geogrids and prefabricated vertical drains that provide confinement to the ballast layer in addition to rapid radial drainage, assures a more resilient long-term performance of the ballast and formation layer [1, 2, 3, 4]. Two field trials were employed for validating the numerical analysis. The first field trial was conducted on a section of a fully instrumented railway track along the coastal town of Bulli, in New South Wales, Australia. The main objectives were to study the benefits of a geocomposite (i.e. combination of biaxial geogrid and non-woven polypropylene geotextile) installed at the ballast-capping interface, and to evaluate the performance of moderately graded recycled ballast in comparison with traditionally used uniform fresh ballast. The second field trial was carried out at the Sandgate Rail Grade Separation Project located between the regional towns of Maitland and Newcastle, in Eastern New South Wales, to verify the performance of a track built on thick soft estuarine clay, stabilised by prefabricated vertical drains (PVDs).

In this chapter, the field measurements are used for the calibration of constitutive models and their successive implementation in the finite element method, capturing the elasto-plastic deformation characteristics of ballast and reinforced by geogrid.

12.1 USE OF GEOCOMPOSITE UNDER RAILWAY TRACK

The comprehensive knowledge of the complex mechanisms associated with track deterioration is essential in the accurate prediction of a typical track maintenance cycle. Various simplified analytical and empirical design methods have been used in the past to estimate settlements and stress-transfer between the track layers. However, these design methods are based on the linear elasticity approach, and they often give only relatively crude estimates (Doyle, [5], Li and Selig, [6]). Given the complexities of the behavior of the composite track system consisting of rail, sleeper, ballast, sub-ballast

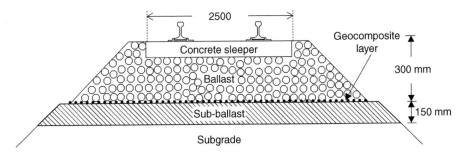

Figure 12.1 Section of ballasted track bed with geocomposite layer (modified after Indraratna et al., [2]).

Table 12.1 Grain size characteristics of fresh ballast, recycled ballast and capping materials (Indraratna and Salim, [1]; Indraratna et al., [2]).

Material type	Particle shape	d_{max} (mm)	d_{min} (mm)	d_{50} (mm)	C_u	C_c
Fresh Ballast	Highly angular	75.0	19.0	35.0	1.5	1.0
Recycled Ballast	Semi-angular	75.0	9.5	38.0	1.8	1.0
Sub-ballast	Semi-rounded	19.0	0.05	0.26	5.0	1.2

and subgrade under repeated (cyclic) traffic loads in the real track environment, the current design techniques used by rail industries worldwide are often over-simplified.

To understand this complex mechanism of track deformations and to evaluate the potential benefits of geosynthetics in rail track, a comprehensive field trial was imperative. An instrumented track was constructed between two turnouts at Bulli town about 10 km north of Wollongong city. The total length of the instrumented track section was 60 m, and it was divided into four sections, each 15 m in length. Two sections were built without the inclusion of a geocomposite layer, while the remaining two sections were built placing a geocomposite layer at the ballast-capping interface (Fig. 12.1). To measure the vertical and horizontal deformations of ballast, settlement pegs and displacement transducers were installed at the sleeper-ballast and ballast-capping interfaces in different track sections. The vertical and horizontal stresses developed in the track bed under repeated wheel loads were measured by pressure cells (230 mm diameter) installed at different locations of selected sections of the fresh ballast.

The overall thickness of granular layer was kept as 450 mm including a ballast layer of 300 mm and a capping layer of 150 mm. The particle size, gradation, and other index properties of ballast used at the Bulli site were in accordance with the Technical Specification TS 3402 [7], which represents sharp highly angular coarse aggregates of crushed volcanic latite basalt. Concrete sleepers were used in the test track. Recycled ballast was acquired from a recycled plant near Sydney. Table 12.1 shows the grain size characteristics of fresh ballast, recycled ballast and the sub-ballast materials used. Electrical Friction Cone Penetrometer (EFCP) tests reported that the subgrade soil is a stiff overconsolidated silty clay and had more than sufficient strength to support the train loads. The bedrock is a highly weathered sandstone having weak to medium compressive strength.

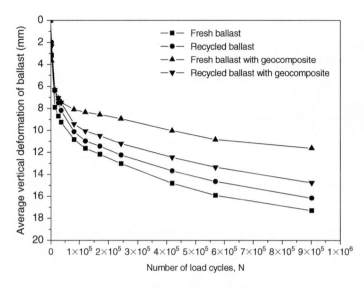

Figure 12.2 Vertical deformation of the ballast layer (modified after Indraratna et al., [2]).

A bi-axial geogrid was placed over the non-woven polypropylene geotextile to serve as the geocomposite layer, which was installed at the ballast-capping interface. While the geogrid reinforces and confines the ballast at the interface, the geotextile layer acts as a drainage medium as well as serving the purpose of 'separator' between the coarse ballast and finer sub-ballast. The technical specifications of geosynthetics used at the site are described in details elsewhere (Indraratna and Salim, [1]). In summary, in order to investigate the overall performance of the ballast layer, the average vertical deformation was considered by correcting for the vertical displacements of the sleeper-ballast and ballast-capping interfaces. The vertical displacement at each interface was obtained by taking the mean of measurements taken beneath the rail and at the edge of sleeper. The values of average vertical ballast deformation are plotted against the number of load cycles (N) in Figure 12.2. In the recycled ballast, the vertical displacements are smaller compared to the case of fresh ballast. The better performance of this moderately-graded recycled ballast is partly attributed to less breakage as they are often less angular, thereby preventing corner breakage due to high contact stresses. Moreover, compared to uniform gradations, a well-graded particle size distribution provides better particle interlock and less particle movement. The geocomposite inclusion induces a decrease in average vertical deformation of recycled ballast at a large number of cycles. The load distribution capacity of ballast layer is improved by the placement of a flexible and resilient geocomposite layer resulting in a substantial reduction of settlement under high cyclic loading.

12.1.1   Finite element analysis

An elasto-plastic constitutive model of a composite multi-layer track system including rail, sleeper, ballast, sub-ballast and subgrade is adopted here. Numerical simulations

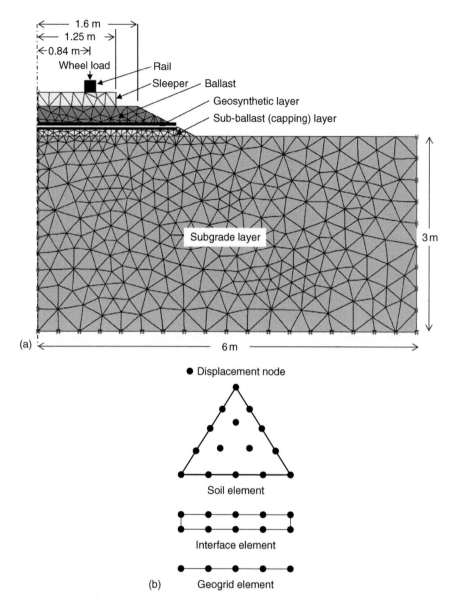

Figure 12.3 (a) Finite element mesh discretisation of a rail track and (b) 15-node continuum soil, 10-node Interface and 5-node geogrid element.

are performed using a two-dimensional plane-strain finite element analysis (PLAXIS) to predict the track behaviour with and without geosynthetics. The FEM software PLAXIS has been popular, and has demonstrated its success in the numerical analysis of numerous geotechnical problems. A typical plain strain track model is simulated in the Finite Element discretization as shown in Figure 12.3a.

The subgrade soil and the track layers are modeled using 15-node linear strain quadrilateral (LSQ) elements. For representing geogrid elements, 5-node line (tension) elements are adopted. Since it is also necessary to model the interaction between granular media and geogrid, special 10-node interface elements are used. Figure 12.3b shows the details of these elements. The 15-node isoparametric element provides a fourth order interpolation for displacements. The numerical integration by the Gaussian scheme involves twelve Gauss points (stress points). The 3 m high and 6 m wide finite element model is discretised to 794 fifteen-node LSQ elements, 26 five-node line elements for geogrid and, 52 five-node elements at the interface. The mesh generation of PLAXIS version 8.0 used here follows a robust triangulation procedure to form 'unstructured meshes', which are considered to be numerically efficient when compared to regular 'structured meshes'.

The nodes along the bottom boundary of the section are considered as pinned supports, i.e., they are restrained in both vertical and horizontal directions (i.e. standard fixities). The left and right boundaries are restrained in the horizontal directions, representing smooth contact in the vertical direction. The vertical dynamic wheel load is simulated as a line load representing an axle train load of 25 tons with a dynamic impact factor (DIF) of 1.3. DIF of 1.3 represents a typical 25 tonnes axle load travelling at 80 km/h. Figure 2.6 in Chapter 2 shows the relevant values of DIF. The gauge length of the track is 1.68 m. The shoulder width of ballast is 0.35 m and the side slope of the rail track embankment is 1:2. The constitutive models and material parameters are given in Table 12.2. The flow rule adopted in Hardening Soil model is characterised by a classical linear relation, with the mobilised dilatancy angle (ψ_m) given by (Schanz et al., [8]):

$$\sin \psi_m = (\sin \phi_m - \sin \phi_{cv})/(1 - \sin \phi_m \sin \phi_{cv}) \tag{12.1}$$

where ϕ_{cv} is a material constant (the friction angle at critical state) and:

$$\sin \phi_m = (\sigma'_1 - \sigma'_3)/(\sigma'_1 + \sigma'_3 - 2c \cot \phi) \tag{12.2}$$

According to equation (12.1), ψ_m depends on the values of friction, ϕ and dilatancy angles at failure, ψ which control the quantity ϕ_{cv}. Indraratna and Salim [1] described the dependence of particle breakage and dilatancy on the friction angle of ballast. A modified flow rule considering the energy consumption due to particle breakage during shearing deformations is given by (Salim and Indraratna, [9]):

$$\frac{d\varepsilon_v^p}{d\varepsilon_s^p} = \frac{9(M - \eta)}{9 + 3M - 2\eta M} + \frac{dE_B}{pd\varepsilon_s^p}\left(\frac{9 - 3M}{9 + 3M - 2\eta M}\right)\left(\frac{6 + 4M}{6 + M}\right) \tag{12.3}$$

The experimental values of η, p, M and the computed values of $dE_B/d\varepsilon_{sp}$ which are linearly related to the rate of particle breakage $dB_g/d\varepsilon_1$ can be readily used to predict the flow rule. In the present study, a non-associative flow rule with a dilatancy angle $\psi = 12.95°$ is used. The values of stress-dependent stiffness moduli E_{50}^{ref}, E_{oed}^{ref} and E_{ur}^{ref} are obtained from previously published results of large scale drained triaxial compression tests under monotonic loading conditions (Salim and Indraratna, [10]). Figure 12.4 exhibits the evolution of hardening soil model parameters based on the deviator stress response and shear strains, respectively. The hardening soil model showed better agreement with the strain-hardening behaviour of ballast observed in

Table 12.2 Constitutive model and material parameters adopted in FEM analysis.

Material Parameters	Rail	Concrete Sleeper	Ballast	Sub-ballast	Subgrade	Geogrid
	Rail Track component					
Material Model	Linear Elastic	Linear Elastic	Hardening Soil	Mohr-Coulomb	Mohr-Coulomb	Elastic
Type	Non-porous	Non-porous	Drained	Drained	Drained	–
E (MPa)	210,000	10,000	–	80	34.2	–
E_{50}^{ref} (MPa)	–	–	21.34	–	–	–
E_{oed}^{ref} (MPa)	–	–	21.34	–	–	–
E_{ur}^{ref} (MPa)	–	–	64.02	–	–	–
EA (kN/m)	–	–	–	–	–	1198
γ (kN/m^3)	78	24	15.6	16.67	18.15	–
ν	0.15	0.15	–	0.35	0.33	–
ν_{ur}	–	–	0.2	–	–	–
c (kN/m^2)	–	–	0	0	5.5	–
ϕ (degrees)	–	–	58.47	35	24	–
ψ (degrees)	–	–	12.95	0	0	–
Pref (kN/m^2)	–	–	50	–	–	–
m	–	–	0.5	–	–	–
k_0^{nc}	–	–	0.3	–	–	–
R_f	–	–	0.9	–	–	–

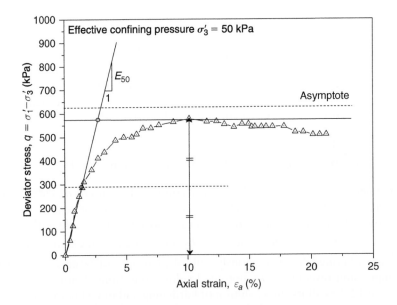

Figure 12.4 Hardening Soil Model: stress-strain relationship for ballast.

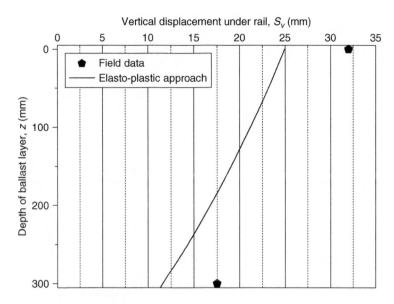

Figure 12.5 Variation of vertical deformation of ballast with the depth.

large scale triaxial tests indicating considerable ballast breakage (Shahin and Indraratna, [11]). Further details of constitutive models used in PLAXIS are given by Brinkgreve [12]. The current formulation of finite element is incapable of conducting postpeak analysis into the strain-softening region. However, such large strains or deformations are not usually permitted in reality, hence the analysis is focused on the peak strength.

12.1.2 Comparison of field results with FEM predictions

In order to validate findings of the finite element analysis, a comparison is made between the elasto-plastic analysis and the observed field data. Figure 12.5 shows the vertical deformation profile predicted by finite element simulations and the measured values of vertical deformation underneath the rail seat at the unreinforced section of the instrumented track. The vertical deformations were monitored at the sleeper-ballast and ballast-capping interfaces using settlement pegs as mentioned earlier. The values predicted by elasto-plastic analysis show slight deviation in contrast to the measured values. This is because the real cyclic nature of wheel loading is not considered and is approximately represented by an equivalent dynamic plain strain analysis. Considering the limitations of elasticity based approaches, this approach of multiplying the static load with a DIF has been employed in practice for a long time (Li and Selig, [6]).

12.2 DESIGN PROCESS FOR SHORT PVDS UNDER RAILWAY TRACK

The Sandgate Rail Grade Separation Project is located at the town of Sandgate between Maitland and Newcastle, in the Lower Hunter Valley of New South Wales (Fig. 12.6).

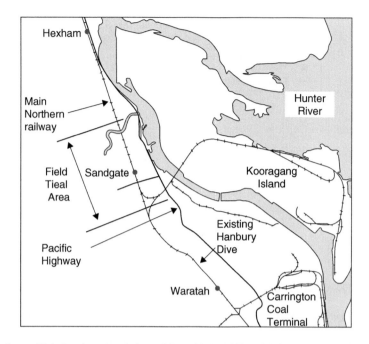

Figure 12.6 Site location (adopted from Hicks, [12] and Indraratna et al., [4]).

The new railway tracks were required to reduce the traffic in the Hunter Valley Coal network. In this section, the rail track stabilised using short prefabricated vertical drains (PVDs) in the soft subgrade soil is presented together with the background of the project, the soil improvement details, design methodology and finite element analysis. The effectiveness of PVDs in improving soil condition has been demonstrated by Indraratna et al. [4]. Preliminary site investigations were conducted for mapping the soil profile along the track. In-situ and laboratory testing programs were carried out to provide relevant soil parameters. Site investigation included 6 boreholes, 14 piezocone (CPTU) tests, 2 in-situ vane shear tests, and 2 test pits. Laboratory testing such as soil index property testing, standard oedometer testing, and vane shear testing were also performed.

A typical soil profile showed that the existing soft compressible soil thickness varies from 4 m to 30 m. The soft residual clay lies beneath the soft soil layer followed by shale bedrock. The soil properties are shown in Figure 12.7. The groundwater level is at the ground surface. The moisture contents of the soil layers are the same as their liquid limits. The soil unit weight varies from 14 to 16 kN/m^3. The undrained shear strength increases from about 10 to 40 kPa. The clay deposit at this site can be considered as lightly overconsolidated (OCR ≈ 1–1.2). The horizontal coefficient of consolidation (c_h) is approximately 2–10 times the vertical coefficient of consolidation (c_v). Based on preliminary numerical analysis conducted by Indraratna et al. [4], PVDs having 8 m length were suggested and installed at 2 m spacing in a triangular pattern Extensive field instrumentations including settlement plates, inclinometers, and vibrating wire

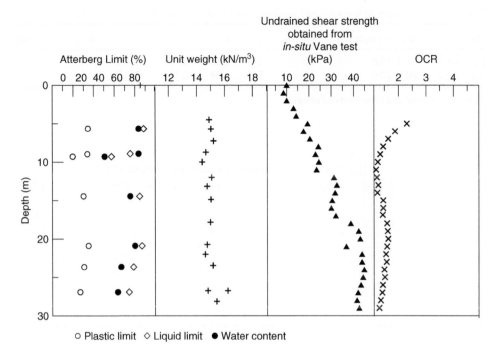

Figure 12.7 Soil properties at Sandgate Rail Grade Separation Project (Indraratna et al., [4]).

piezometers were employed to monitor the track responses. The settlement plates were installed above the surface of the subgrade layer to directly provide a measurement of the vertical subgrade settlement. The main aims of the field monitoring were to:

(a) ensure the stability of track;
(b) validate the design of the track stabilised by PVDs; and
(c) examine the accuracy and reliability of the numerical model through Class A predictions, (i.e. the field measurements were unavailable at the time of finite element modelling).

12.2.1 Preliminary design

Due to the time constraint, the rail track was built immediately after installing PVDs. The train load moving at very low speed was used as the only external surcharge. The equivalent dynamic loading using an impact load factor was used to predict the track behaviour. In this analysis, a static pressure of 104 kPa with an impact factor of 1.3 was applied according to the low train speed (60 km/h) for axle loads up to 25 tonnes, based on the Australian Standards AS 1085.14-1997 [13]. The Soft Soil model and Mohr-Coulomb model were both employed in the finite element code, PLAXIS (Potts, [14], Vermeer and Neher, [15]). The overcompacted surface crust and fill layer were simulated by the Mohr-Coulomb theory, whereas the soft clay deposit was conveniently modelled using the Soft Soil model. The formation was separated into 3 distinct layers,

Table 12.3 Selected parameters for soft soil layer used in the FEM (Indraratna et al., [4]).

Soil layer	Depth of layer (m)	c (kPa)	φ	e	λ	κ	k_h ($\times 10^{-4}$ m/day)
Soft soil-1	1.0–10.0	10	25	2.26	0.131	0.020	1.4
Soft soil-2	10.0–20.0	15	20	2.04	0.141	0.017	1.5

Note: φ Back-calculated from Cam-clay M value.

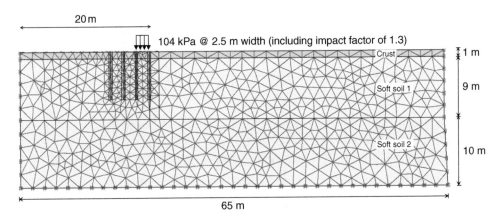

Figure 12.8 Vertical cross section of rail track and foundation (Indraratna et al., [4]).

namely, ballast and fill, Soft soil-1 and Soft soil-2. The soil parameters are given in Table 12.3.

A cross-section of the finite element mesh discritization of the formation beneath the track is shown in Figure 12.8. A plane strain finite element analysis has been employed with liner strain triangular elements with 6 displacement nodes and 3 pore pressure nodes. A total of 4 PVD rows were used in the analysis. An equivalent plane strain analysis with appropriate conversion from axisymmetric to 2-D was adopted to analyse the multi-drain analysis (Indraratna et al., [16]). In this method, the corresponding ratio of the smear zone permeability to the undisturbed zone permeability is given by:

$$\frac{k_{s,ps}}{k_{h,ps}} = \frac{\beta}{k_{h,ps}/k_{h,ax}[\ln(n/s) + k_{h,ax}/k_{s,ax}\ln(s) - 0.75] - \alpha} \qquad (12.4)$$

$$\alpha = \frac{0.67(n-s)^3}{n^2(n-1)} \qquad (12.4a)$$

$$\beta = 2(s-1)[n(n-s-1) + 0.33(s^2+s+1)]/n^2(n-1) \qquad (12.4b)$$

$$n = \frac{d_e}{d_w} \qquad (12.4c)$$

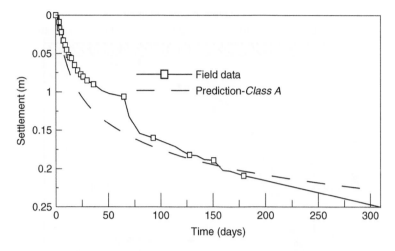

Figure 12.9 Predicted and measured at the centre line of rail tracks (after Indraratna et al., [4]).

$$s = \frac{d_s}{d_w} \tag{12.4d}$$

In the above expressions, d_e = the diameter of unit cell soil cylinder, d_s = the diameter of the smear zone, d_w = the equivalent diameter of the drain, k_s = horizontal soil permeability in the smear zone, k_h = horizontal soil permeability in the undisturbed zone and the top of the drain and subscripts 'ax' and 'ps' denote the axisymmetric and plane strain condition, respectively.

The ratio of equivalent plane strain to axisymmetric permeability in the undisturbed zone can be attained as,

$$\frac{k_{h,ps}}{k_{h,ax}} = \frac{0.67(n-1)^2}{n^2[\ln(n) - 0.75]} \tag{12.5}$$

In the above equation, the equivalent permeability in the smear and undisturbed zone vary with the drain spacing.

12.2.2 Comparison of field with numerical predictions

The field results were released by the track owner (Australian Rail Track Corporation) one year after the finite element predictions. Therefore all predictions can be categorized as Class A (Lambe, [17]). A spacing of 2 m was adopted for Mebra (MD88) vertical drains of 8 m in length. The field data together with the numerical predictions are compared and discussed. The calculated and observed consolidation settlements at the centre line are now presented in Figure 12.9. The predicted settlement matches very well with the field data. The in-situ lateral displacement at 180 days at the rail embankment toe is illustrated in Figure 12.10. As expected, the maximum displacements are measured within the top clay layer, i.e., the softest soil below the 1 m crust.

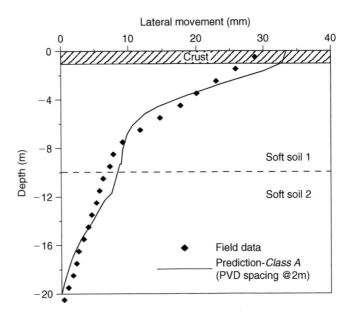

Figure 12.10 Measured and predicted lateral displacement at the embankment toe at 180 days (after Indraratna et al., [4]).

The lateral displacement is restricted to the topmost compacted fill (0–1 m deep). The *Class A* prediction of lateral displacements is also in very good agreement with the field behaviour. The effectiveness of wick drains in reducing the effects of undrained cyclic loading through the reduction in lateral movement is undeniably evident.

REFERENCES

1. Indraratna, B. and Salim, W.: Modelling of particle breakage of coarse aggregates incorporating strength and dilatancy. *Geotechnical Engineering, Proc. Institution of Civil Engineers*, London, 2002, Vol. 155, No. 4, pp. 243–252.
2. Indraratna, B., Nimbalkar, S., Christie D., Rujikiatkamjorn C. and Vinod J.S.: Field Assessment of the Performance of a Ballasted Rail Track with and without Geosynthetics. *Journal of Geotechnical and Geoenvironmental Engineering*, 2010, Vol. 136, No. 7, pp. 907–917.
3. Indraratna B., Nimbalkar, S. and Tennakoon, N.: The Behaviour of Ballasted Track Foundations: Track Drainage and Geosynthetic Reinforcement. *GeoFlorida 2010*, ASCE Annual GI Conference, February 20–24, 2010, (CD-ROM).
4. Indraratna, B., Rujikiatkamjorn, C., Ewers, B. and Adams, M.: Class A prediction of the behaviour of soft estuarine soil foundation stabilised by short vertical drains beneath a rail track. *Journal of Geotechnical and Geoenvironmental Engineering*, 2010, Vol. 136, No. 5, pp. 686–696.
5. Doyle, N.F.: Railway Track Design: A review of current practice. *Occasional paper no. 35, Bureau of Transport Economics*, Commonwealth of Australia, Canberra, 1980.
6. Li, D. and Selig, E.T.: Method for railroad track foundation design, I: Development. *Journal of Geotechnical and Geoenvironmental Engineering*, ASCE, Vol. 124, No. 4, 1998, pp. 316–322.

7. T.S. 3402: Specification for Supply of Aggregates for Ballast. Rail Infrastructure Corporation of NSW, Sydney, Australia, 2001.
8. Schanz, T., Vermeer, P.A. and Bonnier, P.G.: The Hardening Soil Model – Formulation and Verification, Proc. *Plaxis Symposium "Beyond 2000 in Computational Geotechnics"* Amsterdam Balkema, Rotterdam, 1999, 55–58.
9. Salim, W. and Indraratna, B.: A new elasto-plastic constitutive model for coarse granular aggregates incorporating particle breakage. Canadian Geotechnical Journal, 2004, Vol. 41, pp. 657–671.
10. Indraratna, B. and Salim, W.: Deformation and Degradation Mechanics of Recycled Ballast stabilised with Geosynthetics, *J. of Soils and Foundations*, Japanese Geotechnical Society, 2003, Vol. 43 (4), pp. 35–46.
11. Shahin, M.A. and Indraratna, B.: Modeling the mechanical behavior of railway ballast using artificial neural networks. *Canadian Geotechnical Journal*, 2006, Vol. 43, 1144–52.
12. Hicks, M.: Environmental impact statement for the Sandgate Rail Grade Separation, Hunter Valley Region, Australia, 2005.
13. Australia Standards: Railway Permanent Way Material AS 1085.14-1997, Sydney, NSW, Australia, 1997.
14. Potts, D.M.: Guidelines for the use of advanced numerical analysis, London, Thomas Telford, 2002.
15. Vermeer, P.A. and Neher, H.P.: A soft soil model that accounts for creep. In Brinkgreve, R.B.J. (Ed.), *Proc. of the Int. Symp. Beyond 2000 in Computational Geotechnics*, Amsterdam: 1999. pp. 249–261. Rotterdam, A.A. Balkema.
16. Indraratna, B., Rujikiatkamjorn, C. and Sathananthan, I.: Analytical and numerical solutions for a single vertical drain including the effects of vacuum preloading. Canadian Geotechnical Journal, 2005, Vol. 42, pp. 994–1014.
17. Lambe, T.W.: Predictions in soil engineering. *Geotechnique*, 1973, Vol. 23, pp. 149–202.

Chapter 13

Non-destructive Testing and Track Condition Assessment

Regular inspection and maintenance of railway track is always a major task for the rail industry. Ballast fouling is one of the main reasons for track deterioration. Fouling materials come from various sources including ballast breakdown, external materials including coal falling off freight trains, and clay slurry pumped up from subgrade. Ballast fouling will lead to poor drainage in the track and then increase the moisture content of the subgrade posing undrained failure risks. Fouling also reduces the strength and stiffness of the ballast and leads to excess deformation of the track. Highly fouled ballast loses its functions related to drainage, absorbing shocks (impact) and noise levels. Therefore, ballast conditions should be regularly inspected and maintenance should be timely conducted to ensure safe track operations.

Ballasted track is usually monitored by visual inspection at walking speed. Trial pits can be normally excavated at various sections where fouling is anticipated. This method is often cumbersome and inefficient, hence, non-destructive techniques have been recently introduced to monitor the track conditions. In this chapter, two non-destructive techniques, the Ground Penetrating Radar (GPR) and Multichannel Analysis of Surface Wave (MASW) are introduced to evaluate the ballast layer conditions with the aid of a model track built at the University of Wollongong.

13.1 LABORATORY MODEL TRACK

13.1.1 The model track

In order to investigate the actual ground conditions, a full scale railway track containing subgrade, capping layer (sub-ballast), ballast, sleepers and rails was built for conducting non-destructive inspections (Fig. 13.1). The boundary box of the track composed of two layers of plywood to eliminate any reflective radar signals. The internal dimensions of the box were 4.76 m in length, 3.48 m in width and 0.79 m in height. The external layer is 18 mm thick plywood and the inside layer is 12 mm thick water resistant marine plywood. The track can be fully submerged by the aid of a plastic membrane placed between the two layers of plywood. In order to control the moisture condition of the track, perforated pipes were placed at the bottom of the box above the membrane, and timber bracings were used to increase the lateral stiffness.

The track is composed of a 150 mm subgrade of clayey sand, a 150 mm capping layer of road base, and a 490 mm layer of ballast. Additionally, a geotextile and geogrid

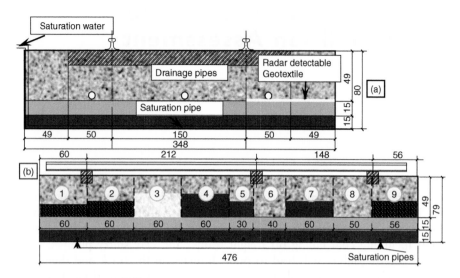

Figure 13.1 Schematic graphs of the model track: (a) traverse direction and (b) longitudinal direction (dimension in cm) (Su et al., [1]).

between the subgrade and capping layer were placed. Radar detectable geotextile was positioned on top of the capping layer at the right side of the box in a longitudinal direction, to test its capability in highlighting the ballast-capping interface (Fig. 13.1). Three drainage pipes were embedded between the capping layer and the ballast, and plastic pipes were also installed in the capping layer in order to measure the moisture content using moisture probes (Fig. 13.2).

13.1.2 Preparation of the ballast sections

As shown in Figure 13.1, the ballast was sub-divided into 9 sections with different fouling conditions. The details of each section are provided in Table 13.1. Different types of fouling material (clayey sand and coal) were used to simulate various fouled ballast. The degree of fouling can be established using the Relative Ballast Fouling Ratio (R_{b-f}) (Indraratna et al., [2]) which is defined by:

$$R_{b-f} = \frac{M_f \times \frac{G_{s-b}}{G_{s-f}}}{M_b} \times 100\% \tag{13.1}$$

where, M_f and M_b, and G_{s-f} and G_{s-b} are the mass and specific gravities of fouling materials and ballast, respectively. This parameter can reflect the influence of specific gravity and particle gradation of fouling material on the degree of ballast fouling.

The same amount of ballast was used in each section. The fouled parts of the sub-sections were prepared using two types of methods. The fouling contents were not very high in sections 1–5, so the fouling materials were added layer by layer while the ballast was being compacted. The thickness of each layer was approximately 40–60 mm.

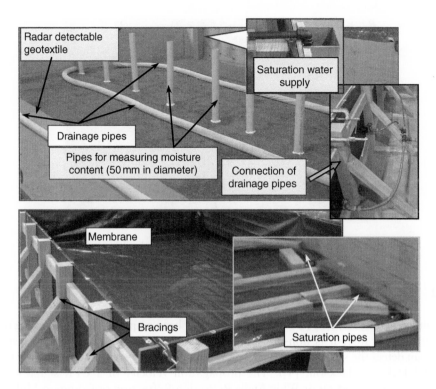

Figure 13.2 Details of the model track box (Su et al., [1]).

Table 13.1 Details for the sub-sections (Su et al., [1]).

	Types of fouling	Thickness of fouled part (cm)	R_{b-f}	Density (ton/m^3)
Section 1	Coal	15	10%	1.675
Section 2	Coal	20	25%	1.807
Section 3	Ballast breakdown	27	25%	2.017
Section 4	Clayey sand	27	25%	2.096
Section 5	Clayey sand	20	10%	1.753
Section 6	Clean	N/A	N/A	1.587
Section 7	Clayey sand	20	50%	1.899
Section 8	Clean	N/A	N/A	1.636
Section 9	Coal	20	50%	1.770

Preparation consisted of placing a layer of clean ballast and, subsequently, a layer of corresponding fouling material calculated according to a given R_{b-f} was spreaded. The ballast and the fouling material were then compacted with a hand operated compactor. The fouling in Sections 7 and 9 was too excessive to be added to the ballast layer by layer, so they were mixed together in a concrete mixer and then compacted layer by layer. The completed track is shown in Figure 13.3.

Figure 13.3 The completed model track with sleepers and rails on it (Su et al., [1]).

13.2 GPR METHOD

Ground penetration radar (GPR) has increasingly been employed for monitoring track conditions, because, it is non-destructive and can monitor the track at high speed. GPR can detect the signal reflections from the layers of sub-structure (Gallagher et al., [3] and Jack and Jackson, [4]). The propagation velocity of a GPR signal can be calibrated using a test pit or Wide Angle Reflection Refraction or Common Mid Point (Clark et al., [5]), and then the thickness of each layer can be calculated based on the propagation velocity and two-way travel time of the radar wave (Hugenschmidt, [6]). Variations of Ballast fouling can be identified from the radargram by the depth of interface (low frequency antennae) or its scattering pattern (high frequency antennae) (Al-Qadi et al., [7]).

13.2.1 Theory background of GPR

GPR is an electromagnetic sounding technique that is used to investigate shallow sub-surface or objects which have contrasting electrical properties (Gallaghera et al., [3] and Daniels, [8]). The GPR operates by transmitting short electromagnetic waves into the subsurface and then recording and displaying the reflected energy. The data obtained from GPR testing is the time domain waveform representing the electromagnetic energy transmitted from the antenna and reflected off subsurface boundaries back to the antenna (Sussmanna et al., [9]). An examination of the reflected radar waveforms enables an interpretation of the material and/or structure under investigation (Clark et al., [10]).

The GPR electromagnetic waves are reflected at interfaces between materials of dissimilar dielectric permittivity. These interfaces include well-defined interfaces, such as the ballast/sub-ballast interface, or undefined interfaces, such as inclusion anomalies and heterogeneities within each layer (Daniels, [8]). Due to the contrast of dielectric permittivity, a portion of the signal energy incident upon the interface will be reflected

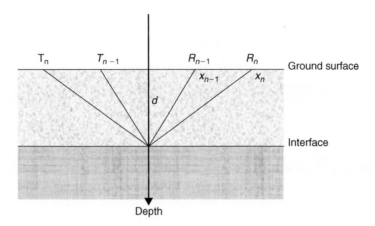

Figure 13.4 Common mid-point measurements (Su et al., [1]).

back and the remaining energy will be transmitted through the interface. The amount of energy reflected from and transmitted through the interface depends upon the extent of the difference in the dielectric properties of the two layers. Knowing the velocity of the wave through the relevant media, the depth is calculated by:

$$d = v\left(\frac{t}{2}\right)$$

(13.2)

where d is the thickness of layer, v the velocity of electromagnetic wave through the layer and t the two-way travel time in this layer.

If the propagation velocity can be measured, or derived, an absolute measurement of depth or thickness can be made. For homogeneous and isotropic materials, the relative propagation velocity can be calculated from (Daniels, [8]):

$$v = \frac{c}{\sqrt{\varepsilon_r}}$$

(13.3)

where ε_r is relative dielectric permittivity of the medium and c speed of light in a vacuum.

In most practical situations the relative permittivity will be unknown. The velocity of propagation must be measured in-situ, estimated by means of direct measurement of the depth to a physical interface or target (i.e. by trial holing or trial pit), or by calculation by means of multiple measurements.

In radar survey, two kinds of velocity measurements can be carried out depending on whether the antenna offset is fixed or can be raised. If the antenna offset can be changed, the common mid-point (CMP) or Wide angle (WA) reflection measurements can be used to calculate the propagation velocity. In the first case, both antennas are simultaneously moved apart at the same speed on either side of the midpoint of the profile. In the second case, one antenna remains stationary while the other is moved along the profile direction (Tillard and Dubois, [11]). Figure 13.4 presents the common

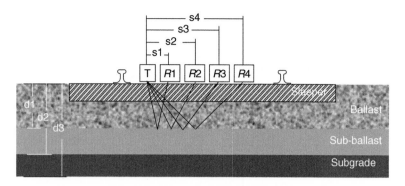

Figure 13.5 A radar system with multi-offset antennae (Su et al., [1]).

mid-point method. In the case of a horizontal reflecting plane in a homogeneous medium, the two-way travel time of the reflected wave can be written as:

$$t^2 = \frac{4x^2}{v^2} + \frac{4d^2}{v^2}$$

(13.4)

where t is the two-way travel time and x offset between antennas, d depth of the reflector and v velocity of radar signal in the medium. Plotting t^2 against x^2 will yield a linear graph of gradient $4/v^2$ and intercept $4d^2/v^2$ and therefore the propagation velocity v and depth d can be determined.

If the antenna offset cannot be varied, the measurement can also be determined using multi-offset method with a multiple pair of antennae or one transmitter and multiple receivers. Figure 13.5 shows a multi-offset antennae system with one transmitter and multiple receivers. With the multi-offset configuration, wave propagation velocity can be calibrated while the system travelling along the track.

13.2.2 Acquisition and processing of GPR data

In order to study the influence of antenna frequency, data were collected using different ground coupled antenna frequencies of 500 MHz, 800 MHz, 1.6 GHz and 2.3 GHz. Before the rails and sleepers were installed, GPR data were collected by pulling the antennae on timber plates placed on the ballast. A wheel encoder was employed to determine the distance the antennae travelled and a X3M control unit and XV11 monitor were used to collect the data for the 500 MHz and 800 MHz antennae (Fig. 13.6). For higher frequency antennae, a CX10 monitor with a combined control unit was used. The horizontal sampling spacing was 0.01 m, while the other acquisition parameters between the antennae were different. Figure 13.7 shows the travelling lines along which the data were collected, including three lines in an X-direction across all the sub-sections and nine lines in a Y-direction, with each line through one section.

After the entire track was completed, more GPR data was acquired using 800 MHz and 1.2 GHz antennae attached to a railway trolley under both dry and wet ballast conditions.

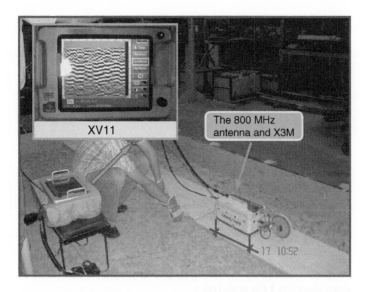

Figure 13.6 Data acquisition using the 800 MHz antenna (Su et al., [1]).

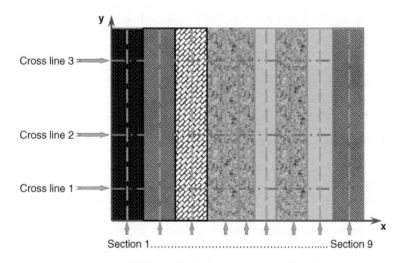

Figure 13.7 Inspection lines (Su et al., [1]).

Raw data were processed to enhance the ratio of signal to noise and highlight the location of the interfaces and texture of radargram. The processing includes band pass filtering, direct current (DC) removal, subtracting mean trace (or background removal), and controlling the gain. The least possible processing should be applied to the raw data to avoid introducing artificial textures into the radargram.

Figure 13.8 shows a comparison between raw and processed radargram from the 500 MHz antenna travelling along line 3. The depth in the radargrams was estimated using a speed of 1.1×10^8 m/s, based on an average dielectric permittivity of the

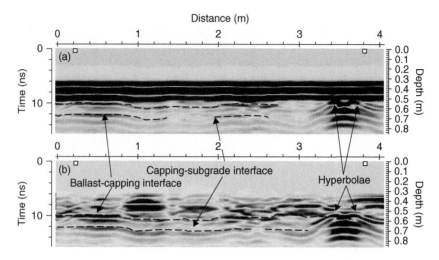

Figure 13.8 Comparison between (a) raw radargram and (b) processed radargram from the 500 MHz antenna along line 3 (Su et al., [1]).

geotechnical materials. Two interfaces and two hyperbolae can be seen on the unprocessed radargram at about 10 nano-seconds but no useful information can be obtained close to the ballast surface due to noise. After applying the DC and background removal methods, there was an obvious improvement in the ratio of signal to noise. Differences from the processed radargram between the textures and patterns can be used to identify the condition of the ballast.

13.2.3   Influence of antenna frequency

The antenna frequency should be determined based on the requirement for both resolution and depth of penetration (Daniels, [8]). Low frequency antennae can penetrate deeper into the ground but they offer only a low resolution, while the high frequency antennae give a high resolution but can only penetrate to a shallower depth. High frequency antennae can monitor ballast condition by providing strong reflections from voids and forming different radargram textures. However, a strong reflection from voids between the ballast also weakens reflections from the existing interfaces and/or foreign objects, which makes them difficult to distinguish. Therefore, four different frequencies were used to discover the optimum frequency for monitoring ballast.

Figure 13.9 shows the processed data collected along Line 3 by directly dragging (a) 500 MHz, (b) 800 MHz, (c) 1.6 GHz and (d) 2.3 GHz antennae over the surface of the ballast. The ballast-capping and capping-subgrade interfaces, and two hyperbolae reflected from two steel pipes (50 mm diameter) could be clearly indicated on the processed radargram of the 500 MHz antenna. Textures for different sub-sections of the track were different but not clear due to the low resolution. The 800 MHz antenna could also identify the interfaces and hyperbolae reflected from the implanted steel pipes. Different textures between clean and fouled sections were noticeable on

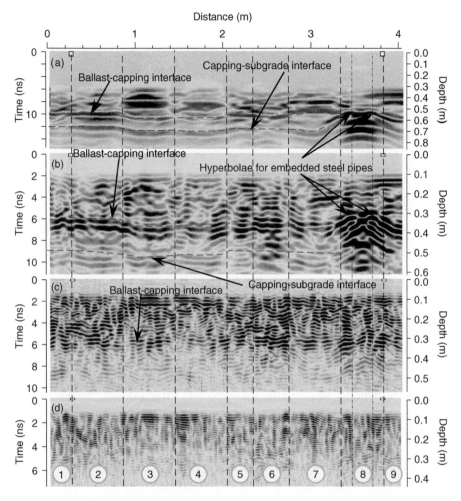

Figure 13.9 Comparison between processed radargrams from antennae (a) 500 MHz, (b) 800 MHz, (c) 1.6 GHz and (d) 2.3 GHz along line 3 (Su et al., [1]).

the radargram. The textures of the clean sections (Sections 6 and 8) were more pronounced than those of the fouled sections (such as Section 7). The 1.6 GHz antenna could not clearly recognise the interfaces between different layers. The hyperbolae were mixed with reflected signals from particles of ballast, and the steel pipe could not be detected. The difference in radargram textures between the clean and fouled sections were comparable to the 800 MHz antenna. It was difficult to observe any interface from the radargram of the 2.3 GHz antenna because of its shallower penetration and interference from signals reflected from ballast particles. The comparison in Figure 13.9 between the four radargrams shows that as the frequency increases, the texture of the radargram become finer, but the ability to distinguish interfaces with the antenna decrease. Of the four frequencies tested here, the 800 MHz antenna gave the clearest image for monitoring the track layers.

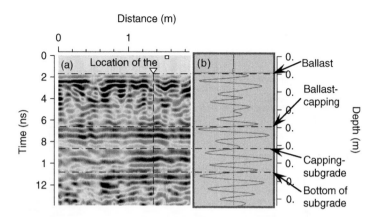

Figure 13.10 (a) Radargram and (b) amplitude profile for 800 MHz antenna along Section 2 (Su et al., [1]).

The radargram and amplitude profile captured by the 800 MHz antenna along Section 2 are shown in Figure 13.10. Interfaces between ballast-capping, capping-subgrade and subgrade-concrete floor are clearly visible on the radargram by continuous reflection bands. From the amplitude profile, a significant increase in amplitude of the reflected signal can be detected at the interfaces owing to the difference in relative dielectric permittivity of the materials. The interfaces can, therefore, easily be located from the radargram and the amplitude profile using image processing tools and a simple mathematical model, respectively. Using the 800 MHz antenna, the textures of the ballast, capping, and subgrade layers can be differentiated indicating its applicability for evaluating the conditions of railway track.

13.2.4 Effect of radar detectable geotextile

Figure 13.11 shows radargrams from the (a) 800 MHz and (b) 1.6 GHz antennae travelling along line 1, in which a layer of Radar detectable geotextile had previously been embedded under the ballast.

Radar detectable geotextile is a type of nonwoven geotextile having a thin aluminum sheet within it. As almost all radar signals are reflected from a metal surface, the interface between the ballast and capping layer shown on the radargram can be clearly shown (Fig. 13.11) which indicates that radar detectable geotextile highlights underground interfaces very effectively. With the existence of radar detectable geotextile, the GPR was able to locate the ballast-capping interface even when the ballast was highly fouled. This is a very useful tool for locating pockets of trapped ballast and deformed capping or subgrade.

The propagation velocity of GPR signal can be estimated by the aid of a certain type of radar detectable geotextile (Carpenter et al., [12]). Figure 13.12 shows a type of radar detectable geotextile with strips of Electric Magnetic (EM) reflective material encapsulated within it. The strips are perpendicular to the rail direction. A reflection hyperbola will be formed when the antennae pass across the reflective strip in the

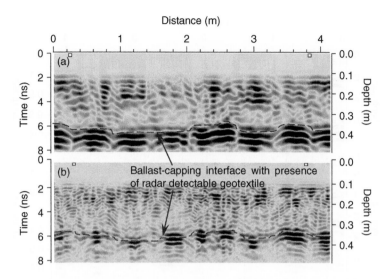

Figure 13.11 Radargram obtained by (a) 800 MHz and (b) 1.6 GHz antennae along line 1 showing the effect of radar detectable geotextile (Su et al., [1]).

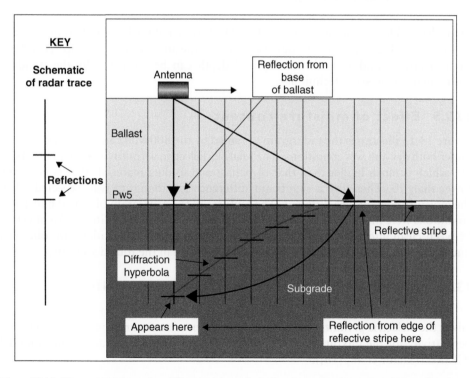

Figure 13.12 The generation of a diffraction hyperbola by the reflective stripe (After Carpenter et al., [12]).

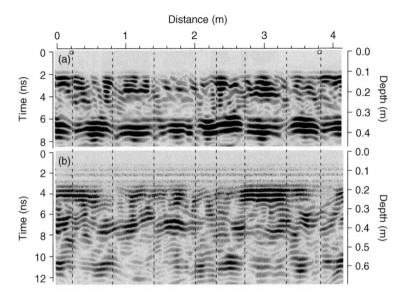

Figure 13.13 Radargram obtained by the 800 MHz antenna along Line 1 under (a) dry and (b) wet conditions (Su et al., [1]).

radar detectable geosynthetic (Fig. 13.12). The formula of the hyperbola is the same as Equation (13.4) except that x is the offset between the antenna and the strip. Therefore, by measuring t and x, both velocity and depth can be estimated by the standard least-squares regression techniques.

13.2.5 Effect of moisture content

Figure 13.13 illustrates the radargram obtained by the 800 MHz antenna along Line 1 under both dry and wet conditions. The relative dielectric permittivity of water is about 80, which is much higher than that of ballast and fouling materials that is normally lower than 10. Therefore a significant difference in radargram was expected when the moisture content was different. The textures of the radargram obtained under wet conditions were much stronger than those obtained under dry conditions. An interface between clean and fouled ballast could even be located on the radargram obtained under wet conditions because of the moisture trapped in the fouled sections.

13.2.6 Applying dielectric permittivity to identify the condition of ballast

Given the dielectric permittivity for each component of a mixture, the relative dielectric permittivity for the mixture can be calculated by the complex refractive index model (CRIM) (Halabe et al., [13]):

$$\sqrt{\varepsilon_T} = \sum_{i=1}^{n} \frac{A_i}{100} \sqrt{\varepsilon_{A,i}} \qquad (13.5)$$

Table 13.2 Calculated propagation velocity and relative dielectric permittivity.

Types of fouling	R_{b-f} (%)	v (m/s)	ε_r
Clean	0	1.43×10^8	4.4
Clean	0	1.42×10^8	4.5
Coal	10	1.38×10^8	4.8
Clayey sand	10	1.31×10^8	5.2
Coal	25	1.29×10^8	5.4
Fine ballast	25	1.23×10^8	5.9
Clayey sand	25	1.25×10^8	5.8
Clayey sand	50	1.18×10^8	6.4
Coal	50	1.20×10^8	6.2

in which, ε_T is the relative dielectric permittivity of the mixture, $\varepsilon_{A,i}$ is the relative dielectric permittivity of component i and A_i is the volumetric percentage of mixture component i.

This model shows that the square root of the dielectric permittivity of a mixture can be determined by multiplying the volumetric percentage of the mixture occupied by the component by the square root of the dielectric permittivity of that component, and subsequently summing the results for all components. Clean ballast consists of particles and air voids between them, but when it becomes fouled, part of the air voids are replaced by fouling particles. The relative dielectric permittivity of the fouled ballast will be greater than the clean ballast because the dielectric permittivity for fouling material is greater than air.

From Figure 13.11, it can be found that the two-way travel time to the ballast-capping interface in different sub-sections is different. However, the depth of the interface is constant along the line. This indicates that the propagation velocity of the radar signal in different sub-sections is different owing to the different relative dielectric permittivity from different conditions of fouling. In the model track, the thickness of the ballast for each sub-section and offset between the transmitter and receiver of a specific shielded antenna were known. The two-way travel time could be obtained from the GPR data so that the propagation velocity of the GPR signal travelling in each sub-section could be calculated using Equation (13.4). The offset between the 800 MHz antenna transmitter and receiver was 0.14 m, so the propagation velocity of the signal for this antenna for each sub-section was determined based on the thicknesses shown in Figure 13.1 and Table 13.1. The corresponding two-way travel time based on the GPR data was acquired along Line 1. The results are summarised in Table 13.2, including the relative dielectric permittivity calculated using Equation (13.3).

Figure 13.14 shows the relationship between relative ballast fouling ratio (R_{b-f}) and the relative dielectric permittivity for ballast fouled with coal, clayey sand, and ballast breakdown, respectively. A significant increase in relative dielectric permittivity can be observed in Figure 13.14 when the degree of fouling increased. For fouling with the same R_{b-f}, the relative dielectric permittivity for ballast fouled with coal was smaller than for clayey sand. The relative dielectric permittivity for coal is smaller

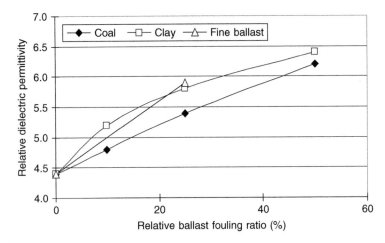

Figure 13.14 Relationship between relative dielectric permittivity and degree of fouling for ballast fouled with different material (Su et al., [1]).

than the soil and rock material. These results indicate that the degree of fouling can be estimated by measuring and calculating the relative dielectric permittivity of fouled ballast, but the type of fouling cannot be differentiated only by a dielectric constant. To achieve this, the texture pattern of the radargram and the amplitude and frequency characteristics of the GPR signal must be analysed and compared. However, the thickness of layers will not be known in a real railway track. In such circumstances, the propagation velocity of the signal can be determined using the CMD or WA methods introduced earlier in Section 13.2.1.

13.3 MULTI-CHANNEL ANALYSIS OF SURFACE WAVE METHOD

A number of geophysical methods have been employed for near-surface characterisation and measurement of shear wave velocity using a wide variety of testing configurations, processing techniques, and inversion algorithms. The most widely used approaches are Spectral Analysis of Surface Waves (SASW) and Multi-channel Analysis of Surface Wave (MASW). The SASW method has been used for sub-surface investigation for several decades (e.g., Nazarian et al., [14], Al-Hunaidi, [15], and Ganji et al., [16]). With this method, the spectral analysis of a surface wave created by an impulsive source and recorded by a pair of receivers is used. The MASW method is a moderately new and improved technique that utilises surface waves from active sources (Park et al., [17] and Xu et al., [18]). The MASW method is more efficient in evaluating shallow sub-surface properties (Park et al., [17] and Zhang et al., [19]). MASW is being increasingly applied to earthquake geotechnical engineering for seismic microzonation and site response studies (Anbazhagan and Sitharam, [20] and Anbazhagan et al., [21]). In particular, it is used to measure the shear wave velocity and dynamic properties, and locate the sub-surface material boundaries and spatial variations of

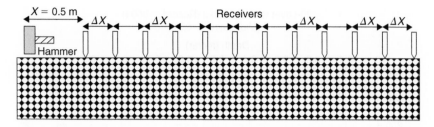

shear wave velocity (Anbazhagan and Sitharam, [22]). MASW can also be used for the characterisation of near surface ground materials (Park et al., [17], 2005, Xia et al., [23] and Kanli et al., [24]).

MASW provides a shear-wave velocity (V_s) profile (i.e. V_s versus depth) by analysing Raleigh-type surface waves on a multi-channel record. An MASW system with a 24-channel SmartSeis seismograph and twelve 10-Hz geophones was used to assess the model track discussed in Section 13.1. At the time of the survey, the ballast was only filled to 270 mm thickness. The seismic waves were generated using a 1-kg sledge hammer and a 70×70 mm aluminium plate with a number of shots. These waves were received by the geophones and further analysed using a software (Anbazhagan et al., [25]).

13.3.1 MASW survey

The MASW survey was conducted by placing 12 geophones parallel to the Y-axis along the Sections 1–9 (Fig. 13.7). The strongest signal was recorded by the receivers when the geophones were placed at 0.25 m (ΔX) interval, and the length between the source to the first receiver was 0.5 m (X). This configuration was applied to all sections. A typical testing arrangement is presented in Figure 13.15. The survey in each section was carried out three times and the seismic signals were recorded every 0.125 ms in a period of 256 ms.

A dispersion curve was initially generated and this is generally displayed as a function of phase velocity versus frequency. Phase velocity can be determined from the linear slope of each component on the swept-frequency record. The accuracy of a dispersion curve can be enhanced by the removal of noise affecting the clarity of important data. High frequency seismic signals were employed to obtain dispersion curves for sections of ballast with a high signal to noise ratio. The frequencies varied from 25 to 60 Hz and had a signal to noise ratio of 80 and above (Fig. 13.16). A typical dispersion curve for a section of ballast is presented in Figure 13.17. An inversion analysis was then carried out by an iterative inversion process that requires the dispersion data to simulate the shear wave velocity (V_s) profile of the medium. A least squares approach allows the process to be automated (Xia et al., [23]) and V_s is updated after each iteration, with Poisson's ratio, density, and model thickness remaining unchanged throughout the inversion. An initial V_s profile should be defined so that V_s at a depth

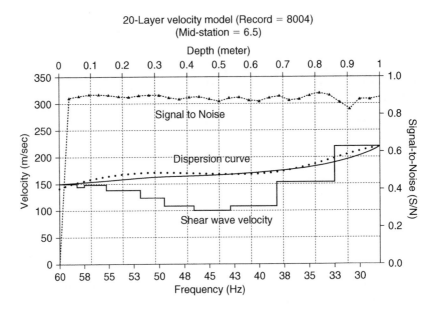

Figure 13.16 Typical velocity of a ballast section (Anbazhagan et al., [25]).

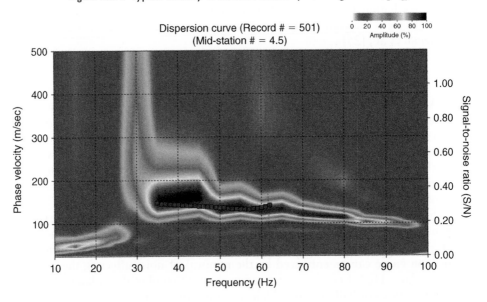

Figure 13.17 Typical dispersion curve of ballast bed (Anbazhagan et al., [25]).

D_f is 1.09 times the measured phase velocity C_f at the frequency where the wavelength λ_f satisfies the following relationship:

$$D_f = a\lambda_f \tag{13.6}$$

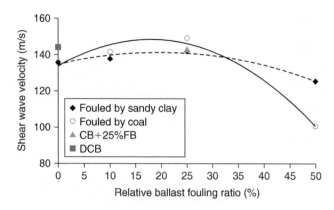

Figure 13.18 Shear wave velocity versus relative ballast fouling ratio (Anbazhagan et al., [25]).

where a is a coefficient that only changes slightly with frequency. A typical shear wave velocity profile obtained for Section 8 is shown in Figure 13.16.

13.3.2 Shear properties of clean and fouled ballast

Shear modulus obtained from seismic survey is widely adopted for site response and seismic microzonation studies. The shear wave velocity for each section of the model track was calculated based on averaging three sets of data having a standard deviation of less than 9. Only four points are available for two types of fouling materials, these points are represented using curves with second order polynomial having a R^2 value of 0.9 and above. The average shear wave velocity of clean ballast was found to (section 6 and 8) vary from 125 to 155 m/s for a density ranging from 1590 to 1660 kg/m^3, which are similar to the shear wave velocity of ballast determined using the resonant column test by Bei [26]. Figure 13.16 shows a typical shear wave velocity for Section 8. The top layer has an average shear wave velocity (V_s) of about 148 m/s which corresponds to clean ballast having a bulk density of 1660 kg/m^3. An average V_s of 135 m/s corresponds to the second layer of clean ballast having a bulk density of 1590 kg/m^3. The average V_s of 115 and 103 m/s corresponds to the capping layer and sub-grade layer below the ballast layer, respectively. Below the sub-grade, the values of V_s increase because of the concrete floor under the model track. In general, the average shear wave velocity of clean ballast is above 125 m/s and fouled ballast is above 80 m/s.

Figure 13.18 shows that initially increase in the degree of fouling increases the shear wave velocity, which is similar to an increase in density due to initial fouling. The shear wave velocity of clean ballast increases when a certain amount of fouling materials is added, after which the velocity of fouled ballast is lower than the clean ballast. With a lower amount of fouling, the shear wave velocity of ballast fouled with coal is slightly greater than when fouled with clayey sand. However, a higher degree of fouling with coal leads to a lower shear wave velocity. The reasons why the shear wave velocity is higher when the amount of coal fouling the ballast is less may be attributed to the size of the particles and specific gravity of the coal. The particles of

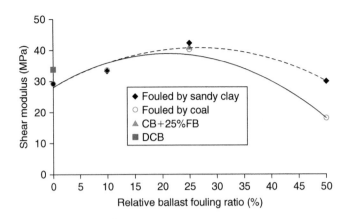

Figure 13.19 Shear modulus versus relative ballast fouling ratio (Anbazhagan et al., [25]).

coal may degrade in the concrete mixer which could lower the shear wave velocity of fouled ballast more than the ballast fouled by clayey sand. The shear wave velocity in Section 3 with ballast fouled by ballast breakdown was similar to the ballast in Section 4 fouled with clayey sand.

The low strain shear moduli of each section were estimated using $G_o = \rho V_s^2$ whilst considering the average shear wave velocity and density of each section. The fouling characteristics and low strain shear modulus of clean and fouled ballast are shown in Figure 13.19.

The shear moduli of clean ballast are approximately 29–34 MPa for the range of density from 1.58 to 1.64 ton/m^3. These values are similar to the shear modulus of fresh ballast given by Ahlf [27] and Suiker et al. [28]. When compared to Sections 6 and 8, the increase in density of clean ballast increases the shear modulus, as expected. If clean ballast is mixed with 25% of fine ballast, the density and compaction of the track bed increases significantly, resulting in higher values of G_{max} to about 41MPa. The shear moduli of ballast fouled by clayey sand vary from 29 to 43 MPa. Whereas, the shear moduli of ballast fouled by coal varies from 17 to 40 MPa. As a result, the lowest shear modulus for Section 9 and the highest value for Section 2 could be observed. Similar patterns can be observed between the sections of ballast fouled by coal and clayey sand due to variations in the specific gravity of fouling materials.

13.3.3 Data interpretation

The shear wave velocity and modulus of fouled ballast increases at the start to reach the maximum values and then begin to decrease. Track maintenance should be carried out based on the degree of fouling, however currently there is no clear criterion to initiate maintenance. This study has shown that after a given degree of fouling, the shear properties of fouled ballast decrease with an increase in the degree of fouling. The optimum fouling point (OFP) represents the highest shear stiffness of fouled ballast, beyond which the shear stiffness decreases drastically. A certain amount of fouling

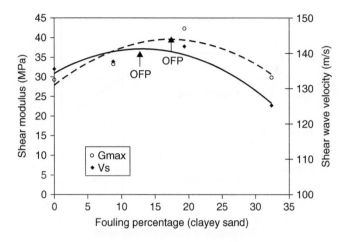

Figure 13.20 Optimum fouling of clayey sand fouled ballast (Anbazhagan et al., [25]).

material can be beneficial towards the track stiffness by optimising the G_{max} of the ballast. To identify the OFP of ballast fouled with clayey sand, the shear wave velocity and modulus with the percentage of fouling are shown in Figure 13.20. The OFP for ballast fouled with clayey sand ranges from 13 to 17% considering both the shear wave velocity and shear modulus. In the field, the ballast density may not vary significantly so the shear wave velocity can be considered to be an ideal parameter for identifying the OFP.

Even though the shear stiffness of fouled ballast decreases after the OFP, it is still greater than the shear stiffness of clean ballast, which means that the track is sufficiently resilient until it reaches a critical fouling point (CFP). Beyond this point, the stiffness and drainage conditions of fouled ballast may not be acceptable and track maintenance will be required. The critical point is a percentage where the shear wave velocity of fouled ballast becomes less than that of clean ballast, and at this point the track shows unacceptable drainage. The permeability of fouled ballast less than 10^{-4} m/s is considered unacceptable based on Selig and Waters [29].

To identify the CFP the shear wave velocity and permeability have been plotted together with respect to the percentage of fouling defined by Selig and Waters [29]. Figures 13.21 and 13.22 show the variation in shear wave velocity and permeability with the percentage of fouling for ballast fouled with clayey sand and coal, respectively. As the fouling of the track bed increases the shear wave velocity, the overall ballast permeability decreases rapidly before approaching OFP. After reaching OFP the permeability decreases slightly. Both figures show that the shear wave velocity of fouled ballast decreases less than that of clean ballast (horizontal line) when the permeability approaches 10^{-4} m/s (vertical line). This point can be defined as the CFP where track maintenance becomes desirable. The critical percentages of fouling for ballast contaminated by clayey sand and coal are about 26% and 16%, respectively.

According to the rail industry, the condition of the track at Bellambi (New South Wales, Australia) was acceptable but relatively poor at Rockhampton (Queensland,

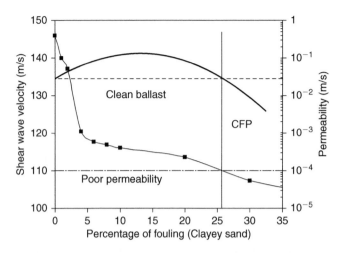

Figure 13.21 Shear wave velocity and permeability of clayey sand fouled ballast (Anbazhagan et al., [25]).

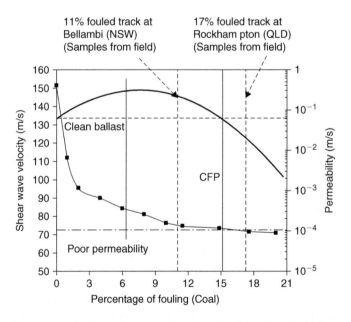

Figure 13.22 Shear wave velocity and permeability of coal fouled ballast (Anbazhagan et al., [25]).

Australia) hence recommended for maintenance. The sample from Bellambi showed that the ballast bed could be categorised as 'moderately clean' based on the percentage of fouling but the sample from Rockhampton was categorised as fouled. The percentage of fouling for these field samples as plotted in Figure 13.22 clarifies that apart from the reduction in shear stiffness (shear wave velocity), the decrease in permeability (drainage) must also be considered simultaneously before maintenance of track is undertaken.

REFERENCES

1. Su, L.J., Rujikiatkamjorn, C. and Indraratna, B.: An evaluation of fouled ballast in a laboratory model track using ground penetrating radar. *ASTM Geotechnical Testing Journal*, Vol. 33, Issue 5, 2010 343–350.

2. Indraratna, B., Rujikiatkamjorn, C. and Su, L.J.: A new parameter for classification and evaluation of railway ballast fouling. *Canadian Geotechnical Journal*, 2010 (Accepted).

3. Gallagher, G.P., Q. Leiper, R. Williamson, M.R. Clark and M.C. Forde: The application of time domain ground penetrating radar to evaluate railway track ballast. *NDT&E International*, Vol. 32, 1999, pp. 463–468.

4. Jack, R. and P. Jackson: Imaging attributes of railway track formation and ballast using ground probing radar. *NDT&E International*, Vol. 32, 1999, pp. 457–462.

5. Clark Max, Michael Gordon, Mike C. Forde: Issues over high-speed non-invasive monitoring of railway trackbed. *NDT&E International*, Vol. 37, 2004, pp. 131–139.

6. Hugenschmidt, J.: Railway track inspection using GPR. *Journal of Applied Geophysics*, Vol. 43, 2000, pp. 147–155.

7. Al-Qadi Imad, L., Wei Xie and Roger Roberts: Scattering analysis of ground-penetrating radar data to quantify railroad ballast contamination. *NDT&E International*, Vol. 41, 2008, pp. 441–447.

8. Daniels, D.J.: Ground Penetrating Radar 2nd Edition. The Institution of Electrical Engierrs, Stevenage, UK, 2004.

9. Sussmanna Theodore R., Ernest, T. Selig and James, P. Hyslip: Railway track condition indicators from ground penetrating radar. *NDT&E International*, Vol. 36, 2003, pp. 157–167.

10. Clark, M.R., R. Gillespie, T. Kemp, D.M. McCann and M.C. Forde: Electromagnetic properties of railway ballast. *NDT&E International*, Vol. 34, 2001, pp. 305–311.

11. Tillard Sylvie and Jean-Claude Dubois: Analysis of GPR data: wave propagation velocity determination. *Journal of Applied Geophysics*, Vol. 33, 1995, pp. 77–91.

12. Carpenter, D., P.J. Jackson and A. Jay: Enhancement of the GPR method of railway trackbed investigation by the installation of radar detectable geosynthetics. *NDT&E International*, Vol. 37 2004, pp. 95–103.

13. Halabe, U.B., Sotoodehnia, A., Maser, K.M. and Kausel, E.: Modeling the Electromagnetic Properties of Concrete. *ACI Materials Journal*, Vol. 90, No. 6, American Concrete Institute, 1993, pp. 552–563.

14. Nazarian, S., Stokoe II, K.H., and Hudson, W.R.: Use of spectral analysis of surface waves method for determination of moduli and thicknesses of pavement systems. *Transport Research Record*, No. 930, 1983, pp. 38–45.

15. Al-Hunaidi, M.O.: Difficulties with phase spectrum unwrapping in spectral analysis of surface waves non-destructive testing of pavements. *Canadian Geotechnical Journal*, Vol. 29, 1992, pp. 506–511.

16. Ganji, V., Gukunski, N., and Maher, A.: Detection of underground obstacles by SASW method-Numerical aspects. *Journal of Geotechnical and Geoenvironmental Engineering*, Vol. 123, No. 3, 1997, pp. 212–219.

17. Park, C.B., Miller, R.D., and Xia, J.: Multi-channel analysis of surface waves. *Geophysics*, Vol. 64, No. 3, 1999, pp. 800–808.

18. Xu, Y., Xia, J., and Miller, R.D.: Quantitative estimation of minimum offset for multi-channel surface-wave survey with actively exciting source. *Journal of Applied Geophysics*, Vol. 59, No. 2, 2006, pp. 117–125.

19. Zhang, S.X., Chan, L.S., and Xia, J.: The selection of field acquisition parameters for dispersion images from multichannel surface wave data. *Pure and Applied Geophysics*, Vol. 161, 2004, pp. 185–201.

20. Anbazhagan, P., and Sitharam, T.G.: Site characterization and site response studies using shear wave velocity. *Journal of Seismology and Earthquake Engineering*, Vol. 10, No. 2, 2008, pp. 53–67.
21. Anbazhagan, P., Sitharam, T.G., and Vipin, K.S.: Site classification and estimation of surface level seismic hazard using geophysical data and probabilistic approach. *Journal of Applied Geophysics*, Vol. 68, No. 2, 2009, pp. 219–230.
22. Anbazhagan, P., and Sitharam, T.G.: Spatial variability of the weathered and engineering bed rock using multichannel analysis of surface wave survey. *Pure and Applied Geophysics*, Vol. 166, 2009, pp. 1–20.
23. Xia, J., Miller, R.D., and Park, C.B.: Estimation of near-surface shear-wave velocity by inversion of Rayleigh wave. *Geophysics*, Vol. 64, No. 3, 1999, pp. 691–700.
24. Kanli, A.I., Tildy, P., Pronay, Z., Pinar, A., and Hemann, L.: Vs^{30} mapping and soil classification for seismic site effect evaluation in Dinar region, SW Turkey. *Geophysics Journal International*, Vol. 165, 2006, pp. 223–235.
25. Anbazhagan, P., Indraratna, B., Rujikiatkamjorn, C. and Su, L.: Using a seismic survey to measure the shear modulus of clean and fouled ballast. *Geomechanics and Geoengineering: an International Journal*, Vol. 5, No. 2, 2010, pp. 117–126.
26. Bei, S.: Effects of railroad track structural components and subgrade on damping and dissipation of train induced vibration. *Doctoral thesis*. The Graduate School, University of Kentucky, Lexington, Kentucky, America, 2005.
27. Ahlf, R.E.: M/W costs: how they are affected by car weight and the track structure. *Railway Track and Structures*, Vol. 71, No. 3, 1975, pp. 34–37.
28. Suiker, A.S.J., Selig, E.T., and Frenkel, R.: Static and cyclic triaxial testing of ballast and subballast. *Journal of Geotechnical and Geoenvironmental Engineering*, Vol. 131, No. 6, 2005, pp. 771–782.
29. Selig, E.T., and Waters, J.M.: Track geotechnology and substructure management. London: Thomas Telford, New York, American Society of Civil Engineers, Publications Sales Department, 1994.

Chapter 14

Track Maintenance

Rail tracks deform both vertically and laterally under cyclic loads resulting from varying traffic loads and speeds, causing deviation from the design geometry. Although these deviations are apparently small, they are usually irregular in nature, deteriorate riding quality and increase dynamic load, which in turn, further worsens the track level and alignment. In order to maintain the design geometry, riding quality and safety levels, rail tracks invariably need maintenance after their construction.

Worldwide, track maintenance is a costly routine exercise. A major portion of the maintenance budget is spent on geotechnical problems [1, 2, 3]. Ballast is the only external constraint applied to the track for holding the running surface geometry [2]. In many countries of the world including the USA, Canada and Australia, hundreds of millions of dollars are spent each year on large terrains of rail track, particularly for maintaining ballast [1, 3]. Effective use of available resources and timely adopting innovative technologies to improve riding quality and safety levels, while minimising maintenance cost still remains a challenging task to the engineers. In this Chapter, the conventional and state-of-the-art machines and methods used in track maintenance are described. An insight is also given on track geotechnology and maintenance of track in cold regions. Further, various techniques are described which can minimise the construction and maintenance challenges in permafrost regions.

14.1 TRACK MAINTENANCE TECHNIQUES

14.1.1 Ballast tamping

Ballast tamping is a traditional method and frequently used all over the world to correct the track geometry. Tamping consists of lifting the track and laterally squeezing the ballast beneath the sleeper to fill the void spaces generated by the lifting operation. The sleepers thus retain their elevated positions.

Figure 14.1 shows a typical tamping machine used for track maintenance and Figure 14.2 gives a closer view of the machine showing tamping tines and lifting rollers. It is a self-propelled machine. The lifting and lining rollers grip the head of rails and can lift the track to a predetermined level. It can also move the rails laterally to re-align the track. Figure 14.3 shows the penetration of ballast layer by the tamping tines.

Ballast tamping is an effective process for re-adjusting the track geometry. However, some detrimental effects, such as ballast damage, loosening of ballast bed and reduced track resistance to lateral displacement and buckling, accompany it. Loosening

Figure 14.1 A typical tamping machine used for track maintenance.

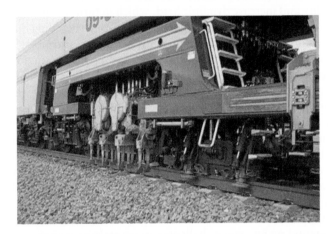

Figure 14.2 A closer view of the tamping machine showing tamping tines and lifting rollers.

Figure 14.3 Tamping tines penetrating the ballast layer.

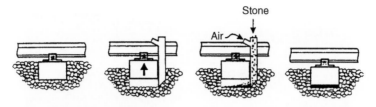

Figure 14.4 Schematic illustration of stoneblowing operation (after Anderson et al., [5]).

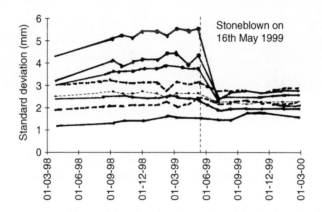

Figure 14.5 Improvement of vertical track profile after stoneblowing (after Anderson et al., [5]).

of ballast by the tamping process causes high settlement in track. Tamping is eventually needed again over a shorter period of time, and in the long run, ballast gradually becomes contaminated (fouled) by fines, which impairs drainage and its ability to hold the track geometry. Eventually fouled ballast will need to be replaced, or cleaned and re-used in track [4].

14.1.2 Stoneblowing

'Stoneblowing' is a new mechanised method of reinstating railway track to its desired line and level [5, 6]. Before the mechanised tamping, track had been re-levelled by 'hand shovel packing', where the sleepers were raised and fine aggregates were shoveled into the voids with minimum disturbance to the well-compacted ballast. The mechanised version of this process is known as 'pneumatic ballast injection' or 'stoneblowing' [5]. The stoneblowing machine lifts the sleeper and blows a predetermined amount of small single size stones into the void beneath the sleeper to create a two layer granular foundation for each sleeper. Figure 14.4 shows schematic operational steps of ballast maintenance by stoneblowing.

Anderson et al. [5] reported the real track data measured in the UK both before and after stoneblowing (Fig. 14.5). They concluded that this technique improves the track profile significantly. Before stoneblowing, the monitored track was deteriorating with time, as revealed by the increasing standard deviation (Fig. 14.5). In contrast,

Figure 14.6 Ballast cleaning machine (after Esveld, [7]).

after the stoneblowing the track quality (represented by the standard deviation) not only improved but was also maintained for a longer period of time.

14.1.3 Ballast cleaning and ballast renewal

As mentioned earlier, when ballast gets excessively fouled (beyond a threshold value), its function is impaired even after using other maintenance techniques (e.g. tamping or stoneblowing). In that case, the contaminated ballast must be cleaned or replaced by fresh ballast. Ballast cleaning and renewal process is a costly and time consuming exercise. It also disrupts traffic flow, and therefore, is not frequently undertaken. Deciding which remedial measure would be appropriate to undertake depends on the site condition and in-situ investigation of foundation materials including subgrade. Traditionally, investigation of track foundation is carried out by a series of cross trenches [4]. However, sinking boreholes using track mounted boring machine will provide further information regarding the foundation condition.

Cleaning the fouled ballast is usually carried out by a track-mounted cleaner, as shown in Figure 14.6. The cleaner digs away the ballast below the sleepers by a chain with 'excavating teeth' attached, conveys it up to a vibrating screen, which separates the dirt (fines) from the coarser aggregates [7]. The dirt is then conveyed away to lineside or spoil wagons for disposal. The cleaned ballast is returned for re-use in track.

The ballast cleaner usually separates the fines from fouled ballast to provide a uniform depth of compacted and clean ballast resting on the smooth cut surface of a compacted subballast layer. However, past experience indicates that the cutter bar is not able to cut the geometrically smooth surface required for the compacted subballast layer due to mechanical vibrations and operator dependent cutting depths [4].

When ballast becomes excessively dirty, it may need to be totally removed rather than on-track cleaning and then replaced with fresh ballast. In these circumstances, the cleaning machine cuts the ballast and conveys it into the wagons. After removing fouled

Figure 14.7 Stockpiles of waste ballast at Chullora (NSW).

ballast, the conveyor/hopper wagons are moved to a discharge side for stockpiling and/or recycling. Figure 14.7 shows a typical large stockpile of waste ballast at a Sydney suburb (Chullora).

To minimise further quarrying for fresh ballast, preserve the environment, and most importantly, to minimise the track construction and maintenance cost, discarded waste ballast can be cleaned and recycled to the track. Laboratory experimental results (presented in Chapters 4 and 5) clearly indicate that recycled ballast when stabilised with appropriate geosynthetics (e.g. geocomposites) can become a viable alternative construction material to the commonly used fresh ballast for track construction and maintenance.

14.2 TRACK GEOTECHNOLOGY AND MAINTENANCE IN COLD REGIONS

The reliable operation of high speed trains in areas where freezing conditions exist, demands the tracks to be free of problems associated with sub-zero temperatures. Soil freezing leads to 'frost heave' producing differential settlement in the track, and subsequent temperature rise induces 'thaw softening' with increased pore water pressures causing reduction in both the bearing capacity and stiffness of the subgrade. In addition, the adverse climatic conditions in combination with high stress levels due to train loading causes rapid ballast degradation thereby presenting considerable serviceability issues.

Freezing of some soils can cause the ground surface to heave by as much as several tens of centimetres. There are three preconditions which are necessary for frost heave to occur, frost–susceptible soil, freezing temperature and availability of water [8]. Also, the frost heave pressure should exceed the weight of track structure above the heaving layer. The overall volume increase can be many times greater than the 9% expansion that occurs when water freezes [9]. When soil freezes, the pore water converts to ice, increasing the resilient modulus, eliminating pore water pressure and decreasing the rate of plastic strain accumulation. Deficiency of the adsorbed water around the fine-grained soil particles or clay lattices only helps towards the growth of ice lenses, eventually pushing the soil particles apart. This segregation results in local cracking and heaving causing uneven ground profile (differential settlement) and significant vertical displacement of track adequate to make the track inoperative [10].

The amount of heaving on a particular railway line will vary with the type of subgrade, drainage conditions, depth of clean ballast, track embankment width, condition of rail cuts among other factors [11]. In the design of track structure, due to the extreme smoothness required for the rail, it is compulsory to investigate the formation characteristics for assessing the risk of frost. Even if the subgrade is non-frost susceptible, unavoidable small amount of frost heave may still occur (under favourable conditions) due to the 9% expansion of pore water on freezing. Assuming the structural layers of the track are saturated with water, the magnitude of this type of heave can be estimated by multiplying the material porosity by the percentage of freeze expansion. However it should be recognised that the composition of the soil and the corresponding water contents vary from place to place, which further leads to differences in heaving over short distances leading to differential heaving [11].

Naturally, frost damage depends on the availability of water in the track domain, making drainage to be of utmost importance as good drainage can minimise the risk of frost damage. For this reason, ballast should be clean and free of soil, coal, remains of plants and other fouling materials. On the other hand, frost also contributes to ballast breakage and increasing the fine content, adding to the fouling of ballast. In fact, the repeated freezing and expansion of the water in grain pores induces tension in the ballast particles which subsequently weakens them, and ultimately causes further degradation. Furthermore, ballast weakened by frost weathering becomes more vulnerable to breakage from traffic loading and during tamping. Nurmikolu [12] studied and reported the progression of frost susceptibility with associated degradation of ballast by means of a series of frost heave tests. The impact of the proportion of fines (that were generated in actual loading environments) on frost susceptibility was evaluated, and it was concluded that the frost susceptibility could be exacerbated with the increase in fines content. In addition, a limit of 1.5% of total fines (i.e. smaller than the 0.020 mm fraction) was proposed based on a frost heave model of crushed rock aggregates.

Before laying the tracks, frost protection works are mandatory in permafrost regions (northern hemisphere in arctic regions). Standards for various countries have been developed by their rail authorities for frost susceptible areas, including Norway, Sweden, Switzerland, Finland, Russia and Canada and some parts of USA. A typical example of negligence of appropriate preventive measures in permafrost areas is the well-known Qinghai-Tibet railway that is the highest (4000 m) and longest (1142 km) rail plateau in the world. It was opened in July 2006, and the settlement and cracking of embankment appeared in some permafrost zones in less than two months has been reported [13].

To prevent frost heaving, a layer of filter and non-frost susceptible material with possible frost insulation boards or a combination of these, is usually placed under the ballast layer in order to prevent frost from penetrating into the frost-susceptible subgrade. For example, in Norway, a layer of peat is normally used as a separator or insulation layer which by virtue of its latent heat of fusion contributes effectively in reducing the frost depth [10]. The thickness of such an insulation layer is determined by the frost penetration depth for a particular area.

Following the effects of frost heave, the subsequent rise in temperature causes the melting of subsurface ice layers (permafrost) causing 'thaw softening'. This not only leads to associated increase in pore pressures and internal seepage, but also causes a reduction in the effective bearing capacity and the stiffness (resilient modulus) of the subgrade. The subsequent live loading will then generate inevitable track settlement. In the permafrost regions, even slight warming due to track engineering activities and passage of trains will promote softening of the subgrade. In addition, change of mean air temperature due to the global warming adds drastically to this effect. The thawing of the underlying permafrost in these countries carry serious budgets for track maintenance and serviceability action plans by the rail organisations.

The use of crushed rock or coarse material (e.g. blockfields, talus, coarse debris) beneath the ballast is sometimes used to lower ground temperatures [14, 15]. The cooling effect is due to continuous exchange of air between the crushed rock and the atmosphere thus increasing evaporation [16]. In addition, thermal-insulation methods (e.g. insulation boards, organic layers) can be adopted in which an insulation layer is put in the embankment so that the heat absorption into permafrost is held back, thus preventing its degeneration. However, some researchers [17] have a view that in long run the permafrost temperature can still rise due to heat accumulation, which would make heat-insulation method inefficient. Shading the surface from solar radiations can assist in cooling the embankment. It avoids the repeated cycles of freeze-thaw which ultimately prevents embankment fill to become loose and weak [18]. This technique is usually used in combination with other methods.

Ventilation ducts or air ducts can also assist in reducing thaw damage to track subgrade [19]. They are installed in the lower portion of the embankment at about half a meter above the original ground surface. During winter, they help to increase the heat loss of underlying soil and in summer they increase the heat absorption [16]. Thermosiphons can also be used to lower the ground temperature due to their excellent heat-transfer ability [17]. A thermosiphon is an airtight vacuum cavity which circulates liquid, having low boiling point (e.g. ammonia, freon), without the need of a mechanical pump. It refers to a method of passive heat exchange based on natural convection. One end of thermosiphon is above the embankment surface and other is embedded into the permafrost layer (Fig. 14.8). Basically, the liquid absorbs heat from the permafrost, becomes less dense, move upwards, condenses and releases heat, thus in turn cooling the permafrost [17].

In summary, a number of techniques are available to minimise the construction and maintenance challenges in permafrost regions, and they include the following or a combination thereof:

- organic insulation layer [10],
- crushed rock layers [14, 15],
- insulation boards [17],
- thermosiphons [17],
- shading boards [18],
- ventilation ducts [16, 19].

The main aim of all the above methods is primarily to maintain the thermal equilibrium of permafrost and to minimise the extent of track maintenance and associated costs.

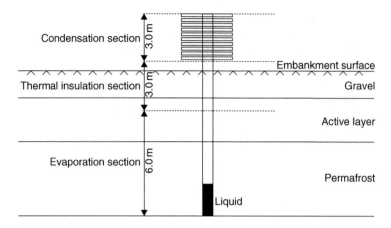

Figure 14.8 Typical thermosiphon (modified after Wen et al., [18]).

REFERENCES

1. Raymond, G.P., Gaskin, P.N. and Svec, O.: Selection and performance of railroad ballast. In: Kerr (ed.): *Railroad Track Mechanics and Technology*. Procedings of a symposium held at Princeton University, 1975, pp. 369–385.
2. Shenton, M.J.: Deformation of railway ballast under repeated loading conditions. In: Kerr (ed.): *Railroad Track Mechanics and Technology*. Procedings of a symposium held at Princeton University, 1975, pp. 387–404.
3. Indraratna, B., Ionescu, D. and Christie, H.D.: Shear behaviour of railway ballast based on large-scale triaxial tests. *Journal of Geotechnical and Geoenvironmental Engineering, ASCE*, Vol. 124, No. 5, 1998, pp. 439–449.
4. Selig, E.T. and Waters, J.M.: *Track Technology and Substructure Management*. Thomas Telford, London, 1994.
5. Anderson, W.F., Fair, P., Key, A.J. and McMichael, P.: The deformation behavior of two layer railway ballast beds. *Proc. 15th Int. Conf. On Soil Mech. Geotech. Engg.*, Istanbul, Vol. 3, 2001, pp. 2041–2044.
6. Key, A.J.: Behaviour of Two Layer Railway Track Ballast under Cyclic and Monotonic Loading. PhD Thesis, University of Shefield, UK, 1998.
7. Esveld, C.: *Modern Railway Track*. MRT-Productions, Netherlands, 2001.
8. Selig, E.T. and Waters, J.M.: *Track Technology and Substructure Management*. Thomas Telford, London, 1994.
9. Harris, J.S.: *Ground freezing in practice*. Thomas Telford, London, UK, 1995.
10. Hartmark, H.: Frost Protection of railway lines. *Engineering Geology*, Vol. 13, 1979, pp. 505–517.
11. Peckover, F.L.: Frost heaving of track-causes and cures. *Proceedings of Technical Conference-American Railway*, Vol. 79, 1978, pp. 143–173.
12. Nurmikolu A.: Degradation and frost susceptibility of crushed rock aggregates used in structural layers of railway track. PhD Thesis, 2005, Tampere University of Technology, Tampere, Finland.
13. Zhu, Z.Y., Ling, X.Z., Chen, S.J., Zhang, F., Wang, L., Wang, Z. and Zou, Z.: Experimental investigation on the train-induced subsidence prediction model of Beiluhe permafrost

subgrade along the Qinghai–Tibet Railway in China. *Cold Regions Science and Technology*, Vol. 62, 2010, pp. 67–75.

14. He, P., Zhang, Z., Cheng, G., Bing, H.: Ventilation properties of blocky stones embankments. *Cold Regions Science and Technology*, Vol. 47, No. 3, 2007, pp. 271–275.

15. Wu, Q., Lu, Z., Tingjun, Z., Ma, W., Liu, Y.: Analysis of cooling effect of crushed rock-based embankment of the Qinghai–Xizang Railway. *Cold Regions Science and Technology*. Vol. 53, No. 3, 2008, pp. 271–282.

16. Cheng, G., Sun, Z., Niu, F.: Application of the roadbed cooling approach in Qinghai–Tibet railway engineering. Cold Regions Science and Technology, Vol. 53, No. 3, 2008, pp. 241–258.

17. Wen, Z., Sheng, Y., Ma, W., Qi, J., Jichun, W.: Analysis on effect of permafrost protection by two-phase closed thermosyphon and insulation jointly in permafrost regions. *Cold Regions Science and Technology*, Vol. 43, No. 3, 2005, pp. 150–163.

18. Yu, Q., Pan, X., Cheng, G., He, N.: An experimental study on the cooling mechanism of a shading board in permafrost engineering. *Cold Regions Science and Technology*, Vol. 53, No. 3, 2008, pp. 298–304.

19. Yu, Q., Niu, F., Pan, X., Bai, Y., Zhang, M.: Investigation of embankment with temperature-controlled ventilation along the Qinghai–Tibet Railway. *Cold Regions Science and Technology* Vol. 53, No. 2, 2008, pp. 193–199.

Chapter 15

Recommended Ballast Gradations

The degradation of ballast is one of the major substructure problems which leads to increased track settlement, increased ballast fouling and reduced drainage. There are several factors affecting ballast deformation and particle breakage, as discussed earlier in Chapter 3. Ballast gradation is a prime factor for the stability, safety and drainage of tracks. A specified ballast gradation must provide the following two key objectives:

- Ballast must have high shear strength to provide increased stability and minimal track deformation. This can be achieved by specifying broadly-graded (well-graded) ballast.
- Ballast must have high permeability to provide adequate drainage, hence readily dissipating excess pore water pressures and increasing the effective stresses. This can be ensured by specifying uniformly-graded ballast.

Clearly, these two objectives are contradictory in terms of required particle size distribution. Higher shear strength of ballast and increased track stability can only be obtained at the expense of ballast drainage capability. The optimum ballast gradation needs a balance between the uniform and broad gradations. Therefore, an attempt was made to find a suitable range of particle size distribution which fulfils the first objective satisfactorily without a significant reduction in the permeability of ballast (i.e. the second objective).

Well-graded ballast gives lower settlement than uniformly-graded aggregates [1, 2]. It has higher shear strength and provides a more stable track with less plastic deformation [3–5]. Well-graded ballast generally attains a higher degree of compaction [2], hence a superior shear strength [4]. An additional advantage of well-graded ballast is that the possibility of inter-mixing between ballast and subballast is low because the voids of well-graded ballast are already filled by the smaller grains [6]. Selig and Waters [7] reported that well-graded distribution extends ballast life and reduces the rate of track settlement. On the other hand, it causes reduced permeability due to smaller void spaces and has high potential of fouling, especially if the source of fouling is from ballast wear. Well-graded ballast is more likely to segregate during transportation and placement, thus making it harder to control in the field [5].

It is expected that both the settlement and degradation of ballast can be significantly reduced by optimising the particle size distribution. In this respect, Lackenby et al. [8] conducted a series of cyclic triaxial tests on ballast varying the particle size distribution. Based on these test results, they recommended a new range of ballast gradation with a uniformity coefficient slightly greater than those specified by the current railway standards (e.g. Australian Standard, AS 2785.7, [9]).

15.1 AUSTRALIAN BALLAST SPECIFICATIONS

The various particle size distributions (PSD) currently used by different rail authorities in Australia (e.g. Rail Infrastructure Corporation (RIC) of NSW, Queensland Rail)

Table 15.1 Railway ballast grading specified by RIC (TS 3402, [10]).

Sieve size (mm)	% passing (by mass) Nominal size: 60 mm
63.0	100
53.0	85–100
37.5	25–65
26.5	0–20
19.0	0–5
13.2	0–2
9.5	0

Table 15.2 Railway ballast grading used by Queensland Rail.

Sieve size (mm)	% passing (by mass) Nominal size: 60 mm
63.0	100
53.0	95–100
37.5	42–64
26.5	4–10
19.0	2–5
13.2	1–4
9.5	0–3
4.75	0

Table 15.3 Railway ballast grading requirements (AS2758.7, [9]).

Sieve size (mm)	% passing (by mass) Nominal size, mm			
	60	60 (steel sleepers)	50	50 (graded aggregates)
63.0	100	100	–	–
53.0	85–100	95–100	100	100
37.5	20–65	35–70	90–100	70–100
26.5	0–20	15–30	20–55	–
19.0	0–5	5–15	0–15	40–60
13.2	0–2	0–10	–	–
9.5	–	0–1	0–5	10–30
4.75	0–1	–	0–1	0–20
1.18	–	–	–	0–10
0.075	0–1	0–1	0–1	0–1

are primarily based on the gradation specified by the Australian Standard for railway ballast [9]. These ballast gradations are presented in Tables 15.1–15.3 for comparison. Although the crib, shoulder and load bearing ballasts play significantly different roles in track, each of these rail authorities specified only one set of PSD for all types of ballast.

The maximum percentage of dust and fine-grained materials (passing 0.075 mm sieve) in ballast is restricted to 1%, as shown in Table 15.3. According to the Australian railway specifications, ballast has been specified to be uniformly-graded for both the upper and lower limits of the gradation range (see Figs. 15.1 and 15.2).

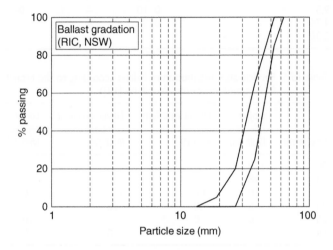

Figure 15.1 Ballast particle size distribution specified by Rail Infrastructure Corporation (RIC), NSW.

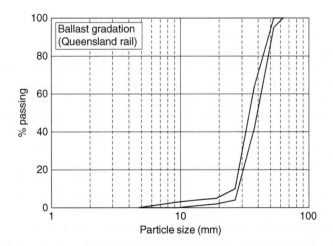

Figure 15.2 Ballast particle size distribution specified by Queensland Rail.

Table 15.4 Ballast gradations (AREMA, [11]).

| Sieve size (mm) | % Passing (by mass) Nominal size square opening, (mm) | | | |
	A: (63.5)	B: (63.5)	C: (50.8)	D: (50.8)
76.2	100	100	–	–
63.5	90–100	80–100	100	100
50.8	–	60–85	95–100	90–100
38.1	25–60	50–70	35–70	60–90
25.4	–	25–50	0–15	10–35
19.1	0–10	–	–	0–10
12.7	0–5	5–20	0–5	–
9.5	–	0–10	–	0–3
4.75	–	0–3	–	–
2.38	–	–	–	–

Table 15.5 Particle size distribution of ballast according to the French Railways (data source: Profillidis, [12]).

Size (mm)	Upper rejection limit (%)	Upper excellent composition limit (%)	Lower excellent composition limit (%)	Lower rejection limit (%)
80	100			
63	98	100		
50	80	86	100	100
40	35	40	76	80
25	0	0	5	10
14			0	0

15.2 INTERNATIONAL RAILWAY BALLAST GRADING

Ballast gradations recommended by the American Railway Engineering and Maintenance-of-way Association [11] are presented in Table 15.4.

In all these gradations, aggregates passing sieve size 200 (i.e. 0.075 mm) should be less than 1% and the limiting value of clay lumps and friable particles is 0.5%. AREMA recommended ballast gradations A, C and D for the mainline tracks, however, gradation B has been included in their recommendation to meet the requirements for other railroads. Gradation B has a uniformity coefficient (C_u) more than 3 and represents relatively more well-graded ballast than other gradations (A, C and D).

Ballast grain size distribution limits specified by the French railways are presented in Table 15.5. In this specification, maximum 2% beyond the limiting values for the particles larger than 63 mm and smaller than 16 mm are accepted. Ballast gradation limits including the excellent composition limits specified by the French Railways are graphically illustrated in Figure 15.3.

A typical composition of ballast grain size used by the British Railways is given in Table 15.6. This gradation represents very uniform ballast, where the uniformity coefficient ($C_u = 1.4$) is much less than most other specifications.

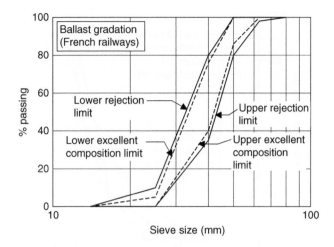

Figure 15.3 Ballast particle size distribution limit curves according to French Railways (modified after Profillidis, [12]).

Table 15.6 Ballast gradation used by the British Railways (after Profillidis, [12]).

Sieve size (mm)	% passing
50	100
28	<20
14	0

15.3 GRADATION EFFECTS ON SETTLEMENT AND BALLAST BREAKAGE

To evaluate the effects of particle size distribution on the deformation and degradation behaviour of ballast, Indraratna et al. [13] conducted cyclic triaxial tests on four different gradations of ballast, as shown in Figure 15.4. Cylindrical ballast specimens were subjected to an effective confining pressure of 45 kPa. To simulate the train axle loads running at high speed, cyclic loading with a maximum deviator stress q_{max} of 300 kPa was applied on the ballast specimens at a frequency of 20 Hz.

Figure 15.5 shows the effects of grain size distribution on the axial and volumetric strains of ballast under cyclic loading. The test results reveal that very uniform to uniform samples give higher axial and volumetric strains. This is attributed to the looser states of the specimens prior to cyclic loading. In contrast, gap-graded and moderately-graded distributions provided denser packing with a higher co-ordination number. Therefore, these gradations provided higher shear strength and thus, decreased the settlement.

Figure 15.6 illustrates the relationship between the uniformity coefficient (C_u) and particle breakage. The test results indicate that ballast breakage decreases as the value of C_u increases, with the exception of the gap graded specimen. The gap-graded ballast

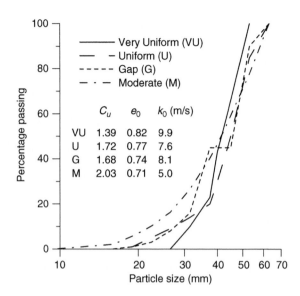

Figure 15.4 Particle size distributions used in cyclic triaxial testing of ballast (after Indraratna et al., [13]).

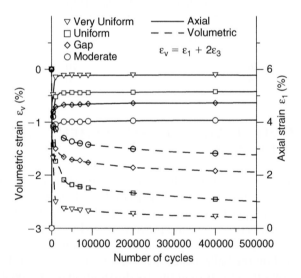

Figure 15.5 Axial and volumetric strain response of different distributions under cyclic loading (after Indraratna et al., [13]).

excluded particle sizes which were found to be highly vulnerable to breakage by the previous researchers [14]. Therefore, the gap-graded specimen shows a smaller amount of breakage than the uniform and very uniform gradations.

As indicated in Figure 15.4, the initial permeability (k_o) for the moderately graded ballast decreased by about 50% from the very uniform distribution. However, in the absence of significant fouling, this permeability of moderately-graded ballast is still

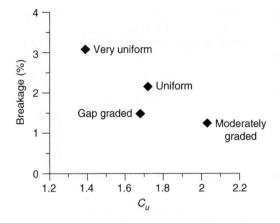

Figure 15.6 Effect of grading on particle breakage (Indraratna et al., [13]).

Table 15.7 Recommended new ballast gradation (after Indraratna et al., [13]).

Sieve size (mm)	% passing (by mass) Nominal size: 60 mm
63.0	100
53.0	85–100
37.5	50–70
26.5	20–35
19.0	10–20
13.2	2–10
9.5	0–5
4.75	0–2
2.36	0–0

considered to be sufficient for track drainage [13]. Moreover, in terms of deformation and resistance to particle breakage, moderately-graded ballast is far superior to uniform gradation, which is used in the current ballast specifications.

15.4 RECOMMENDED BALLAST GRADING

The cyclic test results of ballast varying the gradation indicate that even a modest change in the uniformity coefficient (C_u) substantially affects the deformation and breakage behaviour of ballast. The test results suggest that a distribution similar to the moderate-grading (Fig. 15.4) would give improved track performance. Based on these test findings, Indraratna et al. [13] recommended the following ballast gradation with a uniformity coefficient exceeding 2.2, but not more than 2.6. The recommended ballast gradation, which is relatively more well-graded than the current Australian Standard [9], is presented in Table 15.7. Figure 15.7 graphically illustrates the recommended gradation in comparison with the current Australian Standard.

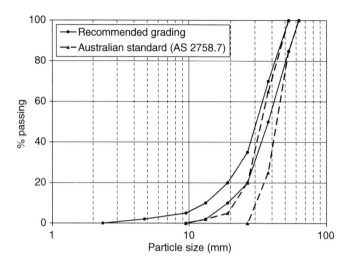

Figure 15.7 Recommended ballast gradation in comparison with current Australian Standard (after Indraratna et al., [13]).

15.5 CONCLUSIONS

The ballast specifications of different countries vary widely with uniformity coefficients ranging from about 1.5–3.0, with a mean in the order of 2 or less. The reasons for the choice of these gradations are not always explained clearly. The gradation of ballast plays a significant role in the strength, deformation, degradation, stability and drainage of tracks. Well-graded ballast gives denser packing, better frictional interlock and hence, lower settlement. However, all ballast specifications demand uniform gradation for free draining. The uniformly-graded ballast gives higher settlement and also more vulnerable to breakage than well-graded ballast.

Recent laboratory test results indicate that the use of slightly broader graded ballast than the current Australian Standard gives considerably lower settlement while not affecting drainage significantly. Moreover, a uniformity coefficient exceeding 2.2 decreases the extent of breakage. From a drainage point of view, this gradation has sufficient permeability and is acceptable for track substructure as long as the ballast is free of fines (fouling) and an appropriate drainage system is constructed along the track. The authors have considered a reasonable balance between the demands for higher strength and free draining in terms of particle size distribution and they recommend a new range of ballast grading with a uniformity coefficient in the order of 2.3–2.6. The proposed new ballast gradation should provide a stronger and more resilient track without causing any significant delay in drainage from the substructure.

REFERENCES

1. Jeffs, T.: Towards ballast life cycle costing. *Proc. 4th International Heavy Haul Railway Conference*, Brisbane, 1989, pp. 439–445.

2. Jeffs, T. and Marich, S.: Ballast characterictics in the laboratory. *Conference on Railway Engineering*, Perth, 1987, pp. 141–147.

3. Raymond, G.P.: Research on railroad ballast specification and evaluation. *Transportation Research Record* 1006, TRB, 1985, pp. 1–8.

4. Raymond, G.P. and Diyaljee, V.A.: Railroad ballast sizing and grading. *J. of the Geotechnical Engineering Division, ASCE*, Vol. 105, No. GT5, 1979, pp. 676–681.

5. Chrismer, S.M.: Considerations of factors affecting ballast performance. *AREA Bulletin AAR Research and Test Dept. Report No. WP-110*, 1985, pp. 118–150.

6. Raymond, G.P.: Railroad ballast prescription: State-of-the-Art. Journal of the Geotechnical Engineering Division, ASCE, Vol. 105, No. GT2, 1979, pp. 305–322.

7. Selig, E.T. and Waters, J.M.: *Track Technology and Substructure Management*. Thomas Telford, London, 1994.

8. Lackenby, J., Indraratna, B. and Khabbaz, H.: The effect of particle size distribution and compaction on ballast degradation. Technical Report No. 4, RAIL-CRC Project No. 6, University of Wollongong, Australia, 2003.

9. AS 2758.7: Aggregates and rock for engineering purposes, Part 7: Railway ballast. *Standards Australia*, NSW, Australia, 1996.

10. T.S. 3402: *Specification for supply of aggregate for ballast*. Rail Infrastructure Corporation of NSW, Sydney, Australia, 2001.

11. AREMA: *Manual for Railway Engineering*. American Railway Engineering and Maintenance-of-way Association, Vol. 1, (Track-Roadway and Ballast), USA, 2003.

12. Profillidis, V.A.: *Railway Engineering*. Avebury Technical, Ashgate Publishing Ltd, UK, 1995.

13. Indraratna, B., Khabbaz, H., Salim, W., Lackenby, J. and Christie, D.: Ballast characteristics and the effects of geosynthetics on rail track deformation. *Int. Conference on Geosynthetics and Geoenvironmental Engineering*, Mumbai, India, 2004, pp. 3–12.

14. Indraratna, B., Salim, W., Ionescu, D. and Christie, D.: Stress-strain and degradation behaviour of railway ballast under static and dynamic loading, based on large-scale triaxial testing. *Proc. 15th Int. Conf. on Soil Mech. and Geotech. Engg*, Istanbul, Vol. 3, 2001, pp. 2093–2096.

Chapter 16

Bio-Engineering for Track Stabilisation

The availability of potential construction sites has continued to decline throughout the world due to over-population in coastal and other metropolitan areas. These circumstances have made engineers to build earth structures, highways, and railways over expansive clays and compressive clay deposits. Following heavy rainfall, seepage beneath the tracks often initiates uneven settlement and potentially hazardous problems if not addressed in a timely manner. The extensive ballast maintenance following heavy rainfall is both costly and time consuming. For example, it has been stated that the cost of maintaining Sydney's rail network was more than two billion dollars in the last decade. Due to the high maintenance costs, the importance of finding appropriate ground improvement techniques to minimise cost can be clearly perceived. Bioengineering aspects of native vegetation are currently being used to improve the soil stiffness, stabilise slopes and control erosion.

Tree roots provide three major stabilising functions: (a) reinforcement of the soil by the roots, (b) dissipation of excess pore pressure through evapo-transpiration and (c) establishing a matric suction that will increase the shear strength. The matric suction induced in the root zone propagates radially and helps stabilise the tracks near the root zone. Figure 16.1 shows schematically, the effect that a single tree located near a railway track has on the ground in its immediate vicinity.

16.1 INTRODUCTION

Various forms of native vegetation are becoming increasingly popular in Australia for improving mechanical and hydrological properties of soft soil. Attempts to quantify these effects have focused on the mechanical strengthening provided by the roots, and the implications of evapo-transpiration on the soil pore water pressure. For instance, the models developed by Chok et al. [2] Operstein & Frydman [3] and Docker & Hubble [4] investigate the reinforcement effect of roots on soil cohesion. The root based soil suction changes have been considered in detail by Indraratna et al. [4] to quantify pore pressure dissipation and induced matric suction incorporating complex inter-relationships among the soil, plant and atmosphere.

The moisture loss from the soil may be classified as: (a) water used for metabolism in plant tissues, and (b) water transpired to the atmosphere. However, as discussed by Radcliffe et al. [6] the required amount of water for photosynthesis or metabolism in plant tissues is negligible compared to the total water uptake by roots. The total

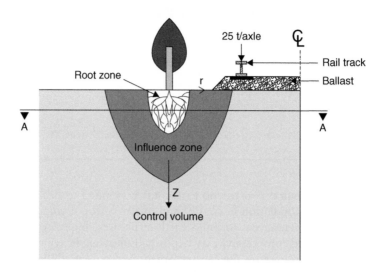

Figure 16.1 Two dimensional vertical view of the root zone and the influence zone of a tree (Fatahi, [1]).

transpiration can then be assumed to be the same as the water uptake through the root zone.

Soil conditions (soil matric suction, and hydraulic conductivity), the vegetation type (root density, the ratio of active roots and leaf area) and atmospheric conditions due to seasons (net solar radiation, temperature, humidity, etc.) influence the rate of root water uptake, hence transpiration. Indraratna et al. [4] formulated a comprehensive equation for calculating the rate of root water uptake considering the interaction between the above features.

16.2 CONCEPTUAL MODELLING

The main variable for estimating the transpiration rate is the root water uptake rate, which is a complex factor because of the considerable variation of the root type and geometry from one species to another. In this section, the key factors, such as soil suction, root distribution and potential transpiration rate are discussed.

16.2.1 Soil suction

Soil suction retards the free water movement towards the root zone and influences the transpiration rate. The root water uptake $(S(x, y, z, t))$ is determined by a combination of the maximum possible root water uptake, S_{max}, and matric suction, ψ:

$$S(x, y, z, t) = S_{max}(x, y, z, t) f(\psi) \tag{16.1}$$

where, $S(x, y, z, t)$ denotes the root water uptake at point (x, y, z) at time t.

In order to calculate $f(\psi)$, the equation suggested by Feddes et al. [7] has been adopted here. The relationship between water uptake and soil suction (Fig. 16.2) as

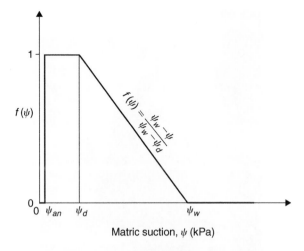

Figure 16.2 Soil suction factor (Indraratna et al., [5]).

suggested by Feddes et al. [7] can be summarised as:

$$
\left.
\begin{cases}
f(\psi) = 0 & \psi < \psi_{an} \\
f(\psi) = 1 & \psi_{an} \le \psi < \psi_d \\
f(\psi) = \dfrac{\psi_w - \psi}{\psi_w - \psi_d} & \psi_d \le \psi < \psi_w \\
f(\psi) = 0 & \psi_w \le \psi
\end{cases}
\right\}
\tag{16.2}
$$

where, ψ_w is the soil suction at wilting point, i.e. the suction limit at which a particular vegetation is unable to draw moisture from the soil. ψ_d is the highest value of ψ and ψ_{an} (soil suction at anaerobiosis point) and it is the lowest value of ψ at $S = S_{max}$, where S_{max} is the maximum rate of root water uptake. An experimental study by Kutilek and Nielsen [8] also confirm the same trend given by Feddes et al. [7] as illustrated in Figure 16.2.

16.2.2 Root distribution

The geometric slope of the root zone is assumed, based on the field observation of typical root cross sections. Trench excavation is one of the appropriate methods to map the root density distribution (Fig. 16.3). The distribution of transpiration within the root zone is a function of the root density, hence,

$$
S(x, y, z, t) = f(\psi)G(\beta)F(T_P)
\tag{16.3}
$$

where, $G(\beta)$ is a function associated with the root density distribution, $F(T_P)$ is a function to consider the potential transpiration distribution, and $\beta(x, y, z, t)$ is the root density.

A traditional agronomical belief implies that the root area of trees below the ground may be as extensive or less than the average canopy above. Some researchers

Figure 16.3 Trench excavation to examine the root density distribution of a native tree (Miram, VIC, Australia).

Docker et al. [4], Dobson and Moffat [9], Sudmeyer [10], Landsberg [11] proposed that the total cross-sectional area of roots, including the depth and distance from the trunk could be determined as an exponential relationship. It is assumed by symmetry that the maximum root density is on a circle with $r = r_0(t)$ at depth of $z = z_0(t)$ and that the root density would decrease exponentially from this maximum value in both vertical and radial directions, thus:

$$\beta(r,z,t) = \beta_{\max}(t)e^{-k_1|z-z_0(t)|-k_2|r-r_0(t)|} \tag{16.4}$$

where, $\beta_{\max}(t)$ is the maximum root density at time t, and k_1 and k_2 are two empirical coefficients depending on the tree root system and type.

16.2.3 Potential transpiration

The potential transpiration is described as evaporation of water from the plant tissues to the atmosphere, assuming that the moisture content of soil is not restricted. The potential transpiration is, therefore, estimated by:

$$T_P = ET_P - E_P \tag{16.5}$$

where, T_P is overall transpiration, ET_P is the potential evapo-transpiration (both plant and soil), and E_P is the potential evaporation from the soil surface.

A combined energy balance and mass balance method can be used to calculate the terms ET_P and E_P. Penman [12], Monteith [13] and Rijtema [14] proposed methods for determining the potential transpiration through potential evapo-transpiration and evaporation. Potential transpiration based on Penman-Brutsaert's model further described by Lai and Katual [15] is given by:

$$T_P = W(R_n - G) + (1 - W)E_A \qquad (16.6)$$

where, T_P is the potential latent heat flux, R_n is the net radiation, G is the soil heat flux, W is a dimensionless weighted function that depends on the slope of the saturation vapor pressure-temperature curve and the psychometric constant, and E_A is the atmospheric drying power function.

The finite element program ABAQUS was employed to evaluate the soil suction generated by transpiration. Equations (16.1)–(16.6) can be typically included as a sub-routine in ABAQUS supplementing the effective stress-based equations.

16.3 VERIFICATION OF THE PROPOSED ROOT WATER UPTAKE MODEL

16.3.1 Case study 1: Miram village (Western Victoria, Australia)

The field investigations were conducted adjacent to an Australian native Black Box tree (Eucalyptus largiflorens) located in Miram village in Western Victoria (Fatahi, [1]). The exact location is shown in Figure 16.4.

The mean daily maximum temperature ranges from 13.7°C in July to 29.7°C in January. The mean monthly rainfall ranges from 20.9 mm in January to 47.7 mm in August, with a mean annual rainfall of 415.3 mm. The mean monthly potential evaporation ranges from 30.45 mm in July to 257.9 mm in January. On an annual basis, the potential evaporation (1483.7 mm/yr) is more than 3 times the average annual rainfall (415.3 mm/yr).

Eucalyptus largiflorens (Black Box) is an Australian native tree, which is very common in the states of New South Wales, Queensland, South Australia, and Victoria. According to Huxley [16] and Genders [17], Eucalyptus largiflorens is an evergreen tree, approximately 10–20 m high and 7.5–15 m in spread with rough bark on trunk and branches. It is a slow growing tree with a relatively shallow root system that thrives under sunny conditions, preferring relatively dry sandy, loamy, and clayey soils. Huxley [16] reported that the Eucalyptus species are deciduous and they continue to grow until the weather becomes too cold in the winter. Deep mulch around the roots prevents the soil from freezing and helps the trees survive very cold conditions. Based on Genders [17], because Eucalyptus largiflorens has shallow roots they should be planted into their permanent positions when small, especially in windy areas.

A mobile drilling rig with 76 mm drill was employed to drill vertical bore holes based on the rotary-dry method. The cores obtained by push-sampling tubes were waxed immediately after extrusion. The geotechnical profile found at the site is shown in Figure 16.5.

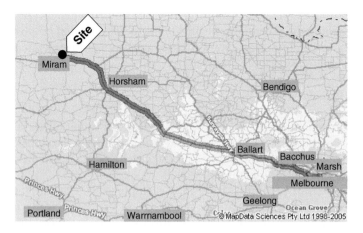

Figure 16.4 Location of the site for the geotechnical investigation, Victoria, Australia (Fatahi, [1]).

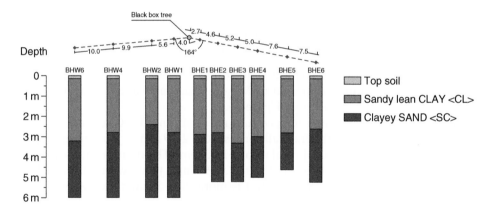

Figure 16.5 Geotechnical section of Miram site (Fatahi, [1]).

The average soil profile can be described as 0.2 m of sandy clay topsoil underlain by brown, firm to hard sandy lean clay to approximately 3 m below the surface. Beneath the clay is a medium dense to dense clayey sand layer approximately 3–6 m deep. The ground-water level is below 6 m. Soil changes are gradual, with no distinct layer boundaries evident below the base of the topsoil.

Two 1 m wide × 35 m long × 3.5 m deep trenches were dug by an excavator to observe the distribution of tree roots and the dimensions of the root zone. An extra trench was excavated between these two trenches to check the relatively homogeneous distribution. Field measurements revealed that the minimum moisture content and matric suction of the top 3 m soil were 9% and 1700 kPa, respectively. Therefore, the wilting point suction of the soil can be estimated to be around 1700 kPa. The

Table 16.1 Parameters of interaction between tree and ground of a Black Box tree at Miram (Fatahi, [1]).

Parameter	Measured Value	Comments
Γ_{max}	6.2%	Measured according to laboratory organic content test
r_{max}	20 m	Estimated from field observation
z_{max}	3 m	Estimated from field observations
r_0	8.5 m	Radial coordinate of the maximum root density point
z_0	1.2 m	Vertical coordinate of the maximum root density point
$\beta_{f\,max}$	659000 m^{-2}	Measured according to organic content
k_1	0.35	Measured according to organic content
k_2	0.55	Measured according to organic content
ψ_w	1700 kPa	Estimated from field measurements
ψ_{an}	4.9 kPa	Clayey soil with air content of 0.04 (Feddes et al., [7])
T_p	80 l/day	Estimated from Indraratna et al. [5]

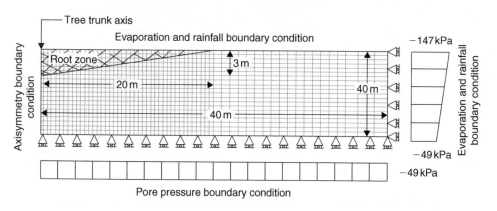

Figure 16.6 Geometry and boundary conditions of the model (Fatahi, [1]).

parameters used in this analysis relating to the interaction between the tree and the atmosphere are given in Table 16.1.

A two dimensional finite element analysis was used to predict the distribution of the soil moisture content and matric suction near a selected Black Box gum tree. The numerical analysis in this case study was based on the basic effective stress theory of unsaturated soils incorporated in the ABAQUS finite element code. The discretised axi-symmetric finite element mesh and specified boundary conditions are shown in Figure 16.6.

Figure 16.7 shows a comparison between the field measurements and the predictions of the numerical model for the volumetric moisture content. The numerical results incorporating the root water uptake model described earlier are in acceptable agreement with the field measurements. According to Figure 16.7, field measurements of moisture content reduction are noticeably different from the finite element predictions close to the tree trunk. This is not surprising as the foliage and the tree trunk alter the uniform distribution of rainfall, and also due to the shade cast beneath the tree canopy, evaporation rate changes as a result of temperature and humidity variations.

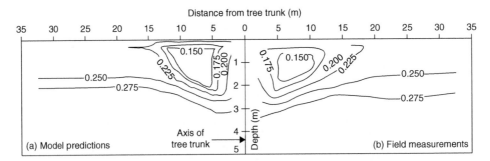

Figure 16.7 Contours of volumetric soil moisture content reduction in vicinity of the tree (a) current numerical analysis results, (b) field measurements in May 2005 (Fatahi, [1]).

Table 16.2 Parameters applied in the finite element analysis (Indraratna et al., [5]).

Parameter	Value	Reference.	Comments
ψ_{an}	4.9 kPa	Feddes et al. [7]	Clay soil with air content of 0.04
ψ_w	1500 kPa	Feddes et al. [7]	$1500 \leq \psi_w \leq 2000$ kPa
ψ_d	40 kPa	Feddes et al. [7]	$40 < \psi_d < 80$ kPa
γ	21 kN/m^3	Powrie et al. [19]	Typical value for Boulder clay
k_s	10^{-10} m/s	Lehane and Simpson [20]	Typical value for Boulder clay
PI	23	Biddle [18]	Measured
e_0	0.60	Powrie et al. [19]	Typical value for Boulder clay
C_c	0.13	Skempton [20]	Typical value for Boulder clay

Consequently, these effects have probably contributed to the disparity between the field data and finite element predictions.

16.3.2 Case study 2: Milton Keynes, United Kingdom

The second case history is related to the results of the field moisture content measured near a single, 14 m high lime tree grown in Boulder Clay near Milton Keynes, UK, as reported by Biddle [18]. Table 16.2 shows the estimated parameters used in the finite element analysis based on the available data in literature. Figure 16.8 illustrates the mesh and element geometry and boundary conditions of the finite element model. A two-dimensional plane strain mesh employing 4-node bilinear displacement and pore pressure elements (CPE4P) was considered. The maximum change in the soil matric suction from the finite element analysis (Fig. 16.9) is found at about 0.5 m depth, which coincides with the same location of the maximum root density.

A comparison between the field measurements and the FEM predictions for moisture content reduction around the lime tree is presented in Figure 16.10. The numerical model is in accordance with the field observations by Biddle [18]. The main differences noted between field data and the predictions are observed at 6–8 m from the trunk. This discrepancy is attributed to the simplicity of the assumed root zone shape. In addition, the foliage prevents uniform distribution of rainfall around the tree. As a

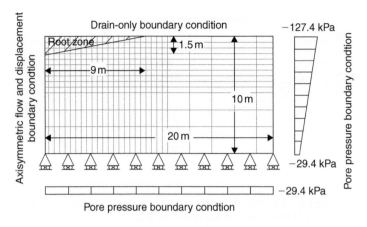

Figure 16.8 The geometry and boundary conditions of case study (Indraratna et al., [5]).

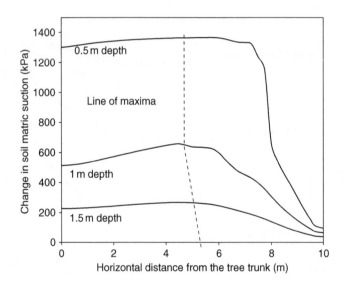

Figure 16.9 Predicted soil matric suction in various depths (Indraratna et al., [5]).

result, moisture content can increase at the canopy edges, thereby further contributing to this disparity.

Figure 16.11 shows the ground settlement at various depths. In this analysis, only the suction related settlement was considered. On the surface, the predicted 80 mm settlement beside the tree trunk decreases to less than 20 mm, at a distance 10 m away from the trunk. As shown in Figure 16.11, the location of the maximum settlement is closer to the trunk at shallower depths, which tends to coincide with the points of maximum change in suction (Fig. 16.10).

It was shown that the numerical analysis incorporating the proposed model could predict the variation of moisture content surrounding the tree trunk. Knowing the

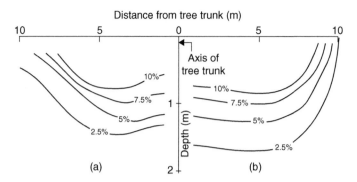

Figure 16.10 Contours of volumetric soil moisture content reduction (%) close to a lime tree: (a) Biddle [18], (b) FEM predictions (Indraratna et al., [5]).

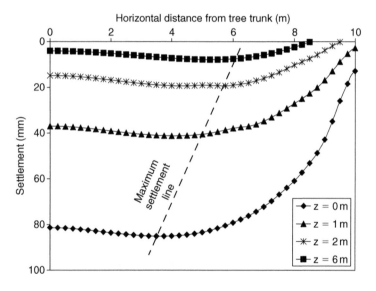

Figure 16.11 Ground settlement at various depths (Indraratna et al., [5]).

moisture content variation, the development of matric suction can be predicted reasonably well using the Soil Moisture Characteristic Curve. Native biostabilisation improves the shear strength of the soil by increasing the matric suction, and also decreases the soil movements. This contribution from trees grown along rail corridors and rail slope is of immense benefit for improving track stability in problematic soil. In other words, native vegetation generating soil suction is comparable to the role of synthetic sub-surface drains with vacuum pressure, in terms of improved drainage (pore water dissipation), and associated increase in shear strength. In addition, the tree roots provide a natural reinforcement effect, which the current model has not simulated thus far.

REFERENCES

1. Fatahi, B.: Modelling of influence of matric suction induced by native vegetation on sub-soil improvement. PhD Thesis, University of Wollongong.
2. Chok Y.H., Kaggwa W.S., Jaksa M.B. and Griffiths D.V.: *Modelling the effects of vegetation on stability of slopes*. 9th Australian New Zealand Conference on Geomechanics, Auckland, 2004, 391–397.
3. Operstein V. and Frydman S.: *The influence of Vegetation on Soil Strength*. Ground Improvement, 2000, 4, 81–89.
4. Docker B.B. And Hubble T.C.T.: Strength and distribution of casuarinas glauca roots in relation to slope stability. Geotechnical Engineering, Swets & Zeitlinger, Lisse, 2001, 745–749.
5. Indraratna, B., Fatahi, B. and Khabbaz, M.: *Numerical Analysis of Matric Suction Effects Induced by Tree Roots*. Geotechnical Engineering, Proc. of Inst. of Civil Engineers, 2006, Vol. 159 No. 2, pp. 77–90.
6. Radcliffe D., Hayden T., Watson K., Crowley P. and Phillips R. E.: *Simulation of soil water within the root zone of a corn crop*. Agronomy Journal, 1980, 72, 19–24.
7. Feddes R.A., Kowalik P.J. and Zaradny H.: *Simulation of field water use and crop yield, Simulation Monograph*. Pudoc, Wageningen, the Netherlands, 1978, 9–30.
8. Kutilek M. and Nielsen D.R.: *Evapotranspiration. Soil Hydrology*, Germany, Catena Verlag, 1994, 195–216.
9. Dobson M.C. and Moffat A.J.: A re-evaluation of objections to tree planting on containment landfills. *Waste Management & Research*, 1995, 13, 579–600.
10. Sudmeyer R.: Tree root morphology in alley system. *Rural Industries Research and Development Corporation*, Australia, 2002, RIRDC Publication No. 02/024, RIRDC Project No. DAW-93A, 2–12.
11. Landsberg J.J.: Tree Water Use and its Implications in Relation to Agroforestry Systems. *Rural Industries Research and Development Corporation* (RIRDC), Australia, 1999, Water and Salinity Issues in Agroforestry No. 5, RIRDC Publication No. 99/37, RIRDC Project No. CSM-4A, 1–24.
12. Penman H.L.: Natural evaporation from open water, bare soil and grass. *Proceedings of the Royal Society of London*, 1948, 193, 120–146.
13. Monteith J.L.: Evaporation and environment. *The State and Movement of Water in Living Organisms*. Academic Press, N.Y., (Fogg, G.E. (Ed.).) 1965.
14. Rijtema P.E.: An analysis of actual evapotranspiration. Agricultural Research Report, Pudoc, Wageningen, 1965, 659–107.
15. Lai C.T. and Katul G.: The dynamic role of root-water uptake in coupling potential to actual transpiration. *Advances in Water Resources*, 2000, 23, 427–439.
16. Hucley, A.J.: *New Royal Horticultural Society Dictionary of Gardening*, MacMillan Press, London. 1992.
17. Genders, R.: *Scented flora of the World*, Robert Hale, London. 1994.
18. Biddle P.G.: Pattern of soil drying and moisture deficit in the vicinity of trees on clay soils. *Geotechnique*, 1983, 33, 2, 107–126.
19. Powrie W., Davies J.N., and Britto A.M.: A cantilever retaining wall supported by a berm during temporary work activities. *ICE conference on retaining structures*. Robinson College, Cambridge, 1992, 418–428.
20. Lehane B.M. and Simpson B.: Modeling glacial till under triaxial conditions using a BRICK soil model. *Canadian geotechnical Journal*, 2000, 37, 1078–1088.
21. Skempton, A.W.: Notes on compressibility of clays. *Quarterly Journal of Geological Society*, London, 1944, 100, 2, 119–135.

Appendices

Appendix A

Derivation of Partial Derivatives of $g(p,q)$ with respect to p and q from a First Order Linear Differential Equation

A simple first order linear differential equation is considered, as given by:

$$\frac{dq}{dp} + pq = 0 \tag{A1}$$

After separating the variables and integrating, the solution of the differential equation (Equation A1) is given by:

$$\ln q + \frac{p^2}{2} + c = 0 \tag{A2}$$

Equation A2 can be re-written in the following form:

$$q - e^{-(p^2/2+c)} = 0 \tag{A3}$$

If Equation A3 represents the function $g = g(p,q)$, then,

$$g(p,q) = q - e^{-(p^2/2+c)} = 0 \tag{A4}$$

Differentiating g with respect to q and p partially gives:

$$\frac{\partial g}{\partial q} = 1 \tag{A5}$$

$$\frac{\partial g}{\partial p} = (-p)\{-e^{-(p^2/2+c)}\} = pq \tag{A6}$$

Derivation of Partial Derivatives of g(p,q) with respect to p and q from a First Order Linear Differential Equation

Determination of Model Parameters from Laboratory Experimental Results

B.I FOR MONOTONIC LOADING MODEL

The current monotonic loading model contains 11 parameters, which can be determined from drained triaxial compression tests with the measurements of particle breakage, as explained below:

The critical state parameters $(M, \lambda_{cs}, \kappa$ and $\Gamma)$ can be evaluated from the critical state line, which is determined from the results of a series of drained triaxial compression tests, as shown in Figs. B1(a)–(b).

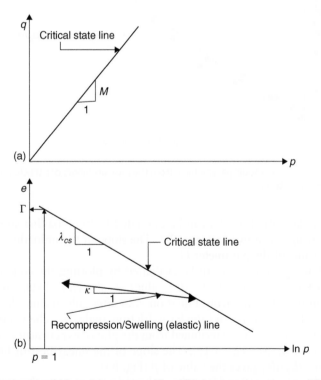

Figure B1 Determination of (a) model parameter M, and (b) the parameters λ_{cs}, κ and Γ from laboratory experimental results.

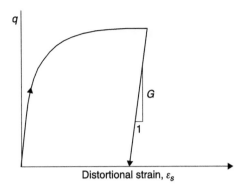

Figure B2 Determination of shear modulus G.

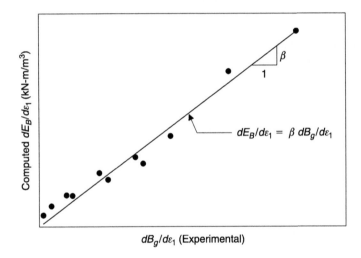

Figure B3 Determination of model parameter β from the measurements of particle breakage in triaxial compression tests.

The elastic shear modulus G, can be evaluated from unloading stress-strain data of triaxial shearing, as shown in Figure B2. The slope of the unloading part of $q - \varepsilon_s$ plot gives the value of the parameter G.

The breakage parameter β can be evaluated by plotting the computed $dE_B/d\varepsilon_1$ values against the experimental $dB_g/d\varepsilon_1$ values, as shown in Figure B3. The values of $dE_B/d\varepsilon_1$ can be computed from Equation 7.23 after substituting the experimental values of $q, p', (1 - d\varepsilon_v/d\varepsilon_1)$, and the basic friction angle, ϕ_f. The values of $dB_g/d\varepsilon_1$ at various strain levels can be determined from the plot of experimental measurements of breakage index B_g (see Fig. 5.19). The slope of the linear best-fit line of the plot $dE_B/d\varepsilon_1$ versus $dB_g/d\varepsilon_1$ gives the value of β (Fig. B3).

The model parameters θ and υ can be evaluated by re-plotting the particle breakage data (B_g) in a modified scale of $\ln \{p_{cs(i)}/p_{(i)}\}B_g$ versus ε_s, as shown in Figs. B4(a)–(b).

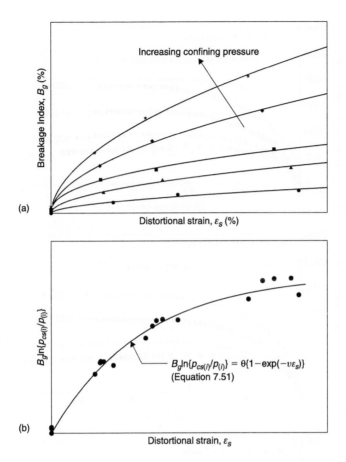

Figure B4   Determination of model parameters θ and υ from breakage measurements, (a) variation of B_g with distortional strain and confining pressure, (b) modelling of particle breakage.

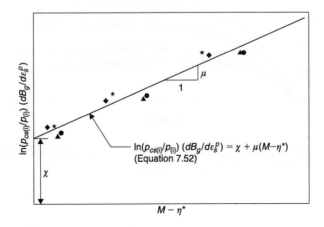

Figure B5   Determination of model parameters χ and μ from triaxial compression tests and breakage measurements.

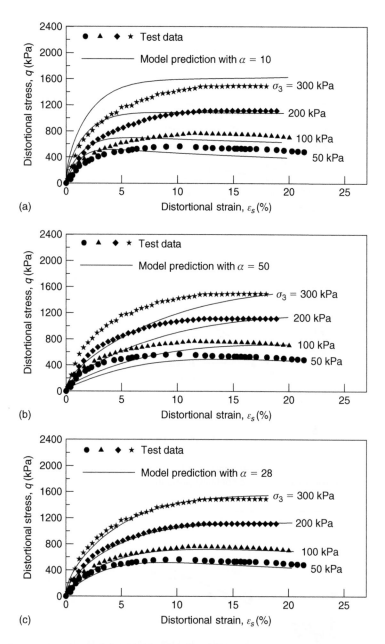

Figure B6 Determination of model parameter α by stiffness matching between analytical predictions and test data using a value of (a) $\alpha = 10$, (b) $\alpha = 50$ and (c) $\alpha = 28$.

Figure B4(b) shows that the variation of particle breakage (B_g) with increasing distortional strain and confining pressure (as shown in Fig. B4a) can be effectively represented by a single function (Equation 7.51), and the coefficients of the exponential function (Equation 7.51) gives the values of θ and υ.

The model parameters χ and μ can be determined by plotting the rate of particle breakage $dB_g/d\varepsilon_s^p$ at various distortional strains and confining pressures in terms of $\ln\{p_{cs(i)}/p_{(i)}\}dB_g/d\varepsilon_s^p$ versus $(M - \eta^*)$, as shown in Figure B5, where $\eta^* = \eta(p/p_{cs})$. The intercept and the slope of the best-fit line of this plot give the values of χ and μ, respectively.

The parameter α can be evaluated by matching the initial stiffness of analytical predictions with a set of experimental results, as shown in Figs. B6(a)–(c). The analytical predictions of stress-strain of ballast using $\alpha = 10$, $\alpha = 50$ and $\alpha = 28$ compared to the test data are shown in Figures B6(a), (b) and (c), respectively. Figure B6(a) shows that the initial stiffness of the stress-strain prediction for $\alpha = 10$ is higher compared to the experimental results. In contrast, $\alpha = 50$ gives the stress-strain predictions with a lower initial stiffness than the test data. Figure B6(c) clearly shows that a value of $\alpha = 28$ gives a very good matching between the analytical predictions and the laboratory measurements.

B.2   FOR CYCLIC LOADING MODEL

The cyclic loading model presented in this study contains additional 4 parameters, which can be evaluated from the laboratory measured data of cyclic stress-strain, as explained in the following:

Figure B7(a) shows a typical stress-strain $(q - \varepsilon_s)$ plot under cyclic loading. The cyclic stress-strain data of Figure B7(a) can be re-plotted as distortional stress versus plastic distortional strain $(q - \varepsilon_s^p)$, as shown in Figure B7(b), by subtracting the elastic component (using Equation 7.41) from the total distortional strain.

The value of the hardening function h (Equation 7.80) at the start of cyclic loading [i.e. at point 'i' in Fig. B7(b)] gives the value of h_i (Equation 7.81). The value of $h_{int(i)}$ (Equation 7.81) for the first reloading 'bc' (Fig. B7b) can be computed by substituting the experimental values of $d\varepsilon_s^p$, p and $d\eta$ for the first incremental load 'bb$_1$'of this reloading into Equation 7.85. The cyclic model parameter ξ_1 can then be evaluated by substituting h_i, $h_{int(i)}$ and the value of ε_v^p at the start of first 'reloading' into Equation 7.81.

Similarly, the values of h_{int} for the following load increments ('b$_1$b$_2$', 'b$_2$b$_3$' etc.) can be computed by substituting the values of $d\varepsilon_s^p$, p and $d\eta$ for the corresponding load increments into Equation 7.85. The value of h (Equation 7.80) at point 'a' gives the value of h_{bound} for the reloading 'bc' (Fig. B7b). The model parameters ξ_2 and γ can be evaluated by a trial and error process after substituting a set of known values of h_{int}, $h_{int(i)}$, h_{bound}, R (Equation 7.83) and ε_v^p for the load increments ('b$_1$b$_2$', 'b$_2$b$_3$', 'b$_3$b$_4$' etc.) of 'bc' into Equation 7.82.

In a similar way, the value of $h_{int(i)}$ for the following reloading 'de' (Fig. B7b) and the values of h_{int} for the load increments ('d$_1$d$_2$', 'd$_2$d$_3$' etc.) of 'de' can be computed. The model parameter ξ_3 can then be evaluated by substituting the values of h_{int}, $h_{int(i)}$, h_{bound}, R, γ and ε_{v1}^p for the load increment 'd$_1$d$_2$' or 'd$_2$d$_3$' into Equation 7.84.

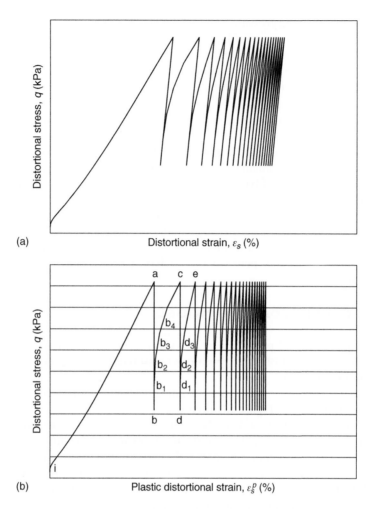

(a)

(b)

Figure B7 Determination of cyclic model parameters ξ_1, ξ_2, ξ_3 and γ from laboratory test data, (a) cyclic stress-strain plot, and (b) cyclic stress-plastic strain plot.

A Pictorial Guide to Track Strengthening, Field Inspection and Instrumentation

Figure C1 Laying the Geogrid over non-woven geotextile to form a geocomposite layer in fully instrumented rail track at Bulli town north of Wollongong City, New South Wales.

Figure C2 Placing of ballast mat over a bridge deck at Bulli.

Figure C3 Automated tamping equipment used at fully instrumented track at Singleton town near Newcastle, New South Wales.

Figure C4 Automated tamping equipment at Singleton.

Figure C5 Installation of pressure cells at Mudies Creek bridge near Singleton.

Figure C6 Installation of pressure cell at the ballast-capping interface at Singleton.

Figure C7 Intrusion of coal fines in to the shoulder ballast in a track near Rockhampton, Queensland.

Figure C8 Migration of coal fines towards bottom of the ballast bed due to vibration and rainwater infiltration, leaving the surface ballast relatively clean – a site near Rockhampton.

Figure C9 Coal and clay sediments fouling the ballast beneath the sleeper at a site in Thirroul, south of Sydney, observed during track maintenance.

Figure C10 Pulverization of ballast plus coal fouling observed near v-crossing close to Thirroul, New South Wales.

Appendix D

Unique Geotechnical and Rail Testing Equipment

Large-scale cyclic triaxial
apparatus

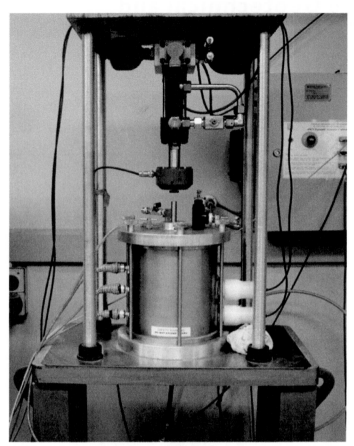

Cyclic subballast filtration
apparatus

Large-scale drop weight
impact testing rig

Model track used for non-destructive testing

Dynamic process simulation
cubicle triaxial chamber

Constant normal stiffness
shear apparatus

Subject Index

Note: Page numbers followed by "*f*" and "*t*" refer to figures and tables, respectively.

Printed and bound by CPI Group (UK) Ltd, Croydon, CR0 4YY

18/10/2024

01776231-0001